Advances in Intelligent Systems and Computing

Volume 1285

The series "Advances in Intelligent Systems and Computing" contains publications on theory, applications, and design methods of Intelligent Systems and Intelligent Computing. Virtually all disciplines such as engineering, natural sciences, computer and information science, ICT, economics, business, e-commerce, environment, healthcare, life science are covered. The list of topics spans all the areas of modern intelligent systems and computing such as: computational intelligence, soft computing including neural networks, fuzzy systems, evolutionary computing and the fusion of these paradigms, social intelligence, ambient intelligence, computational neuroscience, artificial life, virtual worlds and society, cognitive science and systems, Perception and Vision, DNA and immune based systems, self-organizing and adaptive systems, e-Learning and teaching, human-centered and human-centric computing, recommender systems, intelligent control, robotics and mechatronics including human-machine teaming, knowledge-based paradigms, learning paradigms, machine ethics, intelligent data analysis, knowledge management, intelligent agents, intelligent decision making and support, intelligent network security, trust management, interactive entertainment, Web intelligence and multimedia.

The publications within "Advances in Intelligent Systems and Computing" are primarily proceedings of important conferences, symposia and congresses. They cover significant recent developments in the field, both of a foundational and applicable character. An important characteristic feature of the series is the short publication time and world-wide distribution. This permits a rapid and broad dissemination of research results.

**** Indexing: The books of this series are submitted to ISI Proceedings, EI-Compendex, DBLP, SCOPUS, Google Scholar and Springerlink ****

More information about this series at http://www.springer.com/series/11156

Luis M. Bergasa · Manuel Ocaña ·
Rafael Barea · Elena López-Guillén ·
Pedro Revenga
Editors

Advances in Physical Agents II

Proceedings of the 21st International Workshop of Physical Agents (WAF 2020), November 19–20, 2020, Alcalá de Henares, Madrid, Spain

Springer

Editors
Luis M. Bergasa
Electronics Department
University of Alcalá
Madrid, Spain

Manuel Ocaña
Electronics Department
University of Alcalá
Madrid, Spain

Rafael Barea
Electronics Department
University of Alcalá
Madrid, Spain

Elena López-Guillén
Electronics Department
University of Alcalá
Madrid, Spain

Pedro Revenga
Electronics Department
University of Alcalá
Madrid, Spain

ISSN 2194-5357 ISSN 2194-5365 (electronic)
Advances in Intelligent Systems and Computing
ISBN 978-3-030-62578-8 ISBN 978-3-030-62579-5 (eBook)
https://doi.org/10.1007/978-3-030-62579-5

This Springer imprint is published by the registered company Springer Nature Switzerland AG
The registered company address is: Gewerbestrasse 11, 6330 Cham, Switzerland

Preface

This volume of Advances in Intelligent Systems and Computing contains selected papers presented at the 21st International Workshop of Physical Agents (WAF2020), held in Alcalá de Henares, Spain, on November 19–20, 2020. The conference was organized by the University of Alcalá.

WAF2020 is an international forum for fundamental and applied research on artificial intelligence (AI) techniques in applications of control and interaction with physical agents (mobile robots, vehicles, manipulators, etc.)

The conference focuses on a broad range of research challenges in the fields of physical agents, software agents, multiagent systems, human–robot interaction, mobile robots, social robots, cooperating robots, autonomous vehicles, machine learning and robotics, deep-learning and robotics, perception, localization and mapping techniques and autonomous navigation systems, addressing current and future trends in these fields. WAF2020 brings together researchers from academic institutions, leading industrial companies, and government laboratories located around the world for promoting and popularization of the scientific fundamentals of physical agents.

The book was organized in five sections, according to the main conference topics. Each section is devoted to research in the areas of (1) autonomous navigation, localization, and mapping, (2) mobile and social robots, (3) human–robot interaction, (4) perception systems, and (5) deep-learning and robotics.

WAF2020 received 32 contributions. After a thorough peer-review process, the program committee accepted 24 papers, written by authors from five countries. Thank you very much to the authors for their contribution. These papers are published in the present book, achieving an acceptance rate of 75%. Extended versions of selected best papers will be published in *Multimedia Tools and Applications (MTAP) journal* (published by Springer and indexed by ISI-JCR and Scopus).

We would like to take this opportunity to thank members of scientific committee and invited external reviewers for their efforts and expertise in contribution to reviewing, without which it would be impossible to maintain the high standards of peer-reviewed papers.

Thank you very much to our keynote speakers: Pablo Fernández-Alcantarilla (SLAMcore Ltd, UK) and Antonio López-Pena (CVC, Universitat Autonoma de Barcelona, Spain) for sharing their knowledge and experience.

We appreciate the partnership with Springer, EasyChair, and our sponsors Universidad de Alcalá, Red de Agentes Físicos, and RoboCity2030 for their essential support during the preparation of WAF2020.

Thank you very much to WAF2020 team. Their involvement and hard work were crucial to the success of the WAF2020 conference.

November 2020

Luis M. Bergasa
Manuel Ocaña
Rafael Barea
Elena López-Guillén
Pedro Revenga

Organization

Organising Committee

Luis Miguel Bergasa Pascual	University of Alcalá, Spain
Manuel Ocaña Miguel	University of Alcalá, Spain
Rafael Barea Navarro	University of Alcalá, Spain
Elena López Guillén	University of Alcalá, Spain
Pedro Revenga de Toro	University of Alcalá, Spain

Scientific Committee

Agapito Ledezma Espino	University Carlos III, Spain
Angel García Olaya	University Carlos III, Spain
Andrés Rosales	Escuela Politécnica Nacional, Ecuador
Antonio Fernández Caballero	University of Castilla-La Mancha, Spain
Antonio González Muñoz	University of Granada, Spain
Carlos Vázquez Regueiro	University of A Coruña, Spain
Daniel Pizarro Pérez	University of Alcalá, Spain
Diego Viejo Hernando	University of Alicante, Spain
Dimitri Voilmy	Université de Technologie de Troyes, France
Eduardo Molinos	Karlsruhe Institute of Technology, Germany
Eduardo Zalama Casanova	University of Valladolid, Spain
Fernando Fernández Rebollo	University Carlos III de Madrid, Spain
Francisco Martín Rico	University Rey Juan Carlos, Spain
Franziska Kirstein	Blue Ocean Robotics, Denmark
Gustavo Scaglia	Universidad Nacional de San Juan, Argentina
Humberto Martínez Barbera	University of Murcia, Spain
Humberto Rodríguez	Universidad Tecnológica de Panamá, Panamá
Ismael García Varea	University of Castilla-La Mancha, Spain
Iván Armuelles Voinov	University of Panamá, Panamá
Joaquín López Fernández	University of Vigo, Spain

Jorge Dias	University of Coimbra, Portugal
José Daniel Hernández Sosa	University of Las Palmas de Gran Canaria, Spain
José Mª Álvarez	NVIDIA, USA
José María Armingol Moreno	University Carlos III de Madrid, Spain
José María Cañas Plaza	University Rey Juan Carlos, Spain
Juan Adrián Romero Garcés	University of Málaga, Spain
Juan Pedro Bandera Rubio	University of Málaga, Spain
Kailun Yang	Karlsruhe Institute of Technology, Germany
Lluís Ribas i Xirgo	Autonomous University of Barcelona, Spain
Miguel Ángel Cazorla Quevedo	University of Alicante, Spain
Miguel Ángel García García	Autonomous University of Madrid, Spain
Miguel Ángel García Garrido	University of Alcalá, Spain
Miguel García Silvente	University of Granada, Spain
Oscar Andrés Vivas Albán	University of Cauca, Colombia
Pablo Bustos García de Castro	University of Extremadura, Spain
Pablo Fernández Alcantarilla	SLAMcore Ltd., UK
Rashid Mehmood	King Abdul Aziz University, Saudi Arabia
Roberto Iglesias Rodríguez	Universidade de Santiago de Compostela, Spain
Vicente Matellán Olivera	University of León, Spain

Contents

Autonomous Navigation, Localization and Mapping

Defining Adaptive Proxemic Zones for Activity-Aware Navigation 3
Jonatan Ginés Clavero, Francisco Martín Rico, Francisco J. Rodríguez-Lera, José Miguel Guerrero Hernández, and Vicente Matellán Olivera

Lifelong Object Localization in Robotic Applications 18
Cristina Romero-González, Jesus Martínez-Gómez, and Ismael García-Varea

Estimation of Customer Activity Patterns in Open Malls by Means of Combining Localization and Process Mining Techniques 30
Manuel Ocaña, Ángel Llamazares, Pedro A. Revenga, Miguel A. García-Garrido, Noelia Hernández, Pedro Álvarez, Javier Fabra, David Chapela-Campa, Manuel Mucientes, Manuel Lama, Alberto Bugarín, and Jose M. Alonso

Train Here, Drive There: Simulating Real-World Use Cases with Fully-Autonomous Driving Architecture in CARLA Simulator 44
Carlos Gómez-Huélamo, Javier Del Egido, Luis M. Bergasa, Rafael Barea, Elena López-Guillén, Felipe Arango, Javier Araluce, and Joaquín López

DQN-Based Deep Reinforcement Learning for Autonomous Driving ... 60
Óscar Pérez-Gil, Rafael Barea, Elena López-Guillén, Luis M. Bergasa, Pedro A. Revenga, Rodrigo Gutiérrez, and Alejandro Díaz

Mobile and Social Robots

Concept Drift Detection and Adaptation for Robotics and Mobile Devices in Federated and Continual Settings 79
Fernando E. Casado, Dylan Lema, Roberto Iglesias, Carlos V. Regueiro, and Senén Barro

QoS Metrics-in-the-Loop for Better Robot Navigation 94
Renan Salles De Freitas, Adrián Romero-Garcés, Rebeca Marfil, Cristina Vicente-Chicote, Jesús Martínez-Cruz, Juan F. Inglés-Romero, and Antonio Bandera

***DSR$_d$*: A Proposal for a Low-Latency, Distributed Working Memory for CORTEX** 109
Pablo Bustos, Juan C. García, Ramón Cintas, Esteban Martirena, Pilar Bachiller, Pedro Núñez, and Antonio Bandera

Multi-agent System Model of Taxi Fleets 123
Lluís Ribas-Xirgo

Human-Robot Interaction

Can a Social Robot Learn to Gesticulate Just by Observing Humans? 137
Unai Zabala, Igor Rodriguez, José María Martínez-Otzeta, and Elena Lazkano

Evaluation of a Multi-speaker System for Socially Assistive HRI in Real Scenarios 151
Antonio Martínez-Colón, Raquel Viciana-Abad, Jose Manuel Perez-Lorenzo, Christine Evers, and Patrick A. Naylor

Graph Neural Networks for Human-Aware Social Navigation 167
Luis J. Manso, Ronit R. Jorvekar, Diego R. Faria, Pablo Bustos, and Pilar Bachiller

A Toolkit to Generate Social Navigation Datasets 180
Rishabh Baghel, Aditya Kapoor, Pilar Bachiller, Ronit R. Jorvekar, Daniel Rodriguez-Criado, and Luis J. Manso

JBCA: Designing an Adaptative Continuous Authentication Architecture 194
Javier Junquera-Sánchez, Carlos Cilleruelo-Rodríguez, Luis de-Marcos, and José Javier Martínez-Herráiz

Perception Systems

Optimization of a Robotics Gaze Control System 213
Jaime Duque Domingo, Jaime Gómez-García-Bermejo, and Eduardo Zalama

Open-Loop Sidescan Sonar Mosaic and ANN Velocity Estimation 227
José Manuel Bernabé Murcia and Humberto Martínez-Barberá

360° Real-Time 3D Multi-object Detection and Tracking for Autonomous Vehicle Navigation 241
Javier Del Egido, Carlos Gómez-Huélamo, Luis M. Bergasa, Rafael Barea, Elena López-Guillén, Javier Araluce, Rodrigo Gutiérrez, and Miguel Antunes

Towards Fine-Tuning of VQA Models in Public Datasets 256
Miguel E. Ortiz, Luis M. Bergasa, Roberto Arroyo, Sergio Álvarez, and Aitor Aller

Integrating OpenFace 2.0 Toolkit for Driver Attention Estimation in Challenging Accidental Scenarios 274
Javier Araluce, Luis M. Bergasa, Carlos Gómez-Huélamo, Rafael Barea, Elena López-Guillén, Felipe Arango, and Óscar Pérez-Gil

Deep-Learning and Robotics

Embedded Deep Learning Solution for Person Identification and Following with a Robot 291
Ignacio Condés, José-María Cañas, and Eduardo Perdices

Survival Loss: A Neuron Death Regularizer 305
Emilio J. Almazán, Javier Tovar, and Alejandro de la Calle

Reinforcement Learning Experiments and Benchmark for Solving Robotic Reaching Tasks 318
Pierre Aumjaud, David McAuliffe, Francisco Javier Rodríguez-Lera, and Philip Cardiff

Age and Gender Recognition from Speech Using Deep Neural Networks 332
Héctor A. Sánchez-Hevia, Roberto Gil-Pita, Manuel Utrilla-Manso, and Manuel Rosa-Zurera

Monocular 3D Hand Pose Estimation for Teleoperating Low-Cost Actuators 345
Francisco Gomez-Donoso, Félix Escalona, Alejandro Bañuls, Daniel Abellan, and Miguel Cazorla

Author Index 361

Autonomous Navigation, Localization and Mapping

Defining Adaptive Proxemic Zones for Activity-Aware Navigation

Jonatan Ginés Clavero[1(✉)], Francisco Martín Rico[2], Francisco J. Rodríguez-Lera[3], José Miguel Guerrero Hernández[2], and Vicente Matellán Olivera[4]

[1] Escuela Internacional de Doctorado, Rey Juan Carlos University, Móstoles, Spain
j.gines@alumnos.urjc.es
[2] Intelligent Robotics Lab, Rey Juan Carlos University, Fuenlabrada, Spain
[3] Escuela de Ingenierías Industrial e Informática, Universidad de León, León, Spain
[4] Supercomputación Castilla y León, SCAYLE, León, Spain

Abstract. Many of the tasks that a service robot can perform at home involve navigation skills. In a real world scenario, the navigation system should consider individuals beyond just objects, theses days it is necessary to offer particular and dynamic representation in the scenario in order to enhance the HRI experience. In this paper, we use the proxemic theory to do this representation. The proxemic zones are not static. The culture or the context influences them and, if we have this influence into account, we can increase humans' comfort. Moreover, there are collaborative tasks in which these zones take different shapes to allow the task's best performance. This research develops a layer, the *social layer*, to represent and distribute the proxemics zones' information in a standard way, through a cost map and using it to perform a social navigate task. We have evaluated these components in a simulated scenario, performing different collaborative and human-robot interaction tasks and reducing the personal area invasion in a 32%. The material developed during this research can be found in a public repository (https://github.com/IntelligentRoboticsLabs/social_navigation2_WAF), as well as instructions to facilitate the reproducibility of the results.

Keywords: Social robot · Social navigation · Proxemics · Activity-aware · Collaborative navigation

1 Introduction

A human sharing his home with a robot is getting closer every day and is starting to stop being a science-fiction movie thing. So far, domestic robots have a particular purpose, like the vacuum cleaner, but it is expected that in the coming years, these types of robots will have a general-purpose and will be able to solve everyday tasks, as well as naturally interact with humans. It requires robots to

L. M. Bergasa et al. (Eds.): WAF 2020, AISC 1285, pp. 3–17, 2021.
https://doi.org/10.1007/978-3-030-62579-5_1

treat humans in a special way, as they will share space and tasks with them. Humans are letting be another obstacle that robots have to avoid, for example, during navigation. More and more research is being done on *social navigation*. This type of navigation takes into account humans, their social conventions, or their activity, improving their comfort [18]. One of the most used models in social navigation is the proxemics theory [11]. It defines the space around a person as different zones with different radius: intimate, personal, social, and public. In this work, we will focus on the intimate and personal areas. The intimate zone (<0.4 m to the person) is an area that the robot must always respect, so navigation is forbidden in this zone. On the other hand, the personal zone (0.4–1.2 m) is an area where the person interacts with known people or collaborates with others to perform a task. It can be a restricted navigation zone for the robots [1,14,27] or, as described in this article, an adaptive zone. The personal zone can be considered adaptive because, depending on the context, it will be a restricted zone or a cooperation zone where the robot enters to carry out a task with the person. In this way, we make the robot more natural, social, and adaptable. The system can adjust to situations where robots and human beings close or scenarios such as the current pandemic, caused by COVID-19, where the safety distance of 2 m must be respected.

As already mentioned, the proxemics theory does not describe areas with a fixed radius but defines areas that could change according to the context, culture or age, among others. In [28], the authors propose an approach in which the proxemic zones are dynamic and change depending on the spatial context and human intention. Another works have developed methods to follow the social convention of keep on the right when walking in a corridor using this theory as a base [20,24]. However, our proposal is general to any scenario.

Another approach to develop a human-aware navigation is the use of a virtual forces model, the Social Force Model (SFM) [12]. This approach consider the attractive virtual forces, created by the points of interest or the people who want to interact, and repulsive ones, created by the rest of the people or the obstacles. Recent researches use a modern version of the SFM, the Extended Social Force Model [8,25]. In these proposals, they have developed a planner and a controller based on this force model. Our proposal is independent of the planning and control algorithm used so it may be adapted to new and better navigation algorithms.

Another recent approach [15] proposes the use of predefined positions around the person to interact with them. The positions are evaluated based on the user's preferences and choose the best position as the navigation goal. If that position is not reachable, the robot will go to the next best position. As this approximation does not represent the proxemic zones on the map, the robot can invade the intimate or personal zone while moving from one point to another, which would reduce the person's comfort. Our research tries to guarantee the comfort of the humans that interact or collaborate with the robot by establishing the intimate zone as a forbidden navigation zone.

Previous authors' work proposed an initial social layer for the ROS navigation [9]. It takes information about people's moods to adapt their proxemics zones, trying to do not disturb people with a bad mood. The paper at hand adapts this layer to the ROS2 navigation stack and extends the previous work to adapt the proxemics zones according to the context or the collaborative activity conducted by humans and robots.

1.1 Contribution

The main contributions compared to previous works are divided into two, scientific and technical:

- Scientific: A proxemic framework to represent, with adaptive proxemics shapes, human activities, location, culture, or specific situations.
- Scientific: A novel proxemic shape with the addition of what we call the *cooperation zone*. It allows a fluid and natural cooperation between humans and robots in navigation and interaction tasks.
- Technical: The development of open-source ROS2 navigation layers, *people filter*, and *social layer*, brings the scientific community a standard and public framework to represent dynamic proxemics zones in a map and allow them to create complex behaviors using this work as a base.

1.2 Paper Organization

The work presented here presents and discusses both contributions. At first, Sect. 2 defines the framework and how the proxemics zones are built. Section 3 defines the integration with ROS2 and how it works. Section 4 describes the performance of the system using the metrics established by the scientific community and presents an analysis of the experiment results and their implications. Finally, conclusions are presented in Sect. 5.

2 Framework Description

This work proposes a framework for representing the space surrounding a person, the proxemic areas, on a cost map. This representation is fundamental to differentiate humans from the rest of the obstacles, thus enriching the robot's knowledge of its environment. Unlike past research that used proxemic zones based on Gaussian functions of concentric circles [9], in this article, the authors have used Asymmetric Gaussian proxemic zones [13]. These proxemic zones provide us a high adaptation capacity to the context in which the robot is located, varying their size and shape, unlike the used in previous research that only modified their size. The Asymmetric Gaussian are defined by four variables: head (σ_h), side (σ_s), rear (σ_r) and an orientation (Θ). Figure 1 shows a graphic explanation of these parameters.

The high adaptability offered by this type of Gaussian allows us to associate different shapes and sizes of the proxemic zones with different activities of a human's daily life. Thus, associating some values for these variables with human activities, we can build a social map in which people are represented differently based on their activity. For example, people cooking, a person is moving at a certain speed or a person standing in the scenario, Fig. 2.

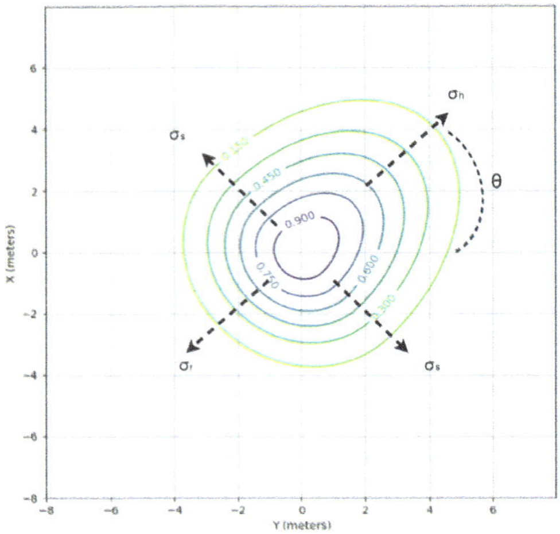

Fig. 1. Asymmetric Gaussian function centered at (0, 0), rotated by $\Theta = \pi/4$, and having variances $\sigma_h = 3.0$, $\sigma_s = 2.0$, and $\sigma_r = 1.8$.

Also, we propose new proxemics shapes oriented to improve in the perform of collaborative tasks between the human and the robot, taking as reference the work of Mead et al. [22]. They show that humans adapt their proxemic zones to interact with a robot. In that way, we have designed proxemic zones that contain a *cooperation zone*, Fig. 3. The robot will occupy the cooperation zone during the collaborative task to keep close to the person. These zones are located within the personal zone but always respecting the intimate zone.

2.1 Asymmetric Gaussian Function as Human Activity Representation

The proposal for represent people and their activities uses the model described in [13]. In this model people generate areas where navigation is forbidden or penalised, using an asymmetric Gaussian function. Let $P_n = \{p_1, p_2 \ldots p_n\}$ be the set of n persons detected in the scenario and $p_i = (x, y, \theta)$ is the pose of the person i.

$$g_{pi}(x,y) = e^{-(A(x-x_i)^2 + 2B(x-x_i)(y-y_i) + C(y-y_i)^2)}$$

With A, B, C:

$$A = \frac{cos(\theta)^2}{2\sigma^2} + \frac{sin(\theta)^2}{2\sigma_s^2}$$

$$B = \frac{sin(2\theta)}{4\sigma^2} - \frac{sin(2\theta)}{4\sigma_s^2}$$

$$C = \frac{sin(\theta)^2}{2\sigma^2} + \frac{cos(\theta)^2}{2\sigma_s^2}$$

where σs, as already mentioned, is the variance on the left and right.

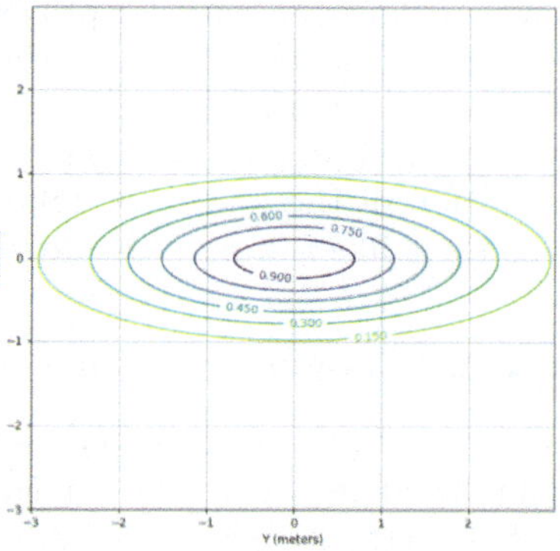

Cooking.
$\sigma_h = 0.5$, $\sigma_s = 1.5$, $\sigma_r = 0.5$

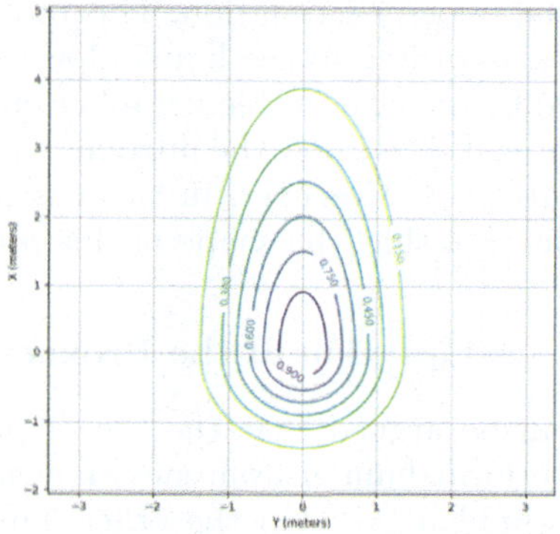

Running.
$\sigma_h = 2.0$, $\sigma_s = 0.7$, $\sigma_r = 0.7$

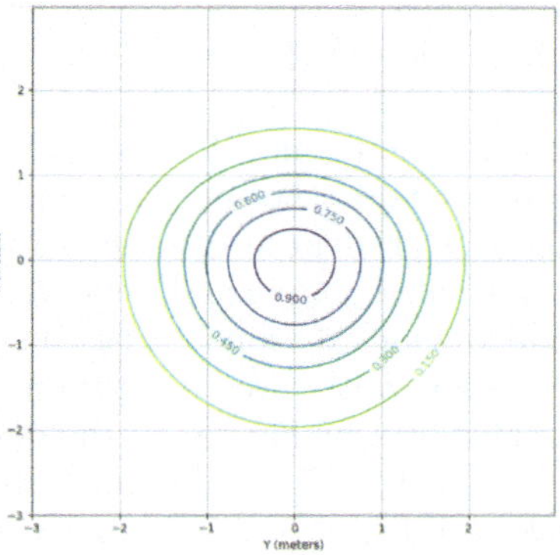

Standing.
$\sigma_h = 0.8$, $\sigma_s = 1.0$, $\sigma_r = 1.0$

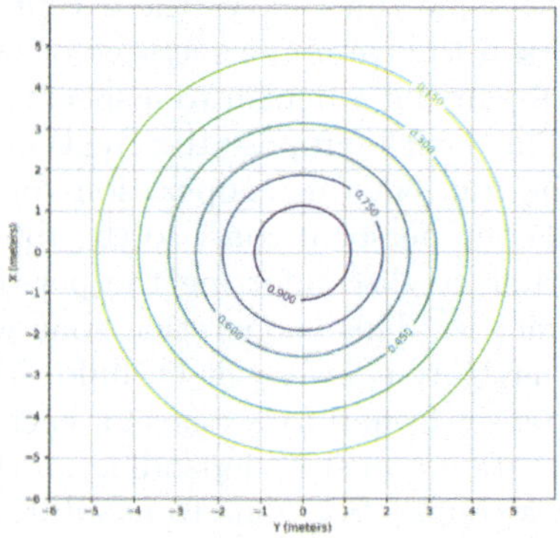

In the bathroom.
$\sigma_h = 2.5$, $\sigma_s = 2.5$, $\sigma_r = 2.5$

Fig. 2. Different proxemic shapes based on the context information.

Using this model allows us to create areas around the people detected with different sizes and shapes. Figure 2 shows four activities' representations. If a person is cooking in the kitchen is expected, he/she is moving from right to left, going from the ceramic stove to the cut zone or the fridge. Using this representation, Fig. 2a, a robot navigating in a domestic environment could pass behind the person, reducing the collision risk and improving his/her comfort. A similar situation is a person moving with a determined velocity in the robot's surroundings, Fig. 2b. The velocity could be estimated, and the proxemic zones will be updated with this estimation, updating the σ_h parameter from the model. Thus, it creates a big zone in a person's front where navigate is forbidden or penalized, avoiding hit with a person in a hurry and with a dynamic size, based on the velocity estimation. This tool also allows us to create different proxemics zones according to the human location. For example, a person in the bathroom could be no comfortable if a robot enters when he/she is in the shower or similar. Figure 2d shows how the proxemics zones have expanded to forbid the navigation in this area. Moreover, in this case, the orientation of the person is unknown. Because of this, the shape of this proxemics zones is like a standard Gaussian.

2.2 Adaptation of the Proxemic Areas, the Cooperation Zone

It can be argued that the use of proxemics zones is the most extent method to perform human awareness navigation, defining areas where the navigation is forbidden [17]. On the other hand, the robots have to behave socially and naturally and perform daily tasks with humans [3]. We can see how these two concepts collide, one restricts navigation around people, and the other promotes human-robot collaboration and interaction. It is necessary to create a mapping or a representation that takes into account the comfort and safety of people and at the same time allows the execution of tasks such as the robot approaching a person to give him a message or offer a information or the robot following or accompanying a person to a specific position.

This article propose the creation of the cooperation zones, Fig. 3. As we see in the Fig. 3d, this cooperation zone is located outside the intimate zone and inside the personal zone, so the robot will keep a prudential distance to avoid colliding or discomforting the person and allowing a comfortable and natural interaction. This cooperation zone will be coded as a free zone on the map, so the navigation in it will be fluid. The cooperation zones are configurable and can be set from 0 to 2 zones for each person. They can be located in the desired orientation and size, depending on the task for which they are designed. In this way, a cooperation zone to facilitate HRI tasks will be located in an orientation equal to the person or a cooperation zone designed to accompany people will be composed of two subzones, one on each side of the person.

3 Building the Social Layer

One of the essential capabilities of a social robot is to move through space. It is necessary to integrate the sensory information obtained by the robot into a map to navigate through space safely and effectively. The previous reason,

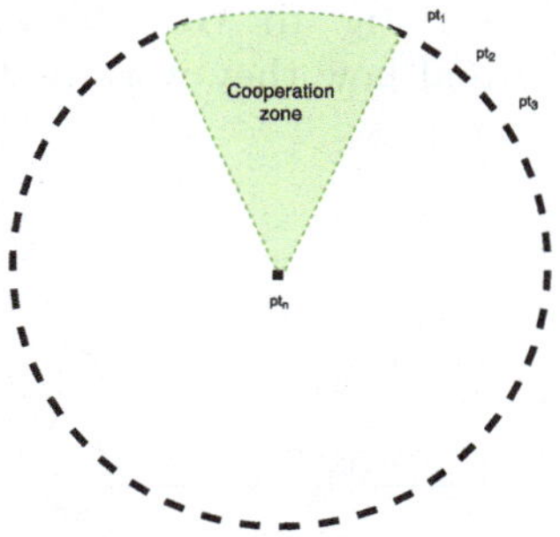

The proxemic base zone is created as a point's vector, excluding the cooperation zone.

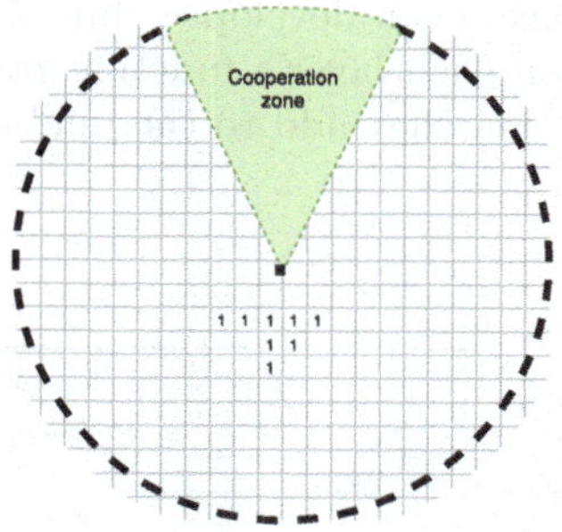

The map cells that are enclosed by the points defined in the previous step are searched.

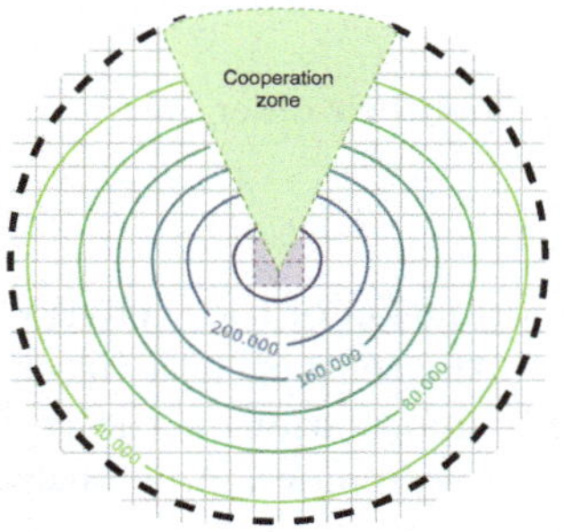

Each cell is evaluated with the Asymmetric Gaussian function defined.

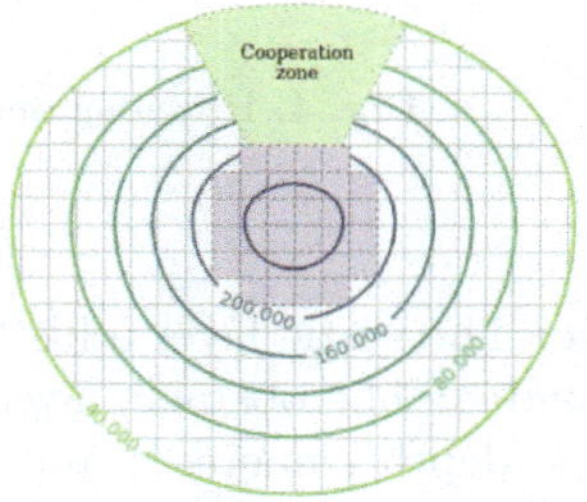

The intimate zone is added to the output of the previou sstep(purplecells). The intimate zone cells will be considered as an obstacle.

Fig. 3. Social layer workflow. From the human position to create its proxemics with a cooperation zone.

coupled with the widespread adoption by industry and the scientific community of ROS/ROS2 as a software development framework for robots, makes the ROS navigation [21] one of the most widely used, robust, and stable packages on the platform. This navigation stack represents the sensory information on two grid maps or cost maps, global and local. The framework chosen for this research is ROS2, the most advanced version of ROS. In this one, the global cost map is used by the planner to calculate the path from robot current position to target, and the local cost map is used by the controller generating movements to follow this path, avoiding unexpected obstacles. Both the global cost map and the local cost map result from the combination of the different layers that compose it [19].

Figure 4 shows how the default cost maps are split into different layers and which layers have been developed during this investigation, Fig. 4b. We will comment briefly on each layer's function to better understand how they work and what information they add to the final map.

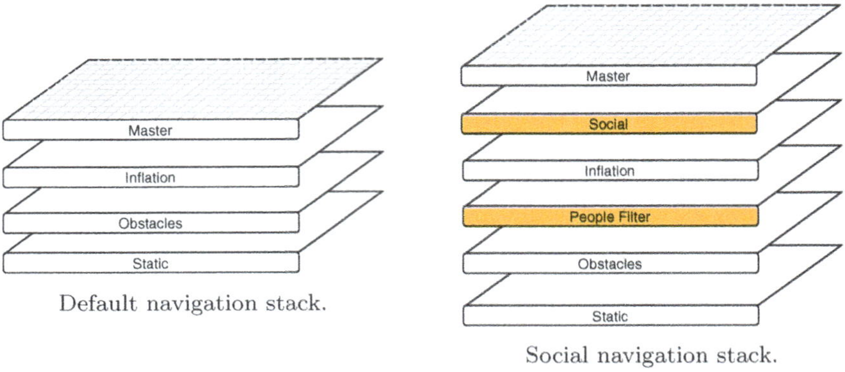

Fig. 4. Previous and propose layers and their order.

- **Static:** The static layer adds the a priori information of the scenario.
- **Obstacles:** The obstacle layer is subscribe to the sensors data (laser, ultrasounds, depth camera...) and materialise this information in the cost map. In addition to adding obstacle information, it also removes previously added information if the object disappears.
- **Inflation:** The inflation layer enlarges each of the obstacles previously added to the map by a radius equal to the robot's radius to ensure the safety of the robot's navigation. Also, the inflation process is made using a Gaussian function, so that the robot's speed is reduced when approaching an obstacle.
- **Master:** It is the final map and will be used by the planner and controller. It contains information about each one of the layers.

We can see how the order in which the layers are placed matters. It must be taken into account when developing a new layer to does not negatively affect the system's overall performance. For example, if we were to place the inflation layer before the obstacle layer, we would only be performing the process of expanding up obstacles to those represented on the scenario's map. The obstacles perceived by the sensors would be seen as points on the map, and it would be complicated to avoid them during navigation.

The layers developed in this research will be briefly presented below.

- **People filter:** It is located after the obstacle layer and has as input the people's position on the map. At this point, the obstacle layer will have included

the information of the people on the map, as they were obstacles. It is necessary to remove this information to have total control of the proxemic areas, so the cells around the people in a radius equal to the radius of their intimate area are marked as free cells.
- **Social:** It is located after the inflation layer and also has as input people's position. This layer adds the information of the people proxemic zones to the map. Section 2 shows in depth how these proxemic zones are.

As mentioned above, the developed layers need people's positions. This information can be obtained from a deep learning system, from a motion capture system, or, as in this article, from a simulator. It does not matter what system is used. The most important thing is that the information is represented correctly with the robot and the map. To do this, we will use the transform tree, tf2 [6], from ROS2. Robot's axes, sensors, map, and how each one is linked to other is represented in the tree. We will include in it the people's positions in the map frame. It will serve us as input for the two proposed layers.

Another essential tool that should be highlighted is the parameters to configure the proxemic zones. Through the ROS2 parameter system, we can set up new proxemic zones or configure, even during the execution of the system, the existing ones. This tool offers us great flexibility, necessary for the easy adapt of the system.

4 Experimental Results

The experiments carried out are aimed at demonstrating the improvement in people's comfort by using the proposed system as opposed to the default navigation system. For this purpose, the metrics already established by the scientific community have been used [16,23], formally described in [27]: d_{min}, average minimum distance to a human during navigation; d_t, distance traveled; τ, navigation time; and Psi, personal space intrusions.

Experiments have also been carried out to show the system's behavior during the execution of two collaborative tasks, escorting and following. ROS2Planning System [26] has been used to implement each of the proposed actions. It is an IA planning framework based on PDDL [7], which uses popf [5] as planner. Using this tool, we can easily decompose complex tasks into a sequence of more uncomplicated actions.

Another fundamental tool for performing the experiments is the pedestrian simulator based on the social force model, PedSim [2,10]. This simulator will provide people's position and orientation in each instant of time.

The tests were performed on a computer with an Intel Core i7-8550U 1.8 GHz processor with 16 Gb of DDR4 RAM and Ubuntu GNU/Linux 18.04 using Gazebo as simulator and ROS2 Eloquent as robot framework.

4.1 Approaching People

In this experiment, we want to compare the behavior of the default navigation system of ROS2 besides the proposed system when performing an approaching

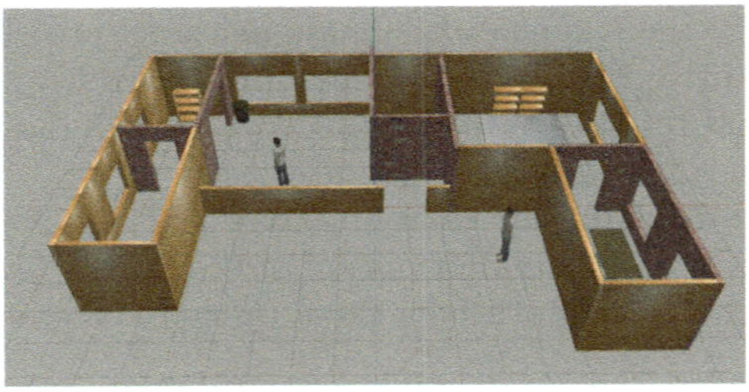

Fig. 5. Domestic scenario from Gazebo simulator.

task. The navigation speed was set at 0.3 m/s [4], the intimate zone radius at 0.4 m, and the personal zone radius at 1.2 m. Figure 6 shows the robot's representation of the person using the two systems. In the first case, we see how the robot does not differentiate people from other obstacles and therefore represents them on the map as such. On the other hand, we can see how the proxemic shape corresponding to the approach action has been established using the proposed system.

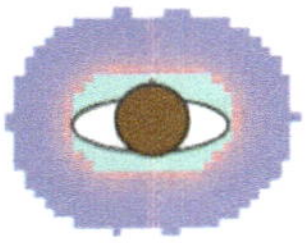

Person like an obstacle.

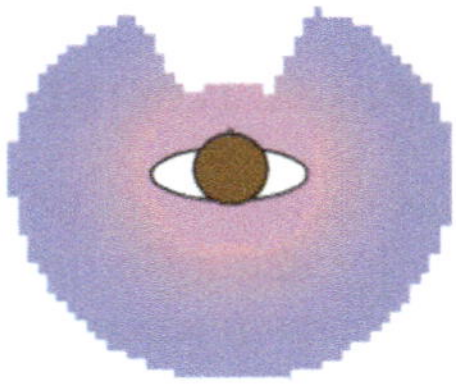

A proxemic zone with the cooperation area for HRI.

Fig. 6. Representations of a person perceived by the robot.

The implementation of the approach action is described below. This action takes as input the position and orientation of the target person and, using an approach similar to that proposed by Koay et al. [15], a set of 3 predefined points are established as possible goals for the navigation system, Fig. 7a. The distance from these points to the human is as follows:

- 1st: Intimate zone radius.
- 2nd: Intimate zone radius plus the robot's radius.
- 3rd: Intimate zone radius plus twice the robot's radius.

These points have an angle equal to the orientation of the person, although in the opposite direction. Once the points have been established, they are evaluated concerning the map in order of proximity to the person. Once the best point is obtained, it is sent to the navigation system. With this implementation, the robot

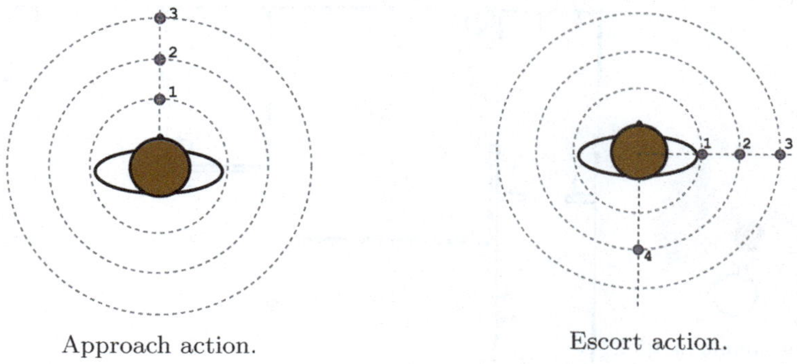

Fig. 7. Navigation predefined points for the actions.

approaches the person from the front, regardless of the representation system we use.

Figure 5 shows the scenario configuration for this experiment. We have the robot in the initial position and the person at a distance of 3.5 m. Each iteration consists of 2 actions, *approach* action, and *return to home* action. During the development of the experiment, the person's position is fixed, and its orientation changes in a random way in each iteration. Table 1 shows the results obtained during the experiment, after the execution of 100 iterations.

Table 1. Social navigation metrics for Approaching Test: for each parameter its mean and standard deviation are provided in parenthesis.

Parameter	ROS2 navigation	Proposed approach
$\tau(s)$	89.4 (22.02)	89.28(17.64)
$d_t(m)$	7.06(1.37)	7.71(1.67)
$d_{min}(m)$	0.68(0.1)	0.57(0.035)
$Psi(Personal)(\%)$	39.86(10.03)	7.83(5.33)
$Psi(Intimate)(\%)$	0(0)	0(0)

We see how there has been a significant decrease in the personal area occupation percentage using the proposed system. Also, the minimum distance to people has been reduced, keeping in 0.54 m–0.60 m shown in [29] as adequate, from the human point of view, to carry out voice interaction tasks or human-robot handovers. The task execution time and distance traveled have been maintained, thus gaining comfort without affecting the system's overall performance.

4.2 Collaborative Actions

This experiment shows the system's behavior during the execution of a collaborative task with a human, the escorting task. The escorting action implementation

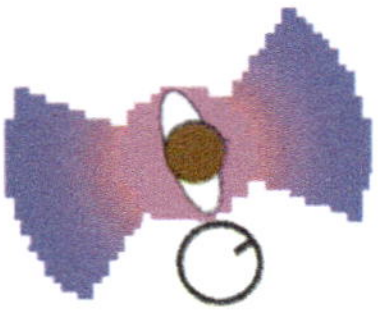

Escort proxemic shape.

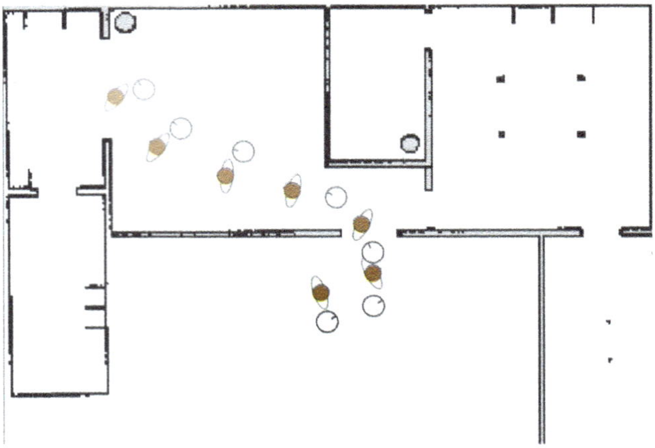

Escorting task path.

Fig. 8. Robot is accompanying a human trough a door.

is very similar to that shown in the previous experiment, Fig. 7b. Now target points are on the right of the person in addition to a point behind the person. This point is useful when, during the task execution, the human passes through a narrow area, such as a door. Figure 8b shows the path taken by the person and the robot during the task. Although the proxemic zone established during the whole execution of the task is the one shown in Fig. 8a, it has been omitted in Fig. 8b to help in the comprehension of the robot's behavior during the task. If one looks in the narrow area of the path, the robot cannot continue next to the person, and it must go now behind them. When there is more space on the stage, the robot recovers its position and stands next to the person.

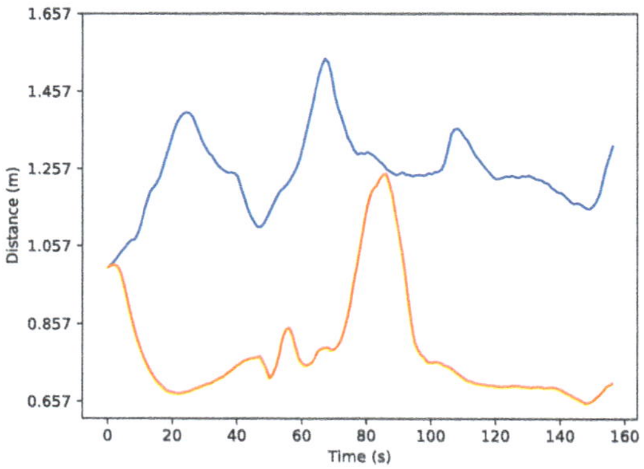

Fig. 9. Mean distance between robot and human during the escorting, (blue) using Lu et al. [20] approach; (orange) using our approach.

Finally, Fig. 9 shows a comparison of the distance between the robot and the human during the escorting task execution using a classical proxemic method, based on concentric circles Gaussian function, and the proposed method. The distance between the person and the robot is always higher. This is due to the robot is always behind the person, since the target points are located in a restricted navigation area, the personal area.

5 Conclusions and Future Works

Perception and context awareness systems attract more and more attention as they provide valuable information about the environment. Accordingly, a domestic robot can create a representation from people different than any other obstacle. It can be useful to perform collaborative navigation tasks between humans and robots or simply to facilitate human-robot interaction without affecting people's comfort.

One of the ways to represent people is through the proxemics theory. It defines the space around a person as different zones with different sizes. These proxemic zones are usually static and regular. The paper at hand proposes the creation of proxemic zones adaptable to the context, human activity, or culture, in order to provide more comfort or safety, and also zones that facilitate the execution of collaborative tasks, with the creation of the so-called *cooperation zone*. It allows performing tasks such as accompanying a person, following them, or approaching them simply and safely. To do this representation, we propose the development of open-source ROS2 navigation layers, *people filter* and *social layer*.

The proposed framework can be tested and used by the scientific community since it is in a public repository, and it offers the necessary resources for the reproducibility of the results.

Future works include the experimentation and study of the solution in a real environment with real participants and the integration in a cognitive architecture that provides to the proposed system more information about the people or their behavior, arriving to learn of their habits or their daily activity automatically.

References

1. Bera, A., Randhavane, T., Prinja, R., Manocha, D.: Sociosense: robot navigation amongst pedestrians with social and psychological constraints. In: 2017 IEEE/RSJ International Conference on Intelligent Robots and Systems (IROS), pp. 7018–7025 (2017). https://doi.org/10.1109/IROS.2017.8206628
2. Billy Okal, T.L.: pedsim_ros (2016). https://github.com/jginesclavero/pedsim_ros/tree/eloquent-dev
3. Breazeal, C.L.: Designing Sociable Robots. MIT Press, Cambridge (2004)
4. Butler, J.T., Agah, A.: Psychological effects of behavior patterns of a mobile personal robot. Auton. Robots **10**(2), 185–202 (2001). https://doi.org/10.1023/A:1008986004181

5. Coles, A.J., Coles, A.I., Fox, M., Long, D.: Forward-chaining partial-order planning. In: Twentieth International Conference on Automated Planning and Scheduling (2010)
6. Foote, T.: TF: the transform library. In: 2013 IEEE International Conference on Technologies for Practical Robot Applications (TePRA), Open-Source Software workshop, pp. 1–6 (2013). https://doi.org/10.1109/TePRA.2013.6556373
7. Fox, M., Long, D.: PDDL2. 1: An extension to PDDL for expressing temporal planning domains. J. Artif. Intell. Res. **20**, 61–124 (2003)
8. Galvan, M., Repiso, E., Sanfeliu, A.: Robot navigation to approach people using G2-spline path planning and extended social force model. In: Advances in Intelligent Systems and Computing. AISC, vol. 1093, pp. 15–27. Springer, Cham (2020). https://doi.org/10.1007/978-3-030-36150-1_2
9. Ginés, J., Martín, F., Vargas, D., Rodríguez, F.J., Matellán, V.: Social navigation in a cognitive architecture using dynamic proxemic zones. Sensors **19**(23) (2019). https://doi.org/10.3390/s19235189
10. Gloor, C.: Pedsim: pedestrian crowd simulation, vol. 5, no. 1 (2016). http://pedsim.silmaril.org
11. Hall, E.T.: The Hidden Dimension, vol. 609. Doubleday, Garden City (1910)
12. Helbing, D., Molnár, P.: Social force model for pedestrian dynamics. Phys. Rev. E **51**(5), 4282–4286 (1995). https://doi.org/10.1103/physreve.51.4282
13. Kirby, R.: Social robot navigation. ProQuest Dissertations and Theses, no. 3470165, p. 232 (2010)
14. Kirby, R., Simmons, R., Forlizzi, J.: Companion: a constraint-optimizing method for person-acceptable navigation. In: RO-MAN 2009 - The 18th IEEE International Symposium on Robot and Human Interactive Communication, pp. 607–612 (2009). https://doi.org/10.1109/ROMAN.2009.5326271
15. Koay, K.L., Syrdal, D., Bormann, R., Saunders, J., Walters, M.L., Dautenhahn, K.: Initial design, implementation and technical evaluation of a context-aware proxemics planner for a social robot. Lecture Notes in Computer Science (including subseries Lecture Notes in Artificial Intelligence and Lecture Notes in Bioinformatics). LNAI, vol. 10652, pp. 12–22. Springer, Heidelberg (2017). https://doi.org/10.1007/978-3-319-70022-9_2
16. Kostavelis, I., Kargakos, A., Giakoumis, D., Tzovaras, D.: Robot's workspace enhancement with dynamic human presence for socially-aware navigation. In: Liu, M., Chen, H., Vincze, M. (eds.) Computer Vision Systems, pp. 279–288. Springer, Cham (2017)
17. Kruse, T., Pandey, A.K., Alami, R., Kirsch, A.: Human-aware robot navigation: a survey. Elsevier. https://www.sciencedirect.com/science/article/pii/S0921889013001048
18. Kruse, T., Pandey, A.K., Alami, R., Kirsch, A.: Human-aware robot navigation: a survey. Robot. Auton. Syst. **61**(12), 1726 – 1743 (2013). https://doi.org/10.1016/j.robot.2013.05.007. http://www.sciencedirect.com/science/article/pii/S0921889013001048
19. Lu, D.V., Hershberger, D., Smart, W.D.: Layered cost maps for context-sensitive navigation. In: IEEE International Conference on Intelligent Robots and Systems, pp. 709–715. Institute of Electrical and Electronics Engineers Inc. (2014). https://doi.org/10.1109/IROS.2014.6942636
20. Lu, D.V., Smart, W.D.: Towards more efficient navigation for robots and humans. In: 2013 IEEE/RSJ International Conference on Intelligent Robots and Systems, pp. 1707–1713 (2013). https://doi.org/10.1109/IROS.2013.6696579

21. Marder-Eppstein, E., Berger, E., Foote, T., Gerkey, B., Konolige, K.: The office marathon: robust navigation in an indoor office environment. In: 2010 IEEE International Conference on Robotics and Automation, pp. 300–307 (2010). https://doi.org/10.1109/ROBOT.2010.5509725
22. Mead, R., Mataric, M.J.: Robots have needs too: people adapt their proxemic preferences to improve autonomous robot recognition of human social signals. In: AISB Convention 2015 (2015)
23. Okal, B., Arras, K.O.: Learning socially normative robot navigation behaviors with Bayesian inverse reinforcement learning. In: 2016 IEEE International Conference on Robotics and Automation (ICRA), pp. 2889–2895. IEEE (2016)
24. Pacchierotti, E., Christensen, H.I., Jensfelt, P.: Evaluation of passing distance for social robots. In: ROMAN 2006 - The 15th IEEE International Symposium on Robot and Human Interactive Communication, pp. 315–320 (2006). https://doi.org/10.1109/ROMAN.2006.314436
25. Repiso, E., Garrell, A., Sanfeliu, A.: Adaptive side-by-side social robot navigation to approach and interact with people. Int. J. Soc. Robot. 1–22 (2019). https://doi.org/10.1007/s12369-019-00559-2
26. Rico, F.M.: ros2_planning_system (2019). https://github.com/IntelligentRoboticsLabs/ros2_planning_system
27. Vega, A., Manso, L.J., Macharet, D.G., Bustos, P., Núñez, P.: Socially aware robot navigation system in human-populated and interactive environments based on an adaptive spatial density function and space affordances. Pattern Recogn. Lett. **118**, 72–84 (2019). https://doi.org/10.1016/j.patrec.2018.07.015
28. Vega-Magro, A., Manso, L.J., Bustos, P., Núñez, P.: A flexible and adaptive spatial density model for context-aware social mapping: towards a more realistic social navigation. In: Proceedings of 15th International Conference on Control, Automation, Robotics and Vision, pp. 1727–1732 (2018)
29. Walters, M.L., Dautenhahn, K., Te Boekhorst, R., Koay, K.L., Syrdal, D.S., Nehaniv, C.L.: An empirical framework for Human-Robot proxemics. In: Adaptive and Emergent Behaviour and Complex Systems - Proceedings of the 23rd Convention of the Society for the Study of Artificial Intelligence and Simulation of Behaviour, AISB 2009, pp. 144–149 (2009)

Lifelong Object Localization in Robotic Applications

Cristina Romero-González(✉), Jesus Martínez-Gómez, and Ismael García-Varea

University of Castilla-La Mancha, Campus Univ. s/n, 02071 Albacete, Spain
{Cristina.RGonzalez,Jesus.Martinez,Ismael.Garcia}@uclm.es

Abstract. One of the most common tasks in assistive robotics is to find some specific object in a home environment. Usually, this task is tackled by adding the objects of interest to a map of the environment as soon as the objects are detected by the vision system of the robot. However, these maps are usually static, and do not take into account the dynamic nature of a home, where anyone could move an object after the robot has seen it. In this paper, we propose a lifelong system to address this problem. The robot takes into account different possible locations for each object, and chooses the more probable one when it is required. We have designed a probability based system that stores possible locations for each object, and updates the probabilities of past locations based on newer detections.

Keywords: Image segmentation · Object detection · Robot vision · Deep learning

1 Introduction

One of the most studied problems in computer vision is object localization. Basically, this task consists in finding the geometrical position of a given object in the environment. This task is not only relevant for computer vision but also for robotics, due to the visual nature of the problem and its importance for manipulation and human-robot interaction applications. Like most computer vision tasks, state-of-the-art solutions are based on deep learning. There are three main methods to approach the object localization problem with CNNs: a) use a sliding window to perform local classifications and then refine the coarse heatmap [1], b) classify objects on the basis of previously detected regions of interest (ROIs) [2], or c) train a model to predict both the class and bounding box of the objects simultaneously [3].

Considering these approaches, object localization could be addressed as a semantic segmentation subtask. For example, in [4] the authors took advantage of state-of-the-art image classifiers to build a 3D map of the environment with the location of the different objects. In that paper the authors translated 2D

L. M. Bergasa et al. (Eds.): WAF 2020, AISC 1285, pp. 18–29, 2021.
https://doi.org/10.1007/978-3-030-62579-5_2

image segmentation solutions to 3D robotic applications in order to exploit the remarkable developments in computer vision with RGB images.

These 3D semantic maps aim at obtaining an accurate segmentation of the 3D space around the robot. As a result, they store, in a grid-based representation, the information about the objects detected with a CNN model within a small volume of space. The smaller the volume the more accurate the 3D map will be, but it will also require more resources to store and update the information in it. Additionally, these maps do not completely address the dynamic nature of robotic applications, where robots and people move around interacting with the environment and the objects in it.

In this paper, we propose a system for long-term object localization, more suitable for robotic applications with limited resources, because it only tracks the interest objects in the environment instead of creating a map of the whole space. Our system is able to consider different locations for each object and assign them a probability based on the last time the object was seen in that position and the output probability from the deep learning model used to detect it. We have run a baseline test in our system in a home-like scenario, which show that this approach could be useful in several robotic applications.

2 Related Work

Most computer vision tasks can be approached with Deep Learning [5] techniques. Convolutional Neural Networks (CNNs) have sufficiently proven their remarkable results in complex tasks, like image classification [6], face recognition [7] or language understanding [8].

In the case of semantic segmentation, the research has been more scarce due to the lack of large datasets with pixel-level annotations. Additionally, these datasets are focused on general scenes and, consequently, the class categories in them are usually focused on large elements (like cars or houses) and rarely have information about the type of objects that indoor robotic applications might require (like mugs or water bottles). However, recently published datasets, like COCO [9] or ADE20K [10], have allowed the development of solutions of scene segmentation. Fully Convolutional networks [11], which represent the state-of-the-art models for semantic segmentation, are able of directly generating pixel-wise predictions. Still, the datasets used to train these models only contain RGB images, so these solutions work in 2D scenarios, while robot vision is by nature 3D.

More specifically, in object localization problems the robot needs to find the 3D pose of an object in the world, not only the position of the object in a captured image. This can be addressed by merging the visual information of a CNN model with the geometric information of RGB-D sensors and the pose of the robot. In [4], the authors proposed a solution based on this approach where the objects in the environment are encoded in a 3D semantic map. In this paper, we are going to exploit the segmentation procedure proposed in that paper in order to create a lifelong object localization system.

3 Segmentation Methods

In this section we will include a brief overview of those steps followed in [4] to detect and calculate the pose of an object in an image that we will exploit in the system presented in this paper.

3.1 Lexical Labels in RGB-D Images

In general, a CNN classifier presents higher confidence values for objects and elements in the environment when the image is focused on them, instead of being a general image of the scene (see Fig. 1). Consequently, to obtain predictions suitable to identify objects that might be of interest in human-robot interaction applications, we do not need a global classification of the whole image depicting the environment, but the categories associated with regions of the image. Besides, we also need to segment the input RGB-D images to determine the position of the detected objects. The same segmentation process can then be used to generate smaller RGB images to be classified with a CNN model and to calculate their geometrical properties.

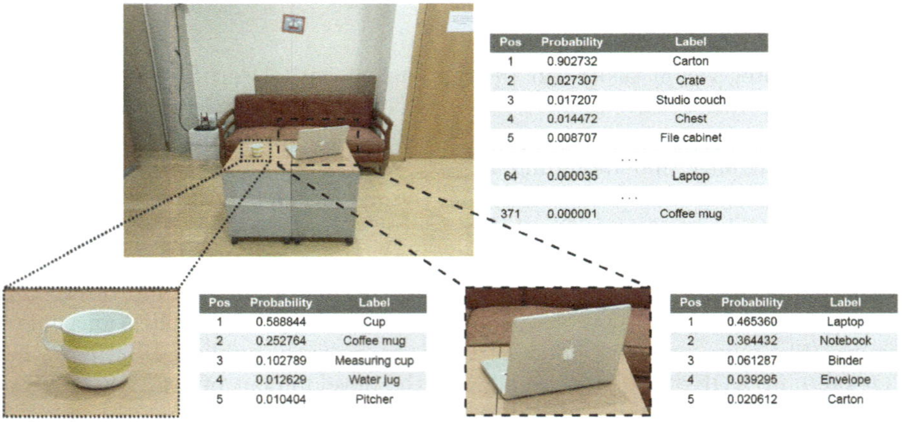

Fig. 1. Example of CNN predictions using the GoogLeNet [12] model for a whole image and for subimages containing objects.

Regarding the classification process, we use a state-of-the-art pretrained CNN classifier to obtain N possible lexical labels and their confidence values, that typically define a probability distribution. In this case, N is the number of output classes from the CNN classifier and it only depends on the training dataset. Formally, given an input RGB-D image $\mathcal{I}$, the first step is to split it in a set of smaller subimages. As we work with RGB-D information, each RGB subimage $I_i^{\mathcal{I}}$ has a corresponding 3D point cloud or subcloud $C_i^{\mathcal{I}}$, and vice versa. So, for each subcloud $C_i^{\mathcal{I}}$ we compute some geometrical information: its 3D centroid (p_i) and bounding cube (defined by its minimum and maximum coordinates, bc_i and BC_i, respectively); and for each subimage $I_i^{\mathcal{I}}$ we obtain the lexical labels, that is, the probability distribution $\mathcal{P}_i^{CNN}$ over the N possible different lexical labels that result from classifying the subimage with a CNN classifier of N classes. Thus, the semantic classification process (f) of a pair of corresponding subimage and subcloud $(I_i^{\mathcal{I}}, C_i^{\mathcal{I}})$ results in the following function:

$$f(I_i^{\mathcal{I}}, C_i^{\mathcal{I}}) = \langle p_i, bc_i, BC_i, \mathcal{P}_i^{CNN} \rangle \quad (1)$$

3.2 Semantic Segmentation

On the basis of the previous step, we need to define the segmentation process to generate the set of subimages to classify. We can take two approaches to this process: to perform a 2D segmentation of the image and calculate the associated 3D subclouds, or to perform a 3D segmentation and classify the associated 2D subimages. Research in 2D segmentation is more extensive, while solutions based on 3D information can benefit from the position of the different pixels besides its intensity value.

In [13], the authors studied how different segmentation methods could affect the quality of generated 3D semantic maps. They evaluated different segmentation methods based on both 2D and 3D information. 2D methods segment the RGB image into subimages that are used to obtain predictions with a pretrained CNN model, while 3D methods use geometrical information from RGB-D images to extract clusters of interest that are then mapped to RGB images in order to obtain the predictions.

The methods considered were a basic sliding window approach and a CNN model for object detection (YOLO [3]), as 2D techniques; and Region Growing Segmentation (RGS [14]) and Constrained Planar Cuts (CPC [15]), as 3D techniques. In Fig. 2) there are some example maps that can be generated with these techniques. 2D segmentation methods are better to identify larger objects (like the table or the couch), meanwhile 3D methods are able to detect smaller objects (like coffee mugs). In general, the best option is YOLO, as the obtained bounding box better adjusts to the object. Also, as a detecting algorithm, YOLO produces fewer false positives than the other methods.

4 Lifelong Object Localization

The use of 3D semantic maps for scene segmentation allows us to have a representation of the whole environment around the robot. However, given the large number of labels that output state-of-the-art image classifiers, this approach might present memory problems while running in large spaces. The alternative that we propose is to use the lexical labels obtained from the 3D semantic segmentation process [13] to keep a list of possible locations for the concerned objects, disregarding the rest of predictions from the CNN model. In other words, while 3D semantic maps answer the question "What objects are here?", our new approach would answer the question "Where I last saw that object?".

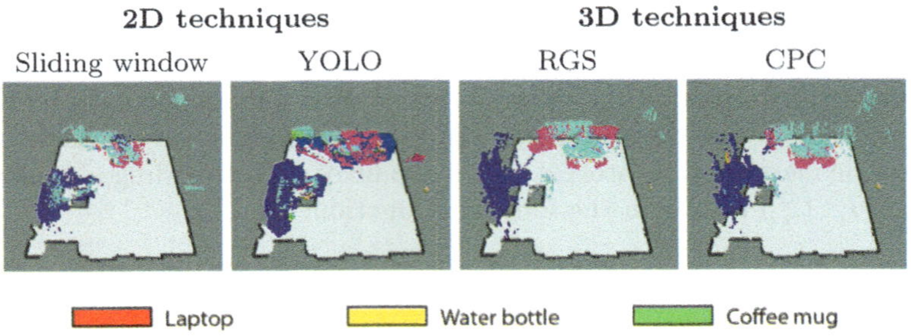

Fig. 2. Example of generated 3D maps with different segmentation techniques.

In this case, given the probability distribution $\mathcal{P}_i^{CNN}$ associated to a pair of subimage and subcloud $(I_i^{\mathcal{I}}, C_i^{\mathcal{I}})$, we only take into account the top M_i (with $M_i < N$) predicted labels. Each of these top predictions is considered a possible detection $(d_{i,m})$ of an object, which includes an associated lexical label $l_{i,m}$, a confidence value $c_{i,m}$ (the probability of the lexical label $l_{i,m}$), a 3D position p_i (the centroid calculated using the subcloud geometrical information), and a timestamp t_i.

$$d_{i,m} = \langle l_{i,m}, c_{i,m}, p_i, t_i \rangle \tag{2}$$

As we assume that the final application will only use a subset of the available object labels in the CNN classifier, we establish a mapping between those labels from the pretrained CNN model and the objects of interest. Consequently, for each object of interest $\mathcal{O}_j$, if the label $l_{i,m}$ can be mapped to the object, we integrate the detection $d_{i,m}$ in a list of possible positions for the object, namely $\mathcal{L}(\mathcal{O}_j)$. This addition is performed in two steps:

1. Short-term validation: We check that the new detection is consistent within a reduced time window, with the aim of adding robustness to false positive predictions from the classifier.
2. Long-term predictions: We add the predictions that are temporarily consistent with the list of possible positions for the object.

In Fig. 3 we can see a basic scheme of the object localization model here proposed. First, we receive an input stream of RGB-D images and we detect the objects in it using a CNN classifier. Then, we will check that the prediction is consistent with the information received in the last seconds. And, finally, we add the position to our database of possible positions for the object.

4.1 Short-Term Validation

The main disadvantage of using CNNs as classifier is the presence of false positive predictions. Usually, the accuracy of these classifiers is evaluated within the context of a multi-label problem, that is, considering the top 5 predictions for a given image, which results in a high number of non-strictly correct labels. Some

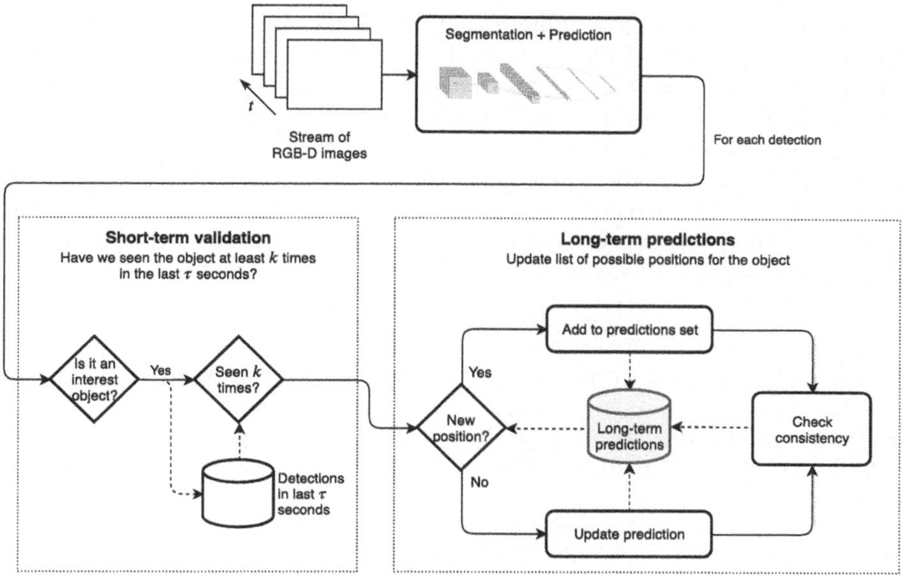

Fig. 3. Object localization model overview. New detections must be validated before they are added to the long-term predictions database.

of these labels might be correlated and thus assigned to the same object (for example, computer and laptop might be considered the same object depending on the final application), or they might be indeed false positives.

To diminish the impact of these multi-label predictions, we add a first step to validate that the obtained labels are consistent in a short span of time. To this end, we maintain a list of short-term predictions, with the predictions of the last τ seconds for each position and object in the environment. Then, if the same object is identified k times in the same position in that τ interval, the position is considered valid and added to the long-term predictions list with the mean confidence value in τ.

The values for τ and k must be carefully chosen based on the predictions per second that we can obtain. This frequency will be directly related to the hardware available on our robot, as the classification time using CNNs is highly dependent on the graphics card, and the segmentation method chosen to obtain the subimages.

4.2 Long-Term Predictions

The long-term predictions are formed by a set of possible positions for each interest object. The positions in the set present a probability value based on the confidence of the CNN output. These probabilities are updated in such a way that they form a probability distribution where, theoretically, the most recently detected position for an object will have a higher associated probability.

Formally, for each object of interest $\mathcal{O}_j$, we define a set of predicted positions $\mathcal{L}(\mathcal{O}_j)$, initially empty $\mathcal{L}(\mathcal{O}_j) = \varnothing$. Each element in this set consists of a tuple with a 3D position p_j and its associated probability $P(p_j)$. Additionally, we establish a learning rate α to integrate the new detections and the previously detected positions for the object.

At run time, those object positions that are time consistent are added to the set. So, given a new detection $d_{i,m}$ of an object $\mathcal{O}_j$ in the position p_i and with confidence value $c_{i,m}$, we calculate the minimum distance between p_i and the stored positions in $\mathcal{L}(\mathcal{O}_j)$, that is,

$$\langle p_s, P(p_s)\rangle = \arg\min_{\forall\langle p,P(p)\rangle\in\mathcal{L}(\mathcal{O}_j)} \|p_i - p\| \tag{3}$$

Then, given a distance threshold ($thres$), we can face the following scenarios:

1. *New position* ($\|p_i - p\| > thres$). The tuple, with the position and its probability (confidence value), is added to the list.

$$\mathcal{L}(\mathcal{O}_j) = \mathcal{L}(\mathcal{O}_j) \cup \{\langle p_i, c_{i,m}\rangle\} \tag{4}$$

2. *Known position* ($\|p_i - p\| \leq thres$). The object $\mathcal{O}_j$ has been previously detected in approximately the same position. So, we combine both detections using the learning rate α as follows:

$$\langle p_s, P(p_s)\rangle \leftarrow \langle \alpha \cdot p_i + (1-\alpha)\cdot p_s, \alpha \cdot c_i + (1-\alpha)\cdot P(p_s)\rangle \tag{5}$$

Finally, after adding/updating the new position, and to maintain consistency in the set of predictions, we might need to adjust the probabilities of the other positions. That is, if we have just added a position p_i to the list (or updated a previous one) with probability $P(p_i)$, the sum of probabilities for the object might be higher than 1. In this case, we would update the probability for the rest of the positions based on the remaining total probability $(1 - P(p_i))$ and the previous probability of each position.

$$P(p_s) \leftarrow \frac{(1 - P(p_i)) \cdot P(p_s)}{\sum_{k\neq i} P(p_k)}, \quad i \neq s \tag{6}$$

where $P(p_s)$ is the stored probability for the position p_s.

4.3 Passive and Active Search

The main goal of this paper is to propose an algorithm to maintain a set of object positions up-to-date. This can be achieved by a) adequately dealing with the detected objects positions discovered while performing another task in a passive search approach, or b) actively searching for a specific object.

In general, when a robot is performing a task, it can be passively observing the environment, which generates a lot of images that are not relevant for the given task and are usually ignored. With this approach, the robot would be using those images to localize objects and incorporate their positions to its knowledge base. For instance, the robot can access its RGB-D camera to discover relevant semantic information while navigating the environment. So, the predictions logic proposed in this paper can be updated as a background process that, ideally, will store the most probable positions for the objects of interest, based on the images passively perceived while performing other tasks.

Additionally, the robot might need to find a specific object in some human-robot interaction applications. In this case, the robot would actively search for the object in the most probable location based on the predictions list. In the worst-case scenario, the object will not be placed in the expected location, due to a false positive prediction or to an external factor (someone moved the object). This will generate an update of the probabilities where the new confidence value for the position will be zero, and the rest of the predictions will be updated according to Eq. (6), increasing their probability and providing a new candidate position. If the object was not found in any of the detected positions, the robot might navigate the environment passively searching for it.

5 Use Case

We have evaluated our proposals in a home-like scenario, a small room with three areas. Its dimensions are 5.95×5.65 m and the overall size of the working scenario is over $33\,\text{m}^2$. The three areas correspond to a dining room (big table and chairs), a kitchen (sink and coffee machine), and a living room (couch, coffee table, and bookshelf). Some pictures of this setup are shown in Fig. 4.

We have placed the following objects in the environment: laptop (x2), coffee mug (x3), water bottle (x2), table (x1), chair (x4), and couch (x1), spread through different "rooms". As robotic platform we have used a PeopleBot robot, with an ASUS Xtion PRO Live camera installed on top of it. We have recorded a sequence of 531 images with the robot following the path shown in Fig. 5, where the robot is able to visualize all the objects in the environment. The images are processed with an nVidia GeForce GTX 970 graphics card.

Given the temporal component of the short-term validation, and considering the segmentation times reported in [13], we only use the two methods that can steadily handle real-time inputs, that is, sliding window and YOLO. We also use this performance information to set the following configuration thresholds, which depend on the frames per second that we are able to process with the available hardware. The short-term memory stores the detections of the last $\tau = 2$ seconds, and whenever an object is detected in approximately the same position $k = 3$ times, it will be added to the long-term predictions. Additionally, we have chosen a learning rate of $\alpha = 0.6$ to stress the significance of more recent detections.

In Fig. 6 we can see the predicted positions for each object at the end of the sequence along with their ground truth centroids. It is important to note that in the visualization with images, there is a small error even in the ground truth positions. These differences are not due to an erroneous position of the centroids, but to the accuracy of the point clouds generated with the RGB-D

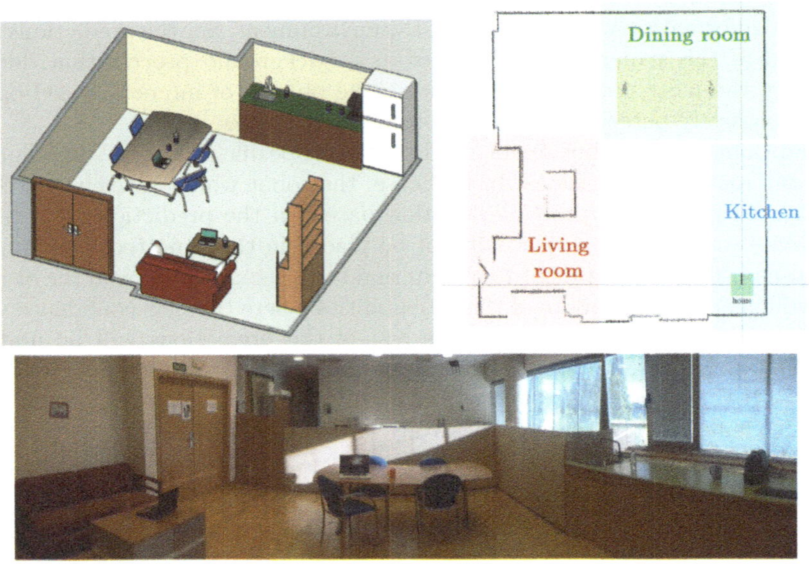

Fig. 4. A 3D model and 2D map of the environment for the experiments (top) and a panoramic picture (bottom).

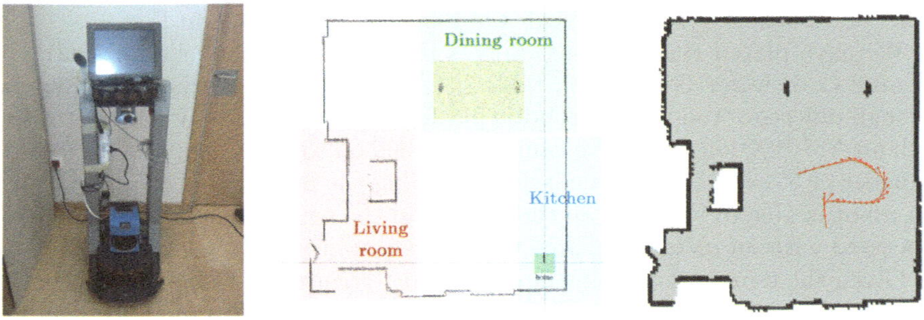

Fig. 5. Robot used in the experiments (left), map with the location of the different rooms in the environment (middle) and the path followed by the robot for the semantic map generation (right), which allows the robot to visualize all the objects in the different rooms.

camera, that becomes more imprecise the farther the objects are. In other words, the images are not in their exact position on the map. We have opted to show the result with these misplaced images because the lifelong localization method here proposed must be robust to these types of errors, and the position should correct itself when visualizing the object from different viewpoints.

Regarding the generated positions, we can observe that sliding window fails to identify some of the smaller objects (coffee mug and water bottle), while YOLO

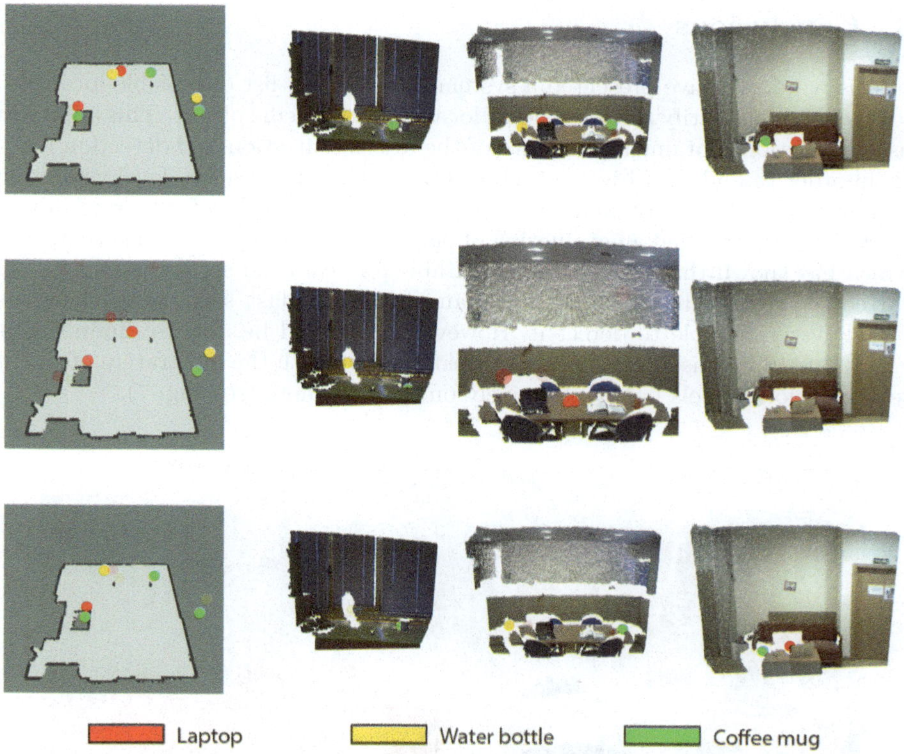

Fig. 6. Comparison of predicted object locations for sliding window segmentation (middle) and YOLO (bottom) against the ground truth centroids (top).

is most robust in this sense. In general, YOLO is the best option to detect and determine the position of the objects in the map, and its use is advisable for our lifelong localization procedure.

In the recorded sequence, the different rooms are visualized in this order: living room, kitchen, living room, and dining room. Theoretically, the objects in the later rooms should have a higher probability (darker color), while the objects in the former rooms should have decreased their probability (lighter color). In Fig. 6 we can observe that both segmentation methods exhibit this behavior. However, given the lower accuracy in the predictions using the sliding window, the probabilities are not high enough in later visualizations of the objects to become the most probable position. For example, the laptop in the dining room is detected last, but the confidence value of the CNN classifier is not high enough to replace the previous detection in the living room. In this regard, YOLO again offers better results because the geometrical features are more accurate and the confidence value for each detection is higher.

6 Conclusions

In this paper we have presented a system to manage a list of possible locations, so a robot can identify the last known location of a specific object. This approach uses predictions obtained with state-of-the-art classification and detection models in order to find the objects of interest, and then, it has a two-level validation system that consistently maintains a list of possible locations for long-term use.

We have performed an evaluation of our system with a prerecorded sequence, which has shown that the system is suitable for real-time applications and can correctly provide the position where an object was last seen, as long as the visual system is able to detect it. However, we should be able to diminish the errors caused by misdetections of the vision system with the integration of recent segmentation models based exclusively on deep learning (see Fig. 7).

Fig. 7. Example of semantic segmentation of an indoors scene with a pretrained deep learning model [16].

Acknowledgments. This work has been partially sponsored by the Spanish Ministry of Science and Innovation under grant numbers TIN2016-77902-C3-1-P and PID2019-106758GB-C33, and by the Regional Council of Education, Culture and Sports of Castilla-La Mancha under grant number SBPLY/17/180501/000493, supported with FEDER funds.

References

1. Tompson, J., Goroshin, R., Jain, A., LeCun, Y., Bregler, C.: Efficient object localization using convolutional networks. In: The IEEE Conference on Computer Vision and Pattern Recognition (CVPR), June 2015
2. Girshick, R.: Fast R-CNN. In: The IEEE International Conference on Computer Vision (ICCV) (2015)
3. Redmon, J., Farhadi, A.: YOLO9000: better, faster, stronger. In: The IEEE Conference on Computer Vision and Pattern Recognition (CVPR), July 2017
4. Romero-González, C., Martínez-Gómez, J., García-Varea, I.: 3D semantic maps for scene segmentation. In: ROBOT 2017: Third Iberian Robotics Conference, Sevilla, Spain, pp. 603–612 (2018)

5. LeCun, Y., Bengio, Y., Hinton, G.: Deep learning. Nature **521**(7553), 436–444 (2015)
6. Krizhevsky, A., Sutskever, I., Hinton, G.E.: Imagenet classification with deep convolutional neural networks. In: Advances in Neural Information Processing Systems 25, pp. 1097–1105 (2012)
7. Taigman, Y., Yang, M., Ranzato, M., Wolf, L.: Deepface: closing the gap to human-level performance in face verification. In: The IEEE Conference on Computer Vision and Pattern Recognition (CVPR), June 2014
8. Collobert, R., Weston, J., Bottou, L., Karlen, M., Kavukcuoglu, K., Kuksa, P.: Natural language processing (almost) from scratch. J. Mach. Learn. Res. **12**(Aug), 2493–2537 (2011)
9. Lin, T.Y., Maire, M., Belongie, S., Hays, J., Perona, P., Ramanan, D., Dollár, P., Zitnick, C.L.: Microsoft coco: common objects in context. In: Computer Vision – ECCV 2014, pp. 740–755. Springer, Heidelberg (2014)
10. Zhou, B., Zhao, H., Puig, X., Fidler, S., Barriuso, A., Torralba, A.: Scene parsing through ade20k dataset. In: Proceedings of the IEEE Conference on Computer Vision and Pattern Recognition (2017)
11. Long, J., Shelhamer, E., Darrell, T.: Fully convolutional networks for semantic segmentation. In: The IEEE Conference on Computer Vision and Pattern Recognition (CVPR), June 2015
12. Szegedy, C., Liu, W., Jia, Y., Sermanet, P., Reed, S., Anguelov, D., Erhan, D., Vanhoucke, V., Rabinovich, A.: Going deeper with convolutions. In: IEEE Conference on Computer Vision and Pattern Recognition (CVPR), pp. 1–9, June 2015
13. Romero-González, C., Martínez-Gómez, J., García-Varea, I.: A review of segmentation methods for 3D semantic mapping. In: Robot 2019: Fourth Iberian Robotics Conference, Porto, Portugal, pp. 620–631 (2020)
14. Rabbani, T., Van Den Heuvel, F., Vosselmann, G.: Segmentation of point clouds using smoothness constraint. Int. Arch. Photogramm. Remote Sens. Spatial Inf. Sci. **36**(5), 248–253 (2006)
15. Schoeler, M., Papon, J., Worgotter, F.: Constrained planar cuts - object partitioning for point clouds. In: The IEEE Conference on Computer Vision and Pattern Recognition (CVPR), June 2015
16. Zhou, B., Zhao, H., Puig, X., Xiao, T., Fidler, S., Barriuso, A., Torralba, A.: Semantic understanding of scenes through the ADE20K dataset. Int. J. Comput. Vision **127**(3), 302–321 (2019)

Estimation of Customer Activity Patterns in Open Malls by Means of Combining Localization and Process Mining Techniques

Manuel Ocaña[1(✉)], Ángel Llamazares[1], Pedro A. Revenga[1], Miguel A. García-Garrido[1], Noelia Hernández[1], Pedro Álvarez[2], Javier Fabra[2], David Chapela-Campa[3], Manuel Mucientes[3], Manuel Lama[3], Alberto Bugarín[3], and Jose M. Alonso[3]

[1] University of Alcalá, Alcalá de Henares, Madrid, Spain
mocana@depeca.uah.es
[2] University of Zaragoza, Zaragoza, Spain
[3] Centro Singular de Investigación en Tecnoloxías Intelixentes (CiTIUS), Universidade de Santiago de Compostela, Santiago de Compostela, Spain

Abstract. In this work we present a method to estimate the activity patterns made by shoppers in open malls based on localization information and process mining techniques. We present our smart phone application for logging information from sensors and a process mining system to discover what kind of activity pattern is made by the shoppers based in the key information of the localization which combine data mining, parallelization and monitoring techniques. Our system has been tested in a challenging scenario such as open mall (mixed classical mall and isolated shops in the street) where location is obtained by means of Global Positioning System (GPS) and a WiFi localization system that is helped by a parallel computing method to speed up the whole process. Activity patterns made by shoppers are obtained by process mining techniques which are applied to the data. The process models obtained provide hints that may help mall's managers to take specific actions considering the activity patterns that customers usually perform while in the mall.

1 Introduction

SOcial Workflows (SOW) coordinate the activities carried out by a group of users [6] who either individually or in cooperation try to achieve a certain objective. The SOW are unstructured flows in which a large number of users who carry out activities of a very diverse nature that they spread over time and typically consume few computing resources. An example of processes that are modelled through these types of workflows (WF) are the marketing campaigns that have as an objective to motivate potential customers in the consumption of a certain

L. M. Bergasa et al. (Eds.): WAF 2020, AISC 1285, pp. 30–43, 2021.
https://doi.org/10.1007/978-3-030-62579-5_3

product or service. However, from a technological point of view, the characteristics of the SOW are: (i) they model unstructured processes where users have multiple options and/or paths, and where they can even carry out additional activities to those initially planned in the WF; (ii) a large number of users; and (iii) depending on the skill or motivation of each user, a same activity can be performed immediately or have a temporary duration much higher (in the order of hours or even days).

These characteristics add a series of challenges when managing and monitoring the execution of SOW. On the one hand, the management of this type of WF requires the handling of a large number of simultaneous executions whose duration extends over time. On the other hand, the analysis of the execution of the WF requires the use of process mining techniques [1] to obtain the real WF followed by the users and, in this way, understanding what really happens in the execution of the process, thereby allowing the improvement of the process and its (dynamic) adaptation to the needs of users. However, for unstructured processes with a huge number of executions, the results of the current process discovery algorithms are WFs that have a structure in spaghetti and therefore very difficult, if not impossible, to interpret. Therefore, it is necessary to develop techniques to extract valuable information about the discovered WF. This is the first axis around which pivots the project Business (Artificial) Intelligence for Social Workflows (BAI4SOW).

On the other hand, the BAI4SOW project adds another dimension to the SOW: the activities to be carried out by users take place in geographic locations (such as an open shopping center, an area of tourist interest, an area where some activity is carried out or, in general, any area of a city) and they are also geolocated activities that users carry out through mobile devices (such as activities of a marketing campaign). Therefore, it is necessary to capture the events related to the execution of this type of activities and, for this, it is required the development of localization techniques and detection of the behavior of users in a given geographic area. This is the second axis of the project BAI4SOW.

In this article, we present the results obtained in BAI4SOW project whose objective is in the development of process mining algorithms for SOW analysis that contain geolocated activities. In addition, given the high number of users who can potentially participate in this type of flow is necessary run the algorithms in grid computing, cluster and cloud infrastructures, minimizing the cost that this computation could have in business clouds like Amazon and Google. This is the third axis of the BAI4SOW project.

The paper is organized as follows, firstly BAI4SOW project is introduced, then localization process with parallel computing is presented, third, test-bed and activity patterns estimation results are discussed.

2 BAI4SOW Global Vision

The main objective of BAI4SOW is the development of intelligent techniques for automatic extraction and user behavior analysis in processes modeled through social WFs. In particular, the target processes are those involving users moving around an exterior area and carrying out the activities of the social WF through their mobile devices. To achieve this objective, a conceptual framework has been proposed composed by the following components (Fig. 1):

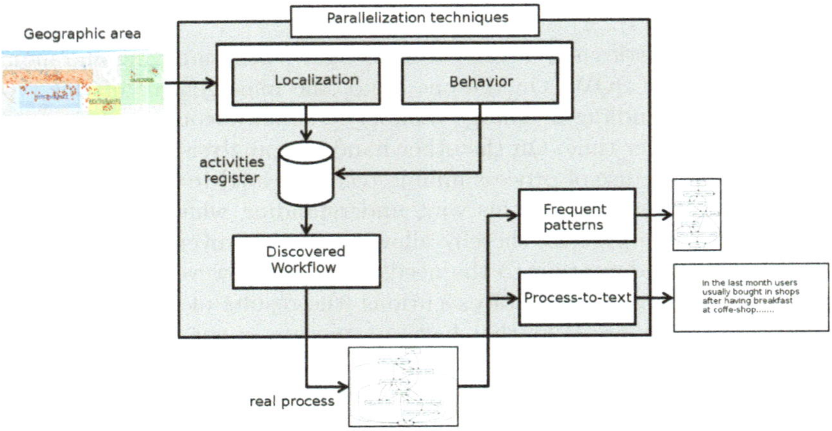

Fig. 1. Conceptual framework of BAI4SOW project.

- *Localization and behavior detection algorithms.* To analyze social WFs is necessary to record the events generated by the activities that each user carries out. For each instance of the process—an execution of the process from start to end—, a trace composed by these events is generated. In those processes involving users moving around a geographic area, this activities' model when a user arrives a certain point (to, for instance, make a purchase or take a photo), sits and moves inside an establishment, etc. Therefore, it is mandatory to *locate* the users in the geographic area where they are moving, as well as to detect users' behavior of interest once they are in a given location. To record the execution of these activities, a set of algorithms have been developed using the information given by the users' mobile devices such as the GPS and Wi-Fi signals sensors.
- *Process Mining Algorithms.* This set of algorithms allow to automatically discover the WF that has been actually executed by the users and, also, to extract relevant information about the process such as: *i) frequent patterns* of activities, which indicate patterns of frequent user behavior; *ii)* the *hierarchical organization* of unstructured and spaghetti WFs, which ease the visualization of the process reducing its complexity at different levels of abstraction; and *iii)* the *linguistic description* of these WFs, which automatically describes in natural language its most relevant characteristics, providing complementary information to the obtained by the users through the visualization of the WFs.

– *Parallelization techniques.* These techniques are used to design a technological infrastructure based on the integration of *grid* and *cloud* resources that support the efficient and large-scale processing of the algorithms developed under the project. This infrastructure integrates cost models to ease the decision of what computation and storage resources are more appropriate in each moment for the execution of the algorithms, considering the possibility of using resources from different infrastructures and suppliers.

It is important to note that this conceptual framework is applicable to SOW containing activities related to the location and behavior of the users taking part in them. However, process mining algorithms and parallelization techniques are directly applicable to any workflow in general, and particularly to unstructured processes with many users taking part in them. The following sections describe each of the components of this conceptual framework.

3 Localization System

As explained in the introduction, knowing the costumer's location is key to obtain heat maps representing the consumer movements and thus, to help mall managers to offer better marketing campaigns or change the shops layout.

In this paper, the method proposed to estimate users' smartphone location is based on the measurement of WiFi signals. WiFi-based localisation was selected because it is a no-installation, no-cost solution: There are WiFi access points (APs) in almost every shop, almost every smartphone has a WiFi interface and there is no need to be connected to WiFi networks to measure WiFi signal, so it is free of charge even for private networks.

Among all existing methods, fingerprint-based WiFi localisation is selected. This kind of methods rely on a two-stage approach: First, during the training stage, the environment is divided in cells and WiFi measurements are collected in each one of them. These measurements are then used to build the localisation system. Then, during the localisation stage, new measurements collected at unknown positions are compared with the stored ones to find the location where they were collected inside the environment.

The proposed method [7] follows this two-stage procedure and, besides, it implements a method to increase the localisation resolution without the need of increasing the number of cells to site-survey. This way, a Support Vector Regresion (SVR) algorithm is used to extrapolate virtual WiFi measurements at positions were no real samples were collected using the real collected information. Figure 2 shows the general architecture of the system.

As it can be seen, during the training stage, real measurements are used to create continuous WiFi surfaces from the collected discrete data using SVR. Using this method, one surface will be created for each existing WiFi AP. The trained WiFi localisation system will be composed of all these surfaces.

Then, during the localisation stage, a new sample collected by each costumer will be searched in these surfaces. This way, the coordinates where difference

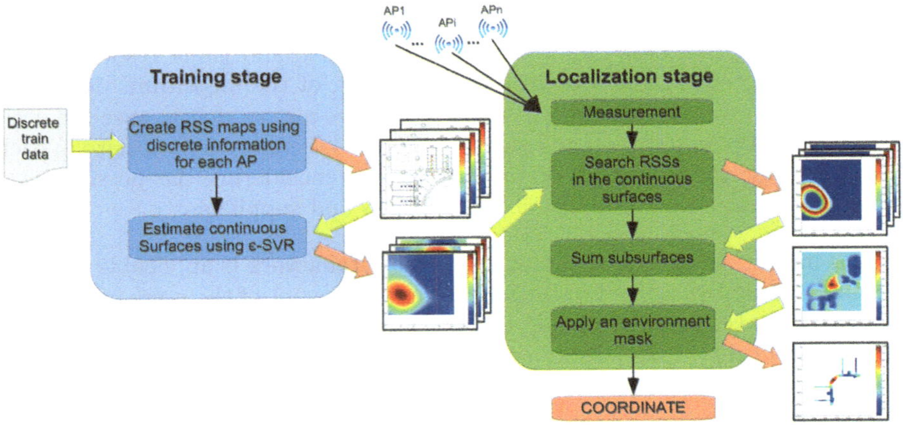

Fig. 2. General architecture of the system [7].

between the Received Signal Strength (RSS) from an AP and the stored RSS for that coordinate is low will be marked with higher scores than the coordinates with greater differences. Finally, the scores for each coordinate will be summed up, to obtain a resulting surface where the coordinate with the highest score is the most likely to be the location of the costumer.

In [7] the method was evaluated for determining the optimal size of cells when it is applied in a medium-size indoor scenario (around 2.500 m^2). The results proved that 15 cm side cells are an efficient option. In this paper, we are interested in evaluating the scalability of the method, from a performance point of view, and its applicability to bigger scenarios. For this reason, 1 cm side cells have been defined in order to increase significantly the number of cells to be processed for the same surface (25 million of cells in the scenario of 2.500 m^2). Besides, all the detected APs (a total of 105) and, as consequence, the 105 surfaces have been used to perform the localization. Two different trajectories were used to test the localisation system, obtaining mean localisation errors around 1,5 m. With this configuration, the execution times are 44.4 h to compute all the surfaces during the training stage and 325 s to determine the current location of a customer. The experiments were conducted using a server with 8 Intel i7-4790K CPUs at 4.00 GHz, a local SSD disk and 32 GB of RAM. Obviously, these performance results show the necessity of accelerating the training and location algorithms in order to be used in real scenarios.

As part of the work, a new parallel version of the algorithms has been programmed. It is based on the master-worker software architecture [3] and has been implemented to be easily executed in cloud-based enviroments. The solution has been deployed in the *Amazon EC2 infrastructure* (specifically, a pool of six *r4.8xlarge* computing instances has been hired to deploy the system programmed) being 16 min and 6.38 s the execution times of the training and location stages, respectively. The speed-up obtained in the calculation of surfaces and the location of customers is proportional to the number of cores available in the hired instances. Besides, the master-worker solution has been programmed as two RESTful services in order to facilitate the integration of the fingerprint-based location method in the BAI4SOW technological infrastructure.

4 Test-Bed and Results

4.1 Smartphone Application Development

In order to build the datasets and to set the ground truth, we have developed a smartphone application (see Fig. 3). This application is used in two ways: firstly the *training dataset* is collected by mean of saving several *paths* at static places, a path consists of frames at key points, and frames are formed by sensors' data and activities at zones and specific places introduced by an expert user; secondly *test dataset* is collected by mean of saving several paths with the user in movement. The activity, zone and place information introduced by the user for the *test dataset* will be used as ground truth in the validation stage.

The application is composed by a main activity and a service. The main activity is an interface to manage the application. The interface has an orange area with the buttons *START* and *STOP* to set the beginning and end of our path, and the name of the file that stores the data logging, as it is shown in the Fig. 3. In the green zone the actual user's selected activity, zone and place information is shown to store it by mean of pushing *SAVE* button. This tuple (activity-zone-place) is selected by pushing the buttons of the three white&blue columns. Finally, in the bottom part of the window, the Google API's recognized activity [5] is shown in order to compare it with our system.

Regarding service, it makes a log of all sensors' data. The advantage of this service is that runs in background, in such way, this allows the use of other applications without losing sensors' data. All the sensor's frames will be synchronized by mean of using the timestamp information.

When application is opened, the steps to follow are the next:

- First it is needed to write the name of the output file and then push the button *START* for initializing the service in background.
- We need to know our real activity for real comparison with Google's Activity Recognition API, which allows detecting users' activity using low power signals. Therefore, in the buttons of the three columns corresponding to place, zone and activity, we must select an option from each column and save the tuple. We will repeat this step every time we change activity.
- The last step is when *STOP* button is pushed, launching data saving in a compressed file.

Fig. 3. Screenshot of the smartphone application developed within the project

4.2 Test-Bed Description

The shopping mall *"C.C. La Dehesa"* is located in Alcalá de Henares (Madrid), this mall has 36.000 m^2 of indoor shopping composed by three big zones: 65 *shops* (orange area), one big *hypermarket* (blue area) and 7 *restaurants* (yellow area). In the shops zone, the ceiling is made by glass, permitting to obtain localization by GPS while in the rest only WiFi localization is available. In the outdoor area, fast food restaurants and shops are located. In this area GPS and WiFi are also available. This test-bed has been selected because it fits perfectly with the definition of open-mall, shops core in indoor-outdoor with mixed localization available.

We have used our app to train and test the BAI4SOW software framework running in three different mobile devices, two smartphones: Samsung Galaxy S6 and S6 edge; and one tablet Samsung Galaxy Tab 2. Table 1 shows the 6 paths that we have considered in order to validate our system. Paths consist of 4 or 5 steps of activities that start in a tuple activity-zone-place or key point and evolve to next one. Paths are executed several times and dates to validate the results obtaining 17 paths for training (17 static places) and 35 for testing (almost 6 times per path in movement). The table also shows the approximated travelled distance at each test's path, that ranges from 250 m to 410 m. For example, **Path 00** consists of the next steps:

Table 1. Paths in shopping mall.

Path	Device	Distance	App waypoint			
			Step	Activity	Zone	Place
00	**S6/S6 E**	**300 m**	1	Eating	Restaurants	Cafetito
			2	Walking	Restaurants	No Place
			3	Walking	Shops	No Place
			4	Shop window	Shops	Poly
01	**S6**	**410 m**	1	Eating	Restaurants	Cafetito
			2	Walking	Restaurants	No Place
			3	Walking	Shops	No Place
			4	Walking	Hypermarket	No Place
			5	Buying	Hypermarket	No Place
02	**Tab2**	**250 m**	1	Eating	Restaurants	Cafetito
			2	Walking	Restaurants	No Place
			3	Walking	Outdoor	No Place
			4	Shop window	Outdoor	McDonalds
			5	Buying	Outdoor	McDonalds
03	**Tab 2/S6**	**300 m**	1	Shop window	Restaurants	Cafetito
			2	Walking	Restaurants	No Place
			3	Walking	Shops	No Place
			4	Shop window	Shops	Poly
04	**Tab 2/S6 E**	**410 m**	1	Shop window	Restaurants	Cafetito
			2	Walking	Restaurants	No Place
			3	Walking	Shops	No Place
			4	Walking	Hypermarket	No Place
			5	Buying	Hypermarket	No Place
05	**S6/S6 E**	**250 m**	1	Shop window	Restaurants	Cafetito
			2	Walking	Restaurants	No Place
			3	Walking	Outdoor	No Place
			4	Shop window	Outdoor	McDonalds
			5	Buying	Outdoor	McDonalds

S6: Samsung Galaxy S6
S6 E: Samsung Galaxy S6 Edge
Tab 2: Samsung Galaxy Tab 2

1. Activity *Eating* (having the breakfast) at *Place* Cafetito in *Restaurants* zone during 20 min.
2. Activity *Walking* along *Restaurants* zone during 2 min.
3. Activity *Walking* along *Shops* zone during 5 min.
4. Activity looking at *Shop window* of Poly *Place* at *Shops* zone during 4 min.

Fig. 4. Open mall CC Dehesa.

This path took us about 300 m of travelled distance and at about 31 min. As you can see, the difference between activities 2 and 3 is the zone, key information that we obtained from the localization system.

The summary of high level activities that we have added to the basic activities recognized by the google's API are:

- Eating: we suppose that person is eating when he/she is localized at a restaurants zone during a higher time than a threshold.
- Walking: accelerometers and localization are key features to determine that person is walking.
- Shop window: this activity means that a person is standing at the front of a shop but lower time than a threshold.
- Buying: localization is key feature to determine that person is buying inside a shop or hypermarket.

These high level activities will help us to elaborate automatic reports in a short future work. As we stated before, localization information is a key feature to determine when a change in activity is detected. This information joined with accelerometers can be crucial to infer the pattern of activities as we will shown in the next sections.

4.3 User Behavioural Analytics Results

In this section we describe how we apply process mining techniques to the activity and localization data obtained and described in the previous sections, aiming to discover and analyze the behaviour of the mall customers. The analysis starts with an event log containing the activities performed by all the customers—formed by multiple executions of the 6 paths described in Table 1. Later, we discover a graph composed by these activities—nodes—and the relations—directed arcs—between them—i.e., a process model. Finally, using this process model, the users behavioral analytics conducted are extracted to understand better how they behave. Among these analytics, a search of commonly executed behavioral activity patterns—sets of activities, and the relations between them—is performed.

Process mining is a research field offering techniques to discover, analyze and enhance processes inside an organization [2]. In this case, we consider the behavior of the customers in the mall as the process to analyze, and the activities that each user carries out while in the mall are the execution of the process. The starting point is an event log storing the activities each customer performs, the instant in which they have been executed, and the customer ID. This event log has been generated with the information collected by the smartphone application, described in Sect. 4.1, which has recorded the activities of the customers inside the mall. With this event log, a process discovery algorithm is applied to discover the process model depicting the behavior of the customers (Fig. 4).

As discovery algorithm we used ProDiGen [8]. ProDiGen is a genetic discovery algorithm searching for complete, precise and simple process models. The result of applying ProDiGen to the recorded event log is depicted in Fig. 5, where the behavior of the customers in the mall can be observed.

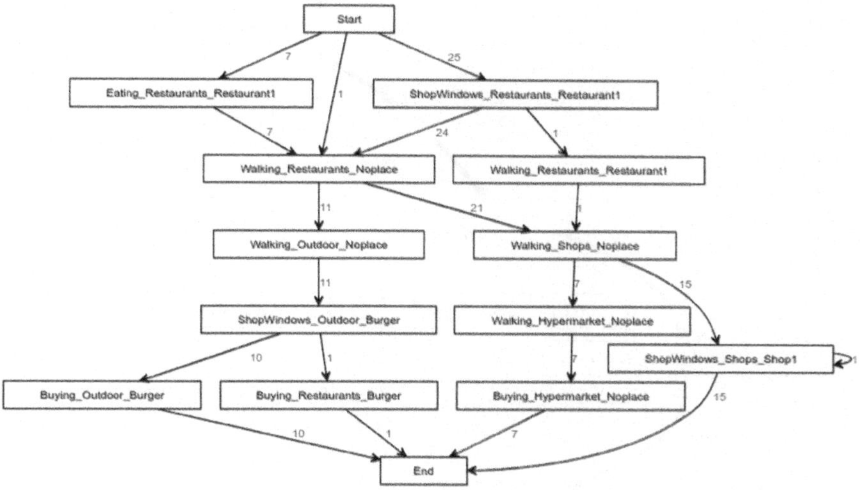

Fig. 5. Process model discovered by ProDiGen [8] depicting the behavior of the customers in the mall.

With this model some information about the behavior of the customers can be extracted: the customers who walk through the restaurants zone, go outdoors, and stop in the front of the shop window of the burger, mostly buy in the outdoor burger. To extract more information, a simplification in the event log can be done removing the paths which can be considered as 'noise', i.e. those executed only by one customer.

Figure 6 shows the process model of the customers behavior without the 'noise' of those single executions, and the relations between the activities of this model colored depending on their frequency. We can see that the most common actions are *i)* start in the shopwindow of Restaurant 1, *ii)* go walking from there through the restaurants zone, and *iii)* walk from the restaurant to the shops zone. Nevertheless, this information does not ensure a causality between these events. For instance, we cannot know if the customers who walk in the shops zone come from activity `Eating_Restaurants_Restaurant1` or from `ShopWindows_Restaurants_Restaurant1`. This information is crucial in the decision making step to select the best decisions to improve the marketing campaign inside the mall.

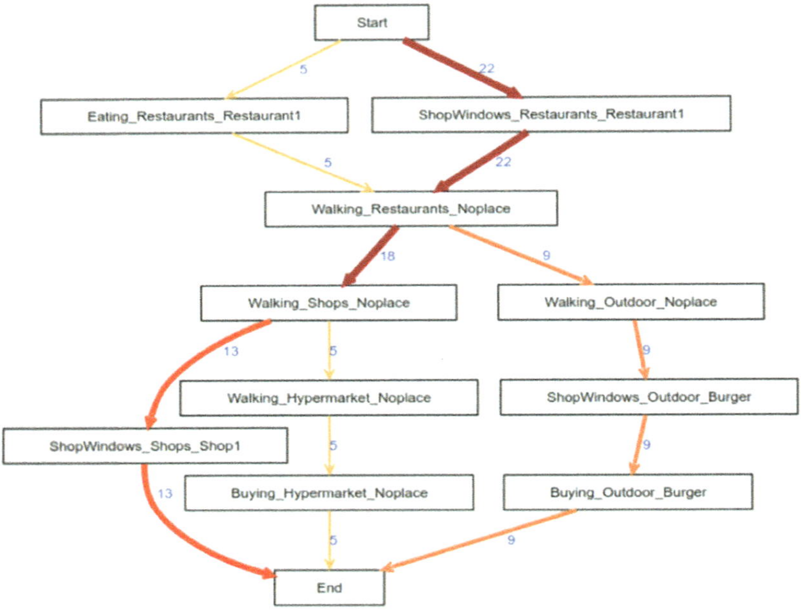

Fig. 6. Process model, discovered with the simplified event log, and with the arcs highlighted depending on their absolute frequency—thicker and a darker color implying a higher frequency.

To obtain this information, WoMine [4] has been used to extract the frequent activity patterns modeling the behavior commonly executed by the users. WoMine is an algorithm that searches, in the process model, for frequent activity

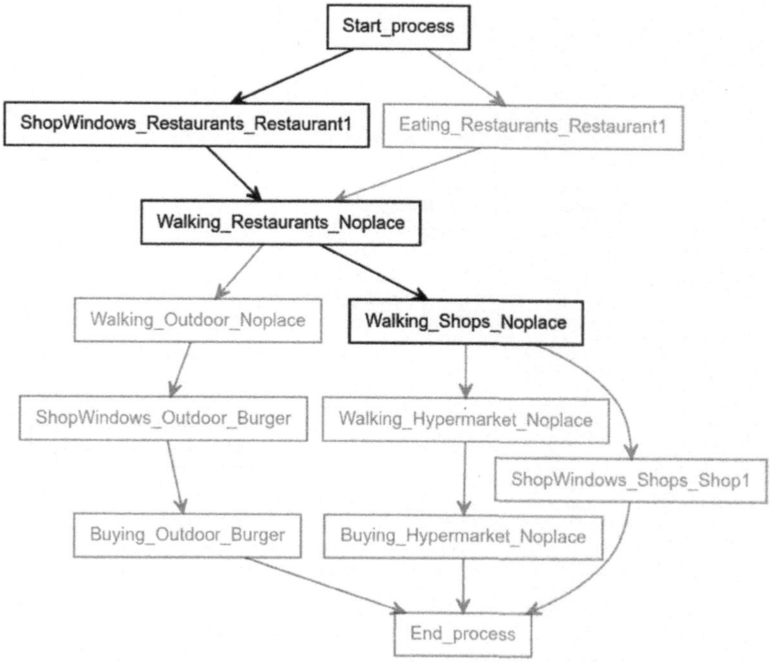

Fig. 7. Frequent pattern (highlighted structure) modeling the behavior performed by a 49% of the customers.

patterns—subgraphs of the process model ensuring all the structure it models has been executed frequently. Figure 7 highlights a behavioral pattern with a frequency of 49%, extracted with WoMine. This behavior has been conducted by the 49% of the recorded customers, who have started in the shopwindow of the Restaurant 1, walking through the restaurants and the shops either ending buying in the hypermarket or passing through the shopwindow of Shop 1. This can hint the need of Restaurant 1 to make changes in order to engage more of the customers who stop at its shopwindow but leave to other areas, without entering in it. Figure 8 depicts another behavioral pattern being executed also by the 49% of the customers. This pattern shows that half of the customers pass walking through the restaurants and shops zones, leaving the mall after staring to the shopwindow of Shop 1. This hints that it might be also interesting to work on the ads at the more frequented zones, Restaurants and Shops, to deal with those customers who leave the mall without consuming anything.

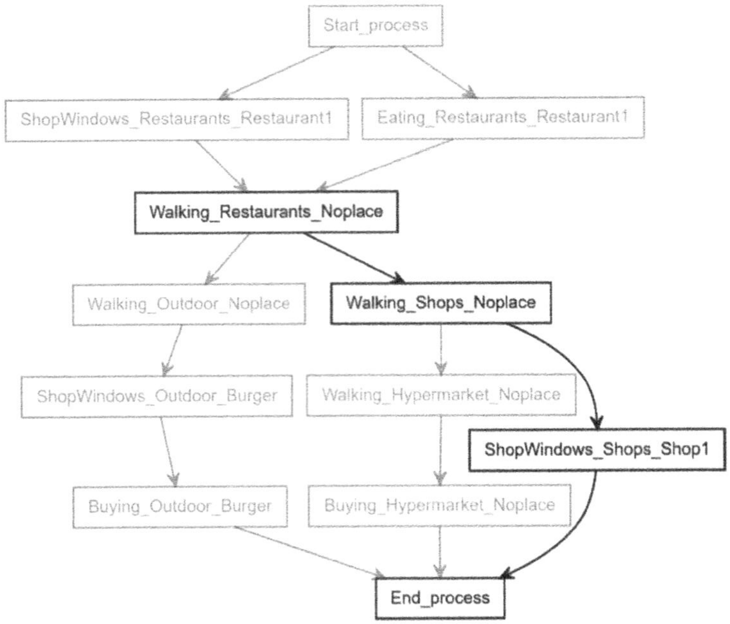

Fig. 8. Frequent pattern (highlighted structure) modeling the behavior performed by a 49% of the customers.

5 Conclusions and Future Works

We have presented a method to estimate the activity patterns made by shoppers in open malls based on localization data. Such data are provided by a specific smartphone application for logging information from sensors and process mining techniques to identify what kind of usual activity patterns are made by the shoppers. We use the localization key information joined with data mining, parallelization and monitoring techniques.

Our system has been tested in a challenging real scenario such as open mall (mixed classical mall and isolated shops in the street) where location is obtained by mean of Global Positioning System (GPS) and a WiFi localization system that is helped by a parallel computing method to speed up the whole process. From these localization data, we apply state of the art process mining techniques to first generate a process model (social workflow) describing the behavior made by the shoppers in the mall. We have also analyzed the most common activity patterns performed by the users. The hints provided by these process mining techniques may help mall's managers to take specific actions considering the customers behaviour and the activities they perform while in the mall.

As future work, we will address the integration of the process-to-text module, which automatically describes in natural language the most relevant characteristics of the customers behaviour, providing complementary information to the one conveyed by the visualization of the process models.

Acknowledgments. This work has been funded by the Cátedra de Ingeniería Avanzada Escriba-no of the UAH under Catedra2017-005, partially supported by the Spanish Ministry of Science, Innovation and Universities and the European Regional Development Fund (ERDF/FEDER program) through grants RTI2018-099646-B-I00, TIN2017-84796-C2-1-R, TIN2017-90773-REDT, RED2018-102641-T, and RYC-2016-19802 (Programa Ramón y Cajal, José M. Alonso). Also by the Galician Ministry of Education, University and Professional Training and the ERDF/FEDER program (ED431F2018/02, ED431C2018/29, ED431G2019/04 grants).

References

1. van der Aalst, W., et al: Process mining manifesto. In: Business Process Management Workshops, pp. 169–194. Springer, Heidelberg (2012)
2. van der Aalst, W.M.P., et al: Process mining manifesto. In: Daniel, F., Barkaoui, K., Dustdar, S. (eds.) Business Process Management Workshops - BPM 2011 International Workshops, Clermont-Ferrand, France, 29 August 2011, Revised Selected Papers, Part I. Lecture Notes in Business Information Processing, vol. 99, pp. 169–194. Springer (2011). https://doi.org/10.1007/978-3-642-28108-2_19
3. Bouziane, H.L., Pérez, C., Priol, T.: Extending software component models with the master-worker paradigm. Parallel Comput. **36**(2), 86–103 (2010). https://doi.org/10.1016/j.parco.2009.12.012. http://www.sciencedirect.com/science/article/pii/S0167819109001343
4. Chapela-Campa, D., Mucientes, M., Lama, M.: Mining frequent patterns in process models. Inf. Sci. **472**, 235–257 (2019). https://doi.org/10.1016/j.ins.2018.09.011
5. Google's Activity Recognition API. https://developers.google.com/location-context/activity-recognition
6. Görg, S., Bergmann, R.: Social workflows-vision and potential study. Inf. Syst. **50**, 1–19 (2015). https://doi.org/10.1016/j.is.2014.12.007. http://www.sciencedirect.com/science/article/pii/S030643791400194X
7. Hernández, N., Ocaña, M., Alonso, J.M., Kim, E.: Continuous space estimation: increasing wifi-based indoor localization resolution without increasing the site-survey effort. Sensors **17**(1), 147–170 (2017). https://doi.org/10.3390/s17010147
8. Vázquez-Barreiros, B., Mucientes, M., Lama, M.: ProDiGen: mining complete, precise and minimal structure process models with a genetic algorithm. Inf. Sci. **294**, 315–333 (2015). https://doi.org/10.1016/j.ins.2014.09.057

Train Here, Drive There: Simulating Real-World Use Cases with Fully-Autonomous Driving Architecture in CARLA Simulator

Carlos Gómez-Huélamo[1(✉)], Javier Del Egido[1], Luis M. Bergasa[1], Rafael Barea[1], Elena López-Guillén[1], Felipe Arango[1], Javier Araluce[1], and Joaquín López[2]

[1] Electronics Department, University of Alcalá (UAH), Madrid, Spain
{carlos.gomezh,javier.egido,juanfelipe.arango,javier.araluce}@edu.uah.es,
{luism.bergasa,rafael.barea,elena.lopezg}@uah.es
[2] Department of Systems Engineering and Automation, University of Vigo, Pontevedra, Spain
joaquin@uvigo.es

Abstract. This work presents the validation of our fully-autonomous driving architecture in the CARLA open-source simulator, by using some challenging driving scenarios inspired on the CARLA Autonomous Driving Challenge (CADC), focusing on our decision-making layer, based on Hierarchical Interpreted Binary Petri Nets (HIBPN). First, our ROS (Robot Operating System) based autonomous driving architecture is introduced. Second, the CARLA simulator is described, outlining the steps conducted to merge our architecture with this simulator and the advantages to create ad-hoc driving scenarios for use cases validation. Finally, the paper validates the architecture by means of some challenging driving scenarios such as: Stop, Pedestrian Crossing, Adaptive Cruise Control (ACC) and Unexpected Pedestrian. Some qualitative (video files) and quantitative (trajectory and linear velocity segmented with its corresponding Petri Net states) results are presented for each use case, validating our architecture in simulation as a preliminary stage before implementing it in our real autonomous electric car.

Keywords: CARLA · Autonomous vehicles · ROS · Simulation · Decision-making · Use cases

1 Introduction

Autonomous vehicles (AV) is one of the most challenging engineering tasks of our era. These AVs are expected to be driven throughout a highly dynamic environment with a reliability greater than human beings and full autonomy. In order to perform this task, an AV must be equipped with robust algorithms and

L. M. Bergasa et al. (Eds.): WAF 2020, AISC 1285, pp. 44–59, 2021.
https://doi.org/10.1007/978-3-030-62579-5_4

multiple sensors in such a way the vehicle is able to perceive the surrounding environment in real-time with a high precision, estimating the position of the objects with a very low error in the 3D/BEV (Bird's Eye View) space. However, making use of the object detection alone is not good enough for an AV to navigate in such complex environments, but the vehicle must also be able to track these obstacles along the scene to estimate their velocity and predict their future positions, also known as scene prediction. In that sense, the perception layer of an intelligent vehicle can be divided into three sequential stages: Detection, Multi-Object Tracking and Scene Prediction stage. The perception layer, based on the combination of different perception sensors and tracking algorithms, must be able to provide the most probable future behaviours, velocity and position of the main objects in the scene. This, combined with the control and mapping/planning layer information, feeds the decision-making layer of the ego-vehicle in order to successfully conduct some common urban use cases such as predicting if a pedestrian will start to cross a certain pedestrian crossing, predict the presence of an adversary vehicle in the next give-way or even when the ego-vehicle is carrying out the Adaptive Cruise Control (ACC), where it must adjust the vehicle speed to maintain a safe distance from ahead vehicles.

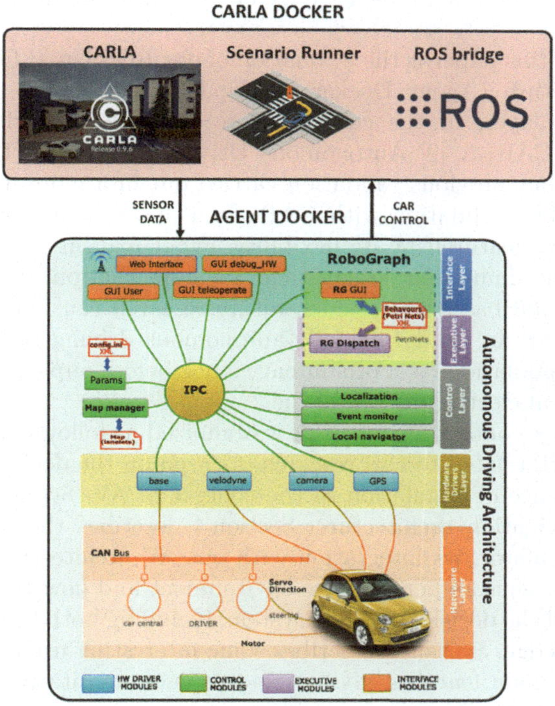

Fig. 1. Our autonomous driving architecture under CARLA simulation. GUI = Graphical User Interface; HW = Hardware; RG = RoboGraph; GPS = Global Positioning System; ROS = Robot Operating System; LiDAR = Light Detection and Ranging; IPC = Inter-Process Communication

In spite of all impressive efforts in the development of autonomous driving technology [8], fully-autonomous navigation in arbitrarily complex environments is still years away. The reason for this is two-fold: Firstly, informed decision-making requires accurate perception. Most of the existing perception systems produce errors at a rate not acceptable for autonomous driving [17]. Secondly, autonomous systems that are operated in complex dynamic environments require intelligent systems that generalizes to unpredictable situations in a timely manner.

An AV must be able to carry out driving decisions based on the offline and online information processed by the vehicle. In that sense, the offline information can be identified as the prior knowledge of the system, such as the topological relations and geographic information of the environment based on high-definition maps [26], the vehicle dynamic and the traffic rules based on behavioural decision-making systems [21]. On the other hand, the online information, also known as the traffic situation, is obtained through the global perception system of the vehicle, which involves different on-board sensors (like Inertial Measurement Unit (IMU)), Light Detection and Ranging (LiDAR), Global Positioning System (GPS), Steering wheel angle and Cameras). Traditionally, autonomous driving systems break down hierarchically into four main components [29]: Route Planning, Executive Layer, Motion Planning and Vehicle Control.

The scope of this paper is the evaluation of our ROS based fully-autonomous driving architecture (Agent Docker in Fig. 1), focusing on the behavioural decision-making layer, in the context of some challenging driving scenarios inspired in the CARLA [9] Autonomous Driving Challenge. This work is the continuation of our previous validation carried out in a more limited simulator (V-REP [14]). The validation with CARLA, a novel open-source autonomous driving simulator, featured by its flexibility, hyper-realism and real-time working, reinforces the simulation design stage. We hope that our distributed system can serve as a solid baseline on which future research can build on to advance the state-of-the-art in validating fully-autonomous driving architectures using hyper-realistic simulation, as a preliminary stage before implementing the architecture in our real electric car prototype.

The remaining content of this work is organized as follows. The next section presents the work related to this research, focused on the decision-making layer and the importance of simulation in the context of AV. Section 3 presents our autonomous navigation architecture. Section 4 describes the main features of the CARLA simulator, its integration with out AV architecture, regarding the way in which the sensors perceive the environment and how this information is processed to feed the decision-making system, including both the traffic situation and prior knowledge. Section 5 describes some interesting use cases, including a table with their main features, giving rise to some quantitative and qualitative results to illustrate the proposed architecture performance. Finally, Sect. 6 deals with the future works and concludes the work.

2 Related Works

As mentioned in the previous section, a fully-autonomous driving architecture (L5 in the J3016 SA [35]) is still years away, mainly due to technical challenges, but also due to social and legal ones [22].

No industry organization has shown a ratified testing methodology for L4/L5 autonomous vehicles. The autonomous driving community gives a simple reason: although some regulations have been defined for these L4/L5 levels, simulation is critical to build safe AV. However, despite the fact that current automotive companies are very good at testing the individual components of the autonomous driving architecture, there is a need to test intelligent vehicles full of advanced sensors [33] and sharing information among them. In this context, artificial intelligence is increasingly being involved in processes such as detecting the most relevant objects around the car (deep learning based multi-object tracking systems), or evaluating the current situation of the vehicle to conduct the safest decision (e.g. deep reinforcement learning applied to behavioural systems). Moreover, it is important to consider the presence of sensor redundancy in order to reach safe navigation in such a way that the different sensors and associated algorithms must be integrated together, required to validate the whole system, not just the individual components.

Regarding urban environment complexity, the system must be tested in countless environments and scenarios, which would escalate the cost and development time exponentially with the physical approach. Regarding this, the use of photo-realistic simulation (virtual development and validation testing) and an appropriate design of the driving scenarios are the current keys to build safe and robust AV. We propose the use of CARLA as the best open-source tool to reach these goals.

On the other hand, behavioural decision-making layer must provide tools to model the sequence of events and action, based on some predefined traffic rules, that can take place in the different traffic scenarios. In terms of AV, different approaches have been proposed to design the decision-making layer, including different heuristic solutions [36] based on identifying a set of driving scenarios or driving contexts (e.g. intersection handling or lane driving), reducing the number of environmental features to which the vehicle must be focused, according to each driving context.

The design of the decision-making layer for AV is challenging due to uncertainty in the knowledge about the driving situation and the state of the vehicle. This uncertainty comes from different sources, such as the estimation of the continuous state of nearby external agents, like other vehicles or pedestrians, whose behaviour is usually unpredictable. Hence, in order to design an optimal decision system uncertainty must be considered. Despite the fact that Partially Observable Markov Decision Processes (POMDP) [20] offer a framework to manage uncertainty, they are not scalable to real-world scenarios because of the associated complexity. Other approaches tackle this layer using simple discrete events systems, which are not enough complex to model real-world driving scenarios. Decision Trees (DT) [11], Hierarchical Finite-State Machines (HFSM) [2] and

Finite-State Machines (FSM) [25] are among the most popular approaches to design a decision making system. Moreover, several teams [21] in the Defense Advanced Research Projects Agency (DARPA) Urban Challenge [7] used some of these approaches.

Nevertheless, for complex problems in which common urban driving scenarios are included, the difficulty in representing the system as FSM lies in dealing with the need to implement the potentially high number of transitions due to the state explosion problem. Behaviours Trees (BT) can get rid of these drawbacks since the transitions among the states are implicit to their control structure formulated as a tree, giving rise to a higher flexibility, maintainability and extensibility with respect to FSMs.

However, BTs and simple discrete event systems share a common problem: A naive implementation usually gives rise to blocking behaviour, which make them unsuitable for autonomous driving. In that sense, concurrency and parallel activities can be easily programmed using Petri Nets (PN). PNs are a powerful tool to design, model and analyze concurrent, sequential and distributed systems. While in a FSM there is always a single current state, in PNs there may be several states that can change the state of the PN. In particular, in this work we model every behaviour as Hierarchical Interpreted Binary Petri Net (HIBPN) [13,14], as shown in Fig. 2(b), where a PN can start/stop another PN depending on its hierarchy.

3 Autonomous Navigation Architecture

Our autonomous navigation framework (Fig. 1) is featured by a modular architecture where individual modules process the information in an asynchronous way. These modules are standalone processes that communicate each other using the ROS Inter-Process Communication (IPC) system. The publish/subscribe concept is used to provide non-blocking communications. These software modules are organized in four layers as following:

- **Hardware drivers layer** Set of programs that implements different hardware devices, such as sensors and actuators.
- **Control layer** Set of programs that implements the vehicle control and navigation functionality. These programs include the path planning (map manager), localization, reactive control (local navigator) and a program to process most of the perception sensors to detect and track relevant events (event monitor).
- **Executive layer** Set of programs that coordinates the sequence of actions required to be executed to perform the current behaviour.
- **Interface layer** Set of processes that interact with the user and enable the communication to other processes for multi-robot applications.

The motion control is broken down into lower-level (LL) reactive control and high-level (HL) planning. First of all, the map manager (control layer) loads the map made up by a sequence of lanelets [4], in which the user must specify the

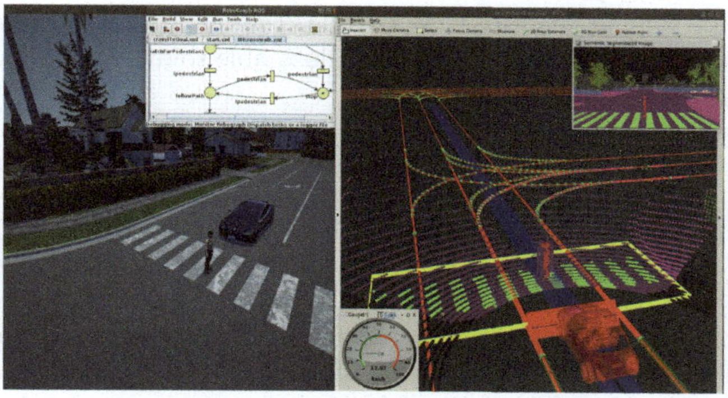

(a)

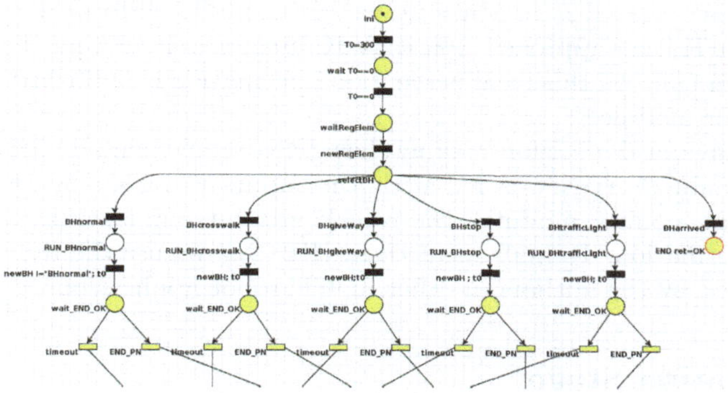

(b)

Fig. 2. (a) Simulation example: on the left, the CARLA simulator illustrates the urban scenario our sensors face. On the right, the RVIZ simulator shows the detection in the coloured point cloud. (b) Start Petri Net of our behaviour decision-making system.

start and goal position of the route. Then, the lanelet path is generated using an A* algorithm [6]. Besides this, the map manager serves other queries from other modules related to the map. For example, it must provide the most relevant lanes around the vehicle, such as intersection lanes or contiguous lanes for Stop and Overtaking use cases respectively, or providing the position of the regulatory elements in the route. Finally, a global path planner calculates a suitable path to be followed by the car. The goal of the local navigation module is to safely follow this path, keeping the vehicle within the driving lane and modifying the vehicle dynamic when was required by the behaviour constraints established by the HL planning. In order to do that, the local navigation system calculates the curvature to guide the car from its current position to a look-at-head position placed in the center of the lane using the Pure Pursuit algorithm [19]. This curvature is used as the reference for the reactive control, where an obstacle avoidance method is implemented based on the Beam Curvature Method (BCM) [12]. Combining both the Pure Pursuit algorithm and BCM, the local navigation system keep the vehicle centered in the driving lane while is able to avoid unknown obstacles which can partially block the lane.

In terms of environment perception, we perform a sensor fusion [14] between LiDAR and camera in order to detect and track the most relevant objects in the scene. First, a semantic segmentation is performed in the RGB image taken from the camera to detect the different obstacles and the driving area. In our real-world vehicle, we use our Efficient Residual Factorized ConvNet (ERFNet) [31] for real-time semantic segmentation, while in simulation, CARLA already provides a sensor including the semantic segmentation of the scene. We merge the 3D LiDAR point cloud and the 2D segmented pixels so as to obtain a coloured point cloud, where the points out of the FoV of the camera are not coloured. Then, we carry out a coloring clustering, obtaining the most relevant objects in the scene, as shown in Fig. 2(a). Multi-Object Tracking is performed by combining the Precision Tracker approach [16], BEV Kalman Filter [18] and Nearest Neighbour algorithm [3]. All this local perception information is published through the event monitor module.

The behavioural decision-making has been implemented using HIBPNs, where the main Petri Net is fed with the inputs provided by the local perception (event monitor module), the vehicle odometry in the map and the map manager information. To implement these HIBPNs, we use the RoboGraph tool [21], employed by the authors in other mobile robot applications.

4 Simulation Stage

As commented in previous sections, virtual testing is becoming increasingly importance in terms of AV. Since the urban environment is highly complex, the navigation architecture must be tested in countless environments and scenarios, which would escalate the cost and development time exponentially with the physical approach. Some of the most used simulators in the field of AV are Microsoft Airsim [34], recently updated to include AV although it was initially designed for drones, NVIDIA DRIVE PX [5], aimed at providing AV and driver assistance functionality powered by deep learning, ROS development studio [24], fully based on the Cloud concept where a cluster of computers allows the parallel simulation of as many vehicles as required, V-REP [30], with an easy integration with ROS and a countless number of vehicles and dynamic parameters and CARLA [9], which is the newest open-source simulator for AV based on Unreal engine. In [14,28] we validated the proposed navigation architecture using the V-REP simulator cause our previous experience in this simulator. However, the paradigms of real-time and high realism could not be deeply analysed due to V-REP is not designed to create high realistic urban driving scenarios, so the analysis of traffic use cases is hard due to the unrealism and slowness of the simulator environment. Since our project is characterized by being open-source, we decided to integrate our autonomous driving architecture with the open-source CARLA simulator. This offers a much more interesting environment in terms of traffic scenarios, perception, real-time and flexibility, which are key concepts for our algorithms. In this work we have used the 0.9.6 version of CARLA, so the CARLA ROS bridge and the Scenario Runner. Additional tools needed for the simulation, were configured according to this version.

In [14] we used the simulation environment, the V-REP ROS bridge and the autonomous navigation architecture in the same host machine, running on Ubuntu 16.04. For this work, we decided to split them into two Ubuntu 18.04 Docker [23] images, the CARLA world and the CARLA ROS bridge in one image and the autonomous navigation architecture in the other one, so as improve the portability, flexibility and isolation of our work.

4.1 Environment

CARLA is implemented as an open-source layer over Unreal Engine 4 (UE4) [32]. This simulation engine provides CARLA an hyper-realistic physics and an

Table 1. Summary of the main features, including inputs, outputs and involved modules, for the main Petri Nets of our work. **MM** = Map Manager; **EM** = Event Monitor; **LN** = Local Navigator; **RGD** = RoboGraph Dispatch; **PC** = Pedestrian Crossing

Petri net	Inputs	Input modules	Outputs	Output modules	Nodes/ transitions
Background	Man/auto	GUI User	Run Selector	RGD	8/9
	goToPoint	GUI User	Stop selector	RGD	
Selector PN	Reg. Element	MM	Run PC	RGD	21/29
	Dist. Reg Element	MM	Stop PC	RGD	
	End Reg. Element	MM	Run GiveWay	RGD	
	Reg. Element	EM	Stop GiveWay	RGD	
	End Reg. Element	EM	Run Stop	RGD	
	FrontCarVel	EM	...	...	
	Odom	Base			
Pedestrian Crossing	NoPedestrian	EM	WatchforPedetrians	EM	10/13
	Pedestrian	EM	SetMaxVel	LN	
	Dist. To PC	MM	StopAtPoint	LN	
	PC Over	MM			
	Force End	RGD			
	Stopped	LN			
Stop	SafetoMerge	EM	CheckSafeMerge	EM	9/12
	NotSafetoMerge	EM	SetMaxVel	LN	
	DistToStop	MM	StopAtPoint	LN	
	StopOver	MM			
	Force End	RGD			
	Stopped	LN			
Adaptive Cruise Control (ACC)	Current Velocity	MM	SetMaxVel	LN	4/6
	FrontCarVel	EM			
	DistToFrontCar	MM			

ecosystem of interoperable plugins. In that sense, CARLA simulates a dynamic world and provides a simple interface between an agent that interacts with the world and itself. In order to support this functionality, CARLA was designed as a server-client system, where the simulation is run and rendered by the server. CARLA environment is made up by 3D models of static objects like infrastructure, buildings or vegetation, as well as dynamic objects such as vehicles, cyclists or pedestrians. They are designed using low-weight geometric models and textures but maintaining visual realism due to a variable level of detail and carefully crafting the materials. On the other hand, the maps provided by CARLA are in OpenDrive [10] format, whilst we use the lanelet approach based on OpenStreetMap (OSM) [15] service. We transform the OpenDrive format into OSM format by using the converter proposed by [1]. Then, based on this lanelet map, we manually include the regulatory traffic information to generate an enriched topological map useful for navigation. Furthermore, the OSM map uses WGS84 coordinates (latitude, longitude and height) whilst the simulator provides Cartesian coordinates (UTM) relative to an origin (usually the center of the map). Then, geometric transformations between both systems are calculated using the libraries implemented in the ROS geodesy package.

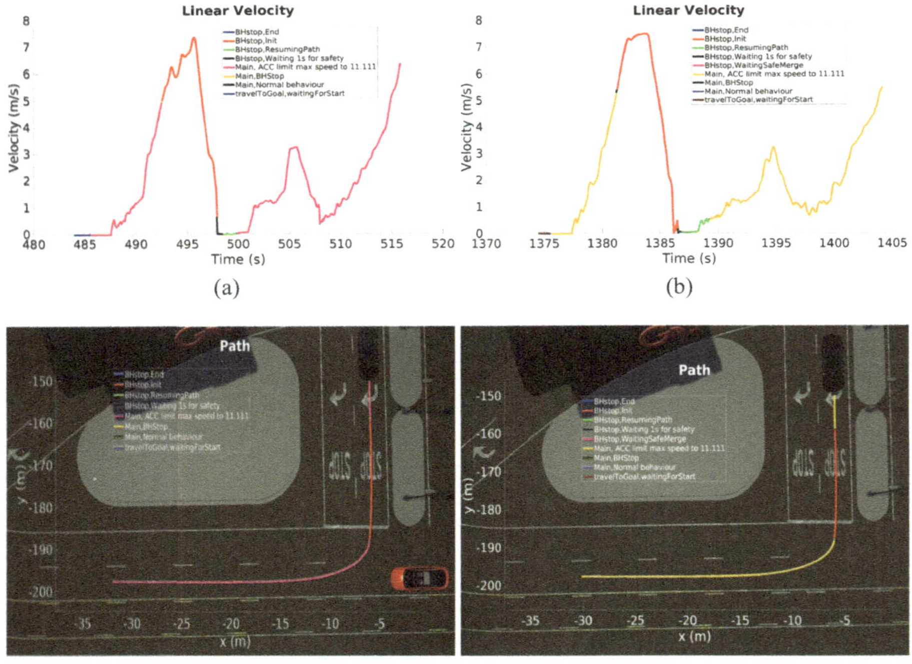

Fig. 3. First and second row represent, respectively, the linear velocities and described trajectory projected onto the corresponding CARLA scenario for Stop with car detection (a, c) and Stop with no car detection (b, d) behaviours

4.2 Vehicle

In order to link the vehicle in CARLA and its corresponding model in RVIZ (Fig. 2(a)), we modify the ROS bridge associated to the CARLA simulator. The CARLA ROS bridge is a ROS package that aims at providing a bridge between CARLA and ROS, being able to send the data captured by the on-board sensors and other variables associated to the vehicle in the form of topics and parameters understood by ROS. In that sense, we modify some parameters of the speed and acceleration PID, the sensors position and orientation, the Ackermann control node and the waypoint publisher to adjust the CARLA ego-vehicle parameters to our project. The reference system of the vehicle is centered on the rear axle, and the orientation of the frame is based on the LiDAR frame, that is, X-axis pointing to the front, Y-axis to the left and Z-axis above. The angular velocity at Z is positive according to the right hand rule, so the right turns are considered negative. The speed commands generated by the simulator are interpreted by the local navigation system (in particular, the low-level controller), giving rise the corresponding accelerations and turning angle.

4.3 Sensors

From the sensors perspective, the agent sensor suite can be modified in a flexible way. Most common sensors in CARLA world are LiDAR, GPS and RGB cameras as well as their corresponding pseudo-sensors that provide semantic segmentation and groundtruth depth. Moreover, camera parameters include 3D orientation and position with respect to the car coordinate system, field-of-view and depth of field. The original configuration of the on-board sensors of the ego-vehicle had a 800×600 RGB camera sensor placed at the front of the vehicle with a FoV of 100°, a 32-channels LiDAR, a GPS sensor, a collision sensor and a lane invasion sensor. In our case, the position of the LiDAR and camera sensors are manually configured in order to obtain the same frames distribution that in our real-world vehicle. Based on the bridge, the LiDAR information can be published in PointCloud2 ROS format with the Z axe pointing up, Y left and the X inwards. Image messages are published with the Z axe inwards, the Y down and the X pointing right in the image plane. We take advantage of the semantic segmentation provided by CARLA so as to colour the LiDAR point cloud and perform the coloring clustering, placing the pseudo-sensor at the same position and orientation that the RGB camera sensor, both with a resolution of 1280×720, since it is the resolution of the ZED camera equipped in our real-world electric car. Moreover, the position of the GPS sensor is displaced to the center of the rear axle, required by the local navigation system. The remaining sensors keep unmodified. As observed, CARLA provides a straightforward way to add or remove sensors from the vehicle or even modify their parameters, to adjust the simulation to the real-world as best as possible.

5 Experimental Results

One of the best advantages of CARLA is the possibility to create ad-hoc urban layouts, useful to validate the navigation architecture in challenging driving scenarios. This code can be downloaded from the Scenario Runner repository, associated to the CARLA GitHub. This repository offers an execution engine for CARLA and traffic scenario definition. The scenarios are inspired in the CARLA Autonomous Driving Challenge (CADC), selected from the NHTSA pre-crash typology [27]. The features of these scenarios can be defined through a Python interface, such as the presence of additional obstacles, start condition for the adversary (e.g. pedestrian or vehicle), position and orientation of the dynamic obstacles in the environment, weather, etc. In this work we study the linear velocity and the odometry of the ego-vehicle throughout the following use cases: Stop with car detection, Stop with no car detection, Pedestrian Crossing, Adaptive Cruise Control (ACC) and Unexpected Pedestrian. A video demo for each of them can be found in the following play list Simulation Use Cases[1]. Due to size constraints, we refer the readers to [14] for a deeper explanation of the behaviour that the ego-vehicle must carry facing to the corresponding traffic scenario. Nevertheless, Table 1 shows the main HIBPNs features used to manage the logic of these use cases(inputs, input modules, outputs, output modules, number of nodes and number of transitions).

The resulting graphics are segmented using different colours, corresponding to the current node of the associated PN. Both Stop use cases are inspired in the Traffic Scenario 09 (Right turn at an intersection with crossing traffic) of the CADC, demonstrating a similar behaviour since the logic performed by the PNs is the same. It is appreciated that the ego-vehicle waits 2.3 s in front of the Stop line in the use case of Stop with no detection (Fig. 3(b), Fig. 3(d)). This time is approximately the same that the ego-vehicle waits with the presence of an adversary vehicle (Fig. 3(a), Fig. 3(c)). This behaviour is coherent, since the Stop PN presents a transition that requests 0 m/s for the ego-vehicle in addition to 1 extra s to wait for safety, in the same way that a real-world car would do. Regarding the Pedestrian Crossing use case (Fig. 4(a), Fig. 4(c), it is inspired in the Traffic Scenario 04 (Obstacle avoidance with prior action) of the CADC. As observed in Fig. 2(a), in this work we define an additional safety area (Yellow rectangle in RVIZ simulator) around the pedestrian crossing, in order to activate the pedestrian detection, carried out by the event monitor, in all this area, including the sidewalks.

While V-REP does not offer the possibility to configure a variable velocity for the adversary, being the ACC limited to decrease the ego-vehicle velocity till the adversary fixed velocity [14], we take advantage of the possibilities offered by CARLA to configure the behaviour of the adversary in this use case. In that sense, Fig. 6(a) shows the linear velocity under the effects of the ACC in blue, being the linear velocity of the ego-vehicle adjusted to the variable adversary linear velocity. Moreover, Fig. 6(c) shows how the distance is decreased till 22/23

[1] Simulation Use Cases link: https://cutt.ly/prUzQLi.

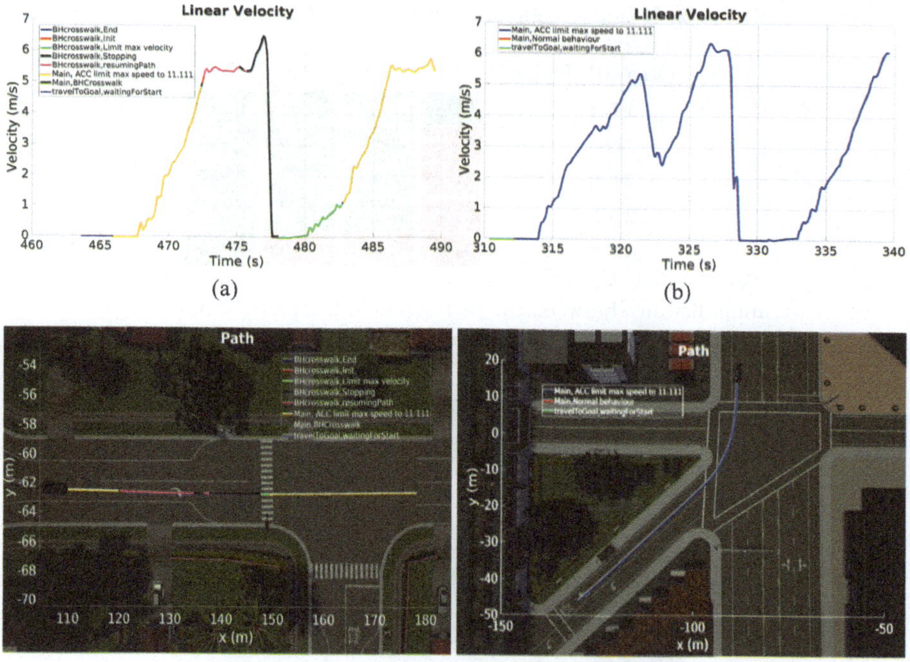

(a) (b)

(c) (d)

Fig. 4. First and second row represent, respectively, the linear velocities and described trajectory projected onto the corresponding CARLA scenario for Pedestrian Crossing (a, c) and Unexpected Pedestrian (b, d) behaviours

m, where the ACC is kept until the traffic scenario is concluded. It can be observed that the ACC behaviour starts at the exact moment in which the adversary vehicle is detected. This new version of our ACC analysis is inspired in the Traffic Scenario 02 (Longitudinal control after leading vehicle's brake) of the CADC.

The last use case that we validate in this paper is inspired in the Traffic Scenario 03 (Obstacle avoidance without prior action) of the CADC. In this traffic scenario, the ego-vehicle suddenly finds an unexpected obstacle on the road and must perform an avoidance maneuver or an emergency break. In our case, we design our own scenario in such a way an unexpected pedestrian jumps to the road, being totally occluded by a bus stop kiosk. As expected, Fig. 4(b) and Fig. 4(d) show that no high-level behaviour is launched because this situation is not included as an use case in the PN. However, our low-level (reactive) control performs an emergency break until the car is stopped in front of the obstacle, and resumes the navigation once the obstacle leaves the driving lane. In this way, we validate the reactive control of our architecture, always in background execution, in those cases in which none of the specific use cases PN are launched.

Table 2. Reactive control analysis on CARLA simulator

NV (km/h)/D (m)	12	16	20	24
14.4	6.75	11.41	15.20	18.73
18	6.17	10.46	13.75	17.16
21.6	5.31	9.07	12.89	16.58
25.2	3.48	8.72	12.15	15.24
28.8	2.66	8.39	11.78	13.12

NV: Nominal Velocity of the ego-vehicle (km/h). D: Minimum distance between the pedestrian and vehicle BEV centroid to allow the pedestrian start its trajectory

Fig. 5. Unexpected Pedestrian scenario. The pedestrian can be found behind the bus stop, so the perception systems detects it at the moment it is entering the road

This strategy shows a good working in unexpected situations due to the high frequent execution of our reactive control.

A robustness analysis of our reactive control is depicted in Table 2. We show some parameters of the obstacle avoidance maneuver for different jumping distances (D(m)) of the unexpected pedestrian and different nominal speeds of the ego-vehicle (NV(Km/h)). For each combination, the distance at which the reactive control detects the obstacle in the road is calculated (m). Green cells indicate no collision has taken place and read that the pedestrian has been overwhelmed. The *D* parameter represents the initial Euclidean distance between the adversary pedestrian and the ego-vehicle BEV centroid, as illustrated in Fig. 5. This distance corresponds with the initial condition of the pedestrian, placed at the sidewalk, 1.5 m away from the road. Whilst in the Pedestrian Crossing use case an additional safety area is considered to detect the presence of pedestrians, in this case the reactive control detects the pedestrian once it is inside the lane. After the detection a braking maneuver is performed. As expected, the faster the vehicle goes, the lower the distance at which the reactive control detects for the first time the pedestrian inside the lane. This is due to the ego-vehicle has travelled a greater distance, increasing the likelihood of colliding with the pedestrian, and the ego-vehicle must perform the emergency break in a shorter distance.

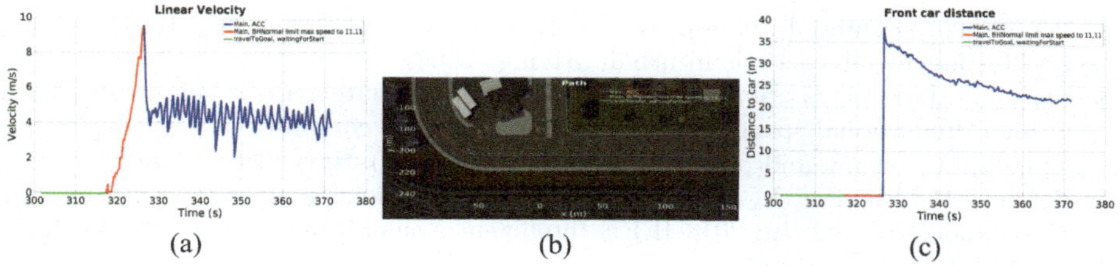

Fig. 6. Analysis of the ACC use case with variable adversary vehicle velocity. (a) represents the ego-vehicle linear velocity, (b) the ego-vehicle odometry projected onto the CARLA world, (c) analysis of the euclidean distance between the ego-vehicle and adversary throughout the use case

6 Conclusions and Future Works

This work presents the validation of our ROS-based fully-autonomous driving architecture, focusing in the decision-making layer, with CARLA, a hyper-realistic, real-time, flexible and open-source simulator for autonomous vehicles. The simulator and its bridge, in charge of communicating the CARLA environment with our ROS-based architecture, on the one hand, and the navigation architecture, on the other hand, have been integrated in two Docker images, in order to gain flexibility, portability and isolation. The decision-making is based on Hierarchical Interpreted Binary Petri Nets (HIBPN), and our perception is based on the fusion of GPS, camera (including semantic segmentation) and LiDAR. The validation has consisted on the study of some traffic scenarios inspired on the CARLA Autonomous Driving Challenge, such as Stop, Pedestrian Crossing, Adaptive Cruise Control and Unexpected Pedestrian, in addition to an analysis of our reactive control in terms of emergency brake without prior action. We hope that our distributed system can serve as a solid baseline on which others can build on to advance the state-of-the-art in validating fully-autonomous driving architectures using virtual testing. As future works, a deep learning based 3D Multi-Object Tracking will be implemented and multiple adversaries will be included in order to get more challenging situations to improve the reliability, effectiveness and robustness of our system, so as to validate the architecture and test it in our real-world electric vehicle.

Acknowledgment. This work has been funded in part from the Spanish MICINN/FEDER through the Techs4AgeCar project (RTI2018-099263-B-C21) and from the RoboCity2030-DIH-CM project (P2018/NMT- 4331), funded by Programas de actividades I+D (CAM) and cofunded by EU Structural Funds.

References

1. Althoff, M., Urban, S., Koschi, M.: Automatic conversion of road networks from opendrive to lanelets. In: 2018 IEEE International Conference on Service Operations and Logistics, and Informatics (SOLI), pp. 157–162. IEEE (2018)

2. Beeson, P., O'Quin, J., Gillan, B., Nimmagadda, T., Ristroph, M., Li, D., Stone, P.: Multiagent interactions in urban driving (2008)
3. Beis, J.S., Lowe, D.G.: Shape indexing using approximate nearest-neighbour search in high-dimensional spaces. In: Proceedings of IEEE computer society conference on computer vision and pattern recognition, pp. 1000–1006. IEEE (1997)
4. Bender, P., Ziegler, J., Stiller, C.: Lanelets: efficient map representation for autonomous driving. In: 2014 IEEE Intelligent Vehicles Symposium Proceedings, pp. 420–425. IEEE (2014)
5. Bojarski, M., Del Testa, D., Dworakowski, D., Firner, B., Flepp, B., Goyal, P., Jackel, L.D., Monfort, M., Muller, U., Zhang, J., et al.: End to end learning for self-driving cars. arXiv preprint arXiv:1604.07316 (2016)
6. Brandes, U.: A faster algorithm for betweenness centrality. J. Math. Sociol. **25**(2), 163–177 (2001)
7. Buehler, M., Iagnemma, K., Singh, S.: The DARPA Urban Challenge: Autonomous Vehicles in City Traffic, vol. 56. Springer, Heidelberg (2009)
8. Dickmanns, E.D., Mysliwetz, B., Christians, T.: An integrated spatio-temporal approach to automatic visual guidance of autonomous vehicles. IEEE Trans. Syst. Man Cybern. **20**(6), 1273–1284 (1990)
9. Dosovitskiy, A., Ros, G., Codevilla, F., López, A., Koltun, V.: CARLA: an open urban driving simulator. arXiv preprint arXiv:1711.03938 (2017)
10. Dupuis, M., Strobl, M., Grezlikowski, H.: Opendrive 2010 and beyond–status and future of the de facto standard for the description of road networks. In: Proceedings of the Driving Simulation Conference Europe, pp. 231–242 (2010)
11. Fernandez, C., Izquierdo, R., Llorca, D.F., Sotelo, M.A.: A comparative analysis of decision trees based classifiers for road detection in urban environments. In: 2015 IEEE 18th International Conference on Intelligent Transportation Systems, pp. 719–724. IEEE (2015)
12. Fernández, J.L., Sanz, R., Benayas, J., Diéguez, A.R.: Improving collision avoidance for mobile robots in partially known environments: the beam curvature method. Robot. Auton. Syst. **46**(4), 205–219 (2004)
13. Fernández, J.L., Sanz, R., Paz, E., Alonso, C.: Using hierarchical binary petri nets to build robust mobile robot applications: robograph. In: 2008 IEEE International Conference on Robotics and Automation, pp. 1372–1377. IEEE (2008)
14. Gómez-Huelamo, C., Bergasa, L.M., Barea, R., López-Guillén, E., Arango, F., Sánchez, P.: Simulating use cases for the UAH autonomous electric car. In: 2019 IEEE Intelligent Transportation Systems Conference (ITSC), pp. 2305–2311. IEEE (2019)
15. Haklay, M., Weber, P.: OpenStreetMap: user-generated street maps. IEEE Pervasive Comput. **7**(4), 12–18 (2008)
16. Held, D., Levinson, J., Thrun, S.: Precision tracking with sparse 3D and dense color 2D data. In: 2013 IEEE International Conference on Robotics and Automation (ICRA), pp. 1138–1145. IEEE (2013)
17. Janai, J., Guney, F., Wulff, J., Black, M.J., Geiger, A.: Slow flow: exploiting high-speed cameras for accurate and diverse optical flow reference data. In: Proceedings of the IEEE Conference on Computer Vision and Pattern Recognition, pp. 3597–3607 (2017)
18. Kalman, R.E.: A new approach to linear filtering and prediction problems (1960)
19. Ko, N.Y., Simmons, R.G.: The lane-curvature method for local obstacle avoidance. In: Proceedings. 1998 IEEE/RSJ International Conference on Intelligent Robots and Systems. Innovations in Theory, Practice and Applications (Cat. No. 98CH36190), vol. 3, pp. 1615–1621. IEEE (1998)

20. Kurniawati, H., Hsu, D., Lee, W.S.: SARSOP: efficient point-based pomdp planning by approximating optimally reachable belief spaces. In: Robotics: Science and Systems, Zurich, Switzerland, vol. 2008 (2008)
21. López, J., Sánchez-Vilariño, P., Sanz, R., Paz, E.: Implementing autonomous driving behaviors using a message driven petri net framework. Sensors **20**(2), 449 (2020)
22. Matthaeia, R., Reschkaa, A., Riekena, J., Dierkesa, F., Ulbricha, S., Winkleb, T., Maurera, M.: Autonomous driving: technical, legal and social aspects (2015)
23. Merkel, D.: Docker: lightweight linux containers for consistent development and deployment. Linux J. **2014**(239), 2 (2014)
24. Michal, D.S., Etzkorn, L.: A comparison of player/stage/gazebo and microsoft robotics developer studio. In: Proceedings of the 49th Annual Southeast Regional Conference, pp. 60–66. ACM (2011)
25. Montemerlo, M., Roy, N., Thrun, S.: Perspectives on standardization in mobile robot programming: the carnegie mellon navigation (CARMEN) toolkit. In: Proceedings 2003 IEEE/RSJ International Conference on Intelligent Robots and Systems (IROS 2003) (Cat. No. 03CH37453), vol. 3, pp. 2436–2441. IEEE (2003)
26. Murciego, E., Huélamo, C.G., Barea, R., Bergasa, L.M., Romera, E., Arango, J.F., Tradacete, M., Sáez, Á.: Topological road mapping for autonomous driving applications. In: Workshop of Physical Agents, pp. 257–270. Springer (2018)
27. Najm, W.G., Smith, J.D., Yanagisawa, M., et al.: Pre-crash scenario typology for crash avoidance research. Technical report, United States. National Highway Traffic Safety Administration (2007)
28. Otero, C., Paz, E., Sanz, R., López, J., Barea, R., Romera, E., Molinos, E., Arroyo, R., Bergasa, L., López, E.: Simulación de vehículos autónomos usando v-rep bajo ros (2017)
29. Paden, B., Čáp, M., Yong, S.Z., Yershov, D., Frazzoli, E.: A survey of motion planning and control techniques for self-driving urban vehicles. IEEE Trans. Intell. Veh. **1**(1), 33–55 (2016)
30. Robotics, C.: V-rep user manual. Ultimo acesso **13**(04) (2015). http://www.coppeliarobotics.com/helpFiles/
31. Romera, E., Alvarez, J.M., Bergasa, L.M., Arroyo, R.: ERFNet: efficient residual factorized convnet for real-time semantic segmentation. IEEE Trans. Intell. Transp. Syst. **19**(1), 263–272 (2018)
32. Sanders, A.: An Introduction to Unreal Engine 4. AK Peters/CRC Press (2016)
33. Schöner, H.: The role of simulation in development and testing of autonomous vehicles. In: Driving Simulation Conference, Stuttgart (2017)
34. Shah, S., Dey, D., Lovett, C., Kapoor, A.: AirSim: high-fidelity visual and physical simulation for autonomous vehicles. In: Field and Service Robotics, pp. 621–635. Springer (2018)
35. Taxonomy, S.: Definitions for terms related to driving automation systems for on-road motor vehicles (J3016). Technical report, Society for Automotive Engineering (2016)
36. Urmson, C., Anhalt, J., Bagnell, D., Baker, C., Bittner, R., Clark, M., Dolan, J., Duggins, D., Galatali, T., Geyer, C., et al.: Autonomous driving in urban environments: boss and the urban challenge. J. Field Robot. **25**(8), 425–466 (2008)

DQN-Based Deep Reinforcement Learning for Autonomous Driving

Óscar Pérez-Gil, Rafael Barea(✉), Elena López-Guillén, Luis M. Bergasa, Pedro A. Revenga, Rodrigo Gutiérrez, and Alejandro Díaz

Electronics Department, Universidad de Alcalá, Madrid, Spain
{o.perezg,rafael.barea,elena.lopezg,luism.bergasa,pedro.revenga, rodrigo.gutierrez,alejando.diazd}@uah.es

Abstract. The goal of this work is to evaluate the task of autonomous driving in urban environment using Deep Q-Network Agents. For this purpose, several approaches based on DQN agents will be studied. The DQN agent learn a policy (set of actions) for lane following tasks using visual and driving features obtained from sensors onboard the vehicle and a model-based path planner. The policy objective is to drive as fast as possible following the center of the lane avoiding collisions and road departures. A dynamic urban simulation environment will be designed using CARLA simulator to validate our proposal. The results show that a DQN agent could be a promising technique for self-driving a vehicle in a urban environment.

Keywords: Autonomous driving · Reinforcement learning · Convolutional neural network · Deep q-network agent · CARLA simulator

1 Introduction

In recent years, autonomous driving plays a pivotal role to solve traffic and transportations problems in urban areas (traffic congestions, accidents, etc) and it is going to change the way of travelling in our world in the future [1]. In the last decade, various challenges, such as the well-known DARPA Urban Challenge and the Intelligent Vehicle Future Challenge (IVFC) have proven that autonomous driving can be a reality in the near future. The teams participating in these events have demonstrated numerous technical frameworks for autonomous driving [2–5]. Nowadays, most self-driving vehicles are geared up with multiple high-precision sensors such as LIDAR and cameras. LIDAR-based detection methods provide accurate depth information and obtain robust results in location, object detection and scene understanding [6] while camera-based methods provide much more detailed semantic information [7]. Lately, Reinforcement Learning (RL) have been used to solve Markov Decision Problems (MDPs). This method tries to calculate the optimal policy of an agent to choose actions in an environment with the goal of maximize a reward function. The results obtained to

L. M. Bergasa et al. (Eds.): WAF 2020, AISC 1285, pp. 60–76, 2021.
https://doi.org/10.1007/978-3-030-62579-5_5

solve computer games or simple decision-making system have been very successful [8,9]. Regarding to autonomous driving, recently, Deep RL approaches have been developed to learn how to drive using sensory system onboard the vehicle [10,11].

In this paper, we are going to focus our study in the performance of Deep Q-Network agents for autonomous driving. The goal is to follow a route as fast as possible avoiding collisions and road departures in a dynamic urban simulation environment. The discrete nature of DQN makes a complex tune for a continuous problem like self-driving, due to the infinite possibles of movement by the car in each step. Studying DQN and the obtained results, we will be able to discuss whether this algorithm is the right one for this navigation purpose. On the other hand, an important part of this work, is the simulator that will be used, CARLA Simulator [12]. CARLA is a hyper-realistic simulator that helps in the development of navigation techniques and in the exportability of the simulated models to real systems. In addition, CARLA provides a bridge with ROS [13] for standardization, and a PythonAPI for easy programming. Both modules allow an easy connection between simulation and the algorithms implemented to control the vehicle. Finally, although working with a real vehicle is not the objective of this work at the moment (safety matters prevent these tests from being performed in real environments without a previous and exhaustive simulation stage), our final goal is that the developed control can be exported to our robotic platform.

2 Related Work

In the last years, several notable approaches to solve autonomous driving challenge have been proposed. These approaches can be classified in three main types: classical or mapping-based, imitation learning (IL) and reinforcement learning (RL).

a) **Classical methods:** Classical autonomous driving systems usually use advanced sensor for environment perception and complex control algorithms for safety navigation in complex scenarios. Typically, these frameworks use a modular architecture where individual modules process information asynchronously. The sensing module captures information from the surroundings using different sensors such cameras, LiDAR, GPS, IMUs, etc. Perception module plays a high-priority role and is responsible for calculating the position of the vehicle and recognizing and detecting the objects present in the surroundings of the environment. Decision module relies on perception module. Once the module understands the behaviour of the scene, it makes the appropriate decision according to the external conditions and the established objective [3,14].
b) **Imitation learning:** This approach tries to learn the optimal policy by following and imitating the expert's decisions. This way, an expert (typically a human) provides a set of driving data [15,16]. Using these labelled data,

the driving policy (agent) is easy to train in a supervised learning. The main advantage of this method is its simplicity and it achieves very good results in end-to-end applications. On the other hand, its main drawback is the difficulty of imitating every potential driving scene being unable to learn behaviors that have not been provided. This drawback causes this approach can be dangerous in some real driving situations that have not been observed.

c) **Reinforcement learning:** This approach is a type of machine learning technique that enables an agent to learn an interactive environment by trial and error using feedback from its own actions and observations.This approach have been successfully tested for solving Markov Decision Problems (MDPs). In recent years, it has been combined with deep learning techniques and have proved its potential to solve autonomous driving problems such as decision making and planning . Deep Reinforcement Learning (DRL) algorithms include: Deep Q-learning Network (DQN), Double-DQN, actor-critic (A2C, A3C), Deep Deterministic Policy Gradient (DDPG) and Twin Delayed DDPG (TD3). All these algorithms have been used to perform autonomous driving tasks, obtaining promising results [17–21].

3 Framework Overview

This section shows the proposed framework for autonomous driving based on DRL (Fig. 1). This approach is based on [17] where a hybrid of a model-based planner and a model-free DRL agent is proposed. The general architecture is based on CARLA simulator, which provides a hyper-realistic and dynamic urban simulation environments. CARLA provides a powerful API that allows the users to control all aspects related to the simulation and permits to configure diverse sensors for environment perception (cameras, LiDAR, GPS, etc). This way, CARLA provides environment and ego-vehicle data. With this data (monocular camera image, and vehicle position and speed), it is possible to obtain visual and driving features. This two set of features can be introduced in a Deep Learning Reinforcement Agent to generate control commands (actions).

4 Method

Autonomous driving tasks can be modeled as a Markov Decision Process (MDP) and a great amount of reinforcement learning algorithms have been developed recently for solving MDP [10,18]. Therefore, Our approach aims to develop a DRL agent that generates autonomous vehicle control actions.

4.1 MDP Formulation

A MPD consists of an agent that observes the state (s_t) of the ego-vehicle (environment state) and generates an action (a_t). This causes the vehicle to move to a new state (s_{t+1}) producing a reward ($r_t = R(s_t, a_t)$ based on the new observation. A Markov decision process is a 4-tuple (S, A, P_a, R_a) where the goal is to

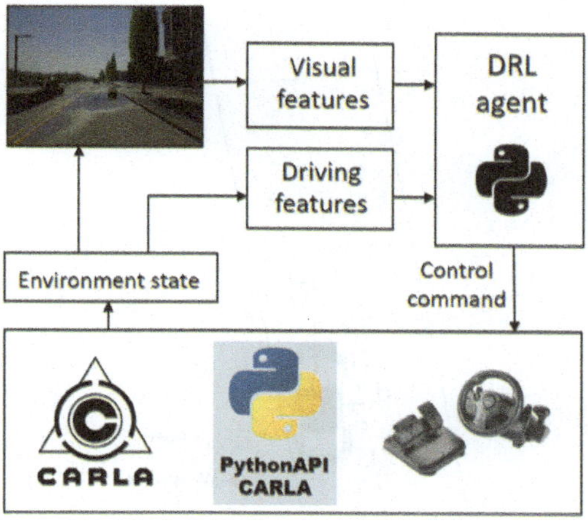

Fig. 1. Framework overview

find a good "policy", that is, a function $\pi(s)$ that the decision maker will choose when is in state s.

a) **State space (S):** This term refers the information which is received from the environment in each algorithm step. In our case, we model s_t as a tuple $s_t = (vf_t, df_t)$ where vf_t is the visual features vector associated to the image I_t or a set of visual features extracted from the image, typically a set of waypoints w_t obtained using a model-based path planner $vf_t = f(I_t, w_t)$. df_t is the driving features vector consisting of an estimation of vehicles speed v_t, distance to the center of the lane d_t and angle between the vehicle and the centre of the lane ϕ_t, $df_t = (v_t, d_t, \phi_t)$. Figure 2 shows the state space where the waypoints are published by CARLA from the planning module.
b) **Action space (A):** To interact with the vehicle available in the simulator, the commands for throttle, steering and brake must be provided in a continuous way. Throttle and brake range is [0, 1] and steering range is [−1, 1]. Therefore, at each step the DRL agent must publish an action $(a_t) = (acc_t, steer_t, brake_t)$ with the commands between the commented ranges.
c) **State transition function** (P_a) is the probability that action a in state s at time t will lead to state s_{t+1} at time t+1. $P_a = P_r(s_{t+1}|s_t, a_t)$.
d) **Reward function** $R(s_{t+1}, s_t, a_t)$: This function generates the immediate reward of translating from s_t to s_{t+1}. The goal in a Markov decision process is to find a good "policy" $\pi(s) = a_t$ that will choose an action given a state. This function will maximize the expectation of cumulative future rewards.

$$E = \sum_{t=0}^{\infty} \gamma^t R(s_t, s_{t+1}) \tag{1}$$

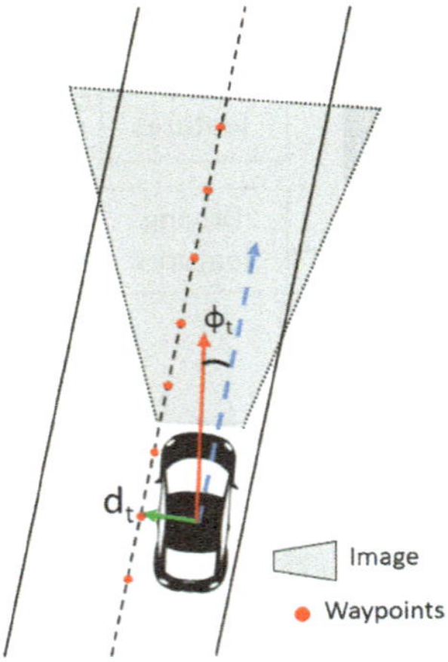

Fig. 2. State space.

4.2 Deep Q-Learning Algorithm for Reinforcement Learning

In recent years, there have been important advances in machine learning. Reinforcement Learning techniques have proved that an agent can learn robust policies in an interactive environment. In this work we are going to focus our effort in DQN [22] to solve our MDP problem.

Q-Learning. This process creates an exact matrix for the agent to maximize its reward in the long run. This approach is only practical for restricted environment, with limited space for observation, due to a high number of states or actions causes a wrong algorithm behaviour. Q-learning is an off-policy, model-free RL based on the Bellman Equation.

The goal of Q-learning is to maximize the Q-value though policy iteration, which runs a loop between policy evaluation and policy improvement. Policy evaluation estimates the value function V (equivalent to the referenced reward function (R)) with the greedy policy which has been obtained from the last policy improvement. On the other hand, policy improvement updates the policy with the action that maximizes V function for each of state.

$$Q(s_t, a_t) = Q(s_t, a_t) + \alpha[r_{t+1} + \gamma maxargQ(s_{t+1}, a_t) - Q(s_t, a_t)] \tag{2}$$

Deep Q-Learning. Q-learning is lack of generality when space of observation increases. Imagine one situation with 10 states and 10 possible actions, we will need a 10 × 10 Q-matrix. If now the number of states increases to 1000, the Q-matrix increases too. To solve this issue, Deep Q-Learning gets rid of the two-dimensional array by introducing a Neural Network.

So now, DQN estimates the Q-value by using a Neural Network, where the state is used as input, and the output is the corresponding Q-value for each action. The difference between D-Learning and Deep Q-Learning lies in:

$$y_j = r_j + \gamma maxargQ(\phi_{j+1}, a'; \theta^-) \quad (3)$$

Where ϕ is equivalent to the state s, while the θ stands for the parameters in the Neural Network. A detailed explanation of DQN algorithm can be found in [22].

5 Deep Q-Network Agents

We have developed various agents that cover a wide variety of model architectures for the Deep Q-Network agents. Models will be first developed in simulation for safety reasons. Therefore, the agent will interact with CARLA [23] and the code will be programmed in Python based on several open-source RL frameworks [17,24] (see Fig. 3) .

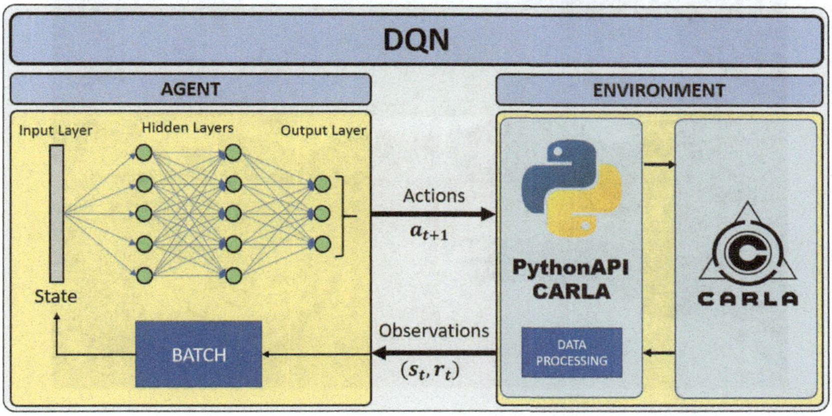

Fig. 3. Reinforcement learning architecture DQN-based.

CARLA is an hyper-realistic simulator which provides a complete set of maps, in addition to the possibility of creating your own maps and import them into it. Figure 10 depicts the map corresponding to "Town01". CARLA also provides a PythonAPI from which different actors (vehicles or pedestrians) in the simulator can be created and controlled by an external python agent. Different types of sensors are allowed to be attached to these actors, many of them will be used in this work, such as collision sensor, odometry, or camera image, being this last the most important for our application.

The cameras can be fixed to a vehicle using its centre as reference point, so that is placed in a position relative to the vehicle's position by using a set of coordinates (x, y, z, pitch, yaw and roll). Moreover, there is a fourth post-processing effect available for some specific cameras like raw depth or semantic segmentation.

Using this PythonAPI, points belonging to the map can be obtained easily, as well as actual vehicle position and speed. These points are called spawn points and they are determined by how the map was created. But what is really important is that CARLA also provides waypoints, which do not depend on how the map was built, and can be used for navigation. The simulator contains a path planning algorithm, both global and local. The global route planner is based on A* algorithm, and is able to built a route between two map points and returning the set of waypoints that joins them, forming a path. Furthermore, thanks to the debugging tools, this path can be drawn in CARLA scene as shown in Fig. 4.

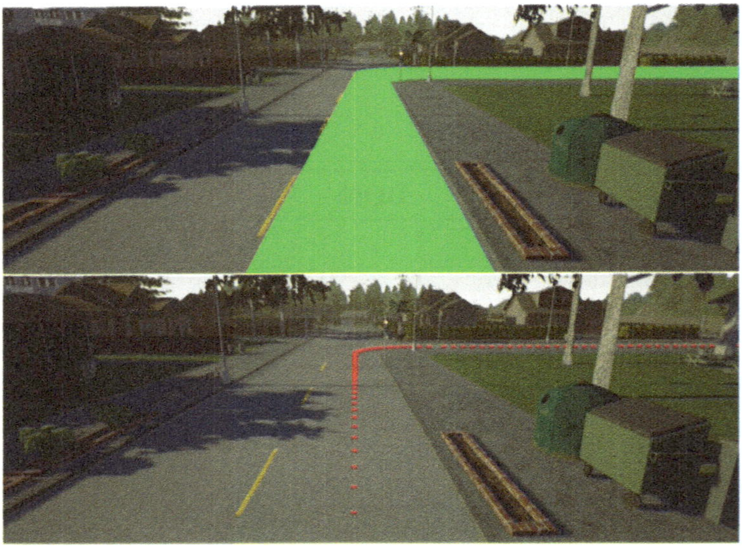

Fig. 4. Route with waypoints.

With all these facilities, some parameters have to be determined for the correct Trajectory with waypoints

a) **Reward function.** The proposed architecture obtains a driving features vector $df_t = (v_t, d_t, \phi_t)$ from the simulator. This vector is composed of the velocity of the vehicle in the direction of its heading v_t, the distance to the center of the lane d_t and the angle regarding the lane direction ϕ_t. Considering that the objective is to go as fast as possible through the center of the lane without leaving the lane and avoiding collisions, the reward function rewards the longitudinal velocity and penalize the transverse velocity and divergence from the center of the lane. This approach is similar to the proposal made in [20] where TORCS (The Open Racing Car Simulator) is used.

$$R = -200 \text{ if collision or lane change or roadway departure} \tag{4}$$

$$R = \sum_t |v_t cos\phi_t| - |v_t sin\phi_t| - |v_t||d_t| \text{ if car in lane} \tag{5}$$

$$R = 100 \text{ if goal position is reached} \tag{6}$$

b) **Control commands (actions).** CARLA needs control commands for steering [−1, 1] and throttle [0, 1]. Brake has not been implemented in this first version because the environment is free of obstacles and the regenerative braking of the vehicle is enough to stop the vehicle. The DQN policy allows generating discrete actions, so it is necessary to simplify the continuous control of actions to a discrete control. Taking this into account, the number of control commands has been simplified to a set of 27 discrete driving actions, discretizing steering angle and throttle position in an uniform way. Table 1 shows the set of control commands where there are 9 steering wheel positions and 3 throttle position.

Table 1. Policy network. 27 classes.

Control commands		
Classes	Steering	Throttle
27	**−1, −0.75, ... 0.75, 1**	**0, 0.5, 1**

5.1 DQN Architectures

This section explains different implemented model architectures. The main difference among them are the features extraction methods. All agents have been programmed using Keras [25], which is a high-level neural networks library, that runs on the top of TensorFlow (open source platform for machine learning) [26].

DQN-CNN Agent. DQN-Convolutional Neural Network is an ambitious agent model, as it covers the whole problem straightforwardly, as shown in Fig. 5. This agent takes data from an image as visual features $vf_t = I_t$ and a set of parameters obtained from the vehicle as driving features $df_t = (v_t, d_t, \phi_t)$. Camera image is connected into a CNN and its output feeds a Fully-connected module, whereas the driving features are directly connected into the Fully-connected module. What is expected to have as CNN output are the road features at that moment, serving as a kind of waypoints for the training of the DRL. Then, both road and driving features are concatenated and introduced on set of Fully-Connected layers from which final action would be obtained. The CNN consist of three convolutional layer with 64 filters of size [7 × 7], [5 × 5] and [3 × 3] respectively. All layers use RELU as activation function. Each layer is followed by an average pooling. The output of the last convolutional layer is flattened and concatenates with driving features vector and fed into 2-fully-connected layers with 300 and 600 hidden units respectively. Finally a classification layer with linear activation outputs the predicted Q value.

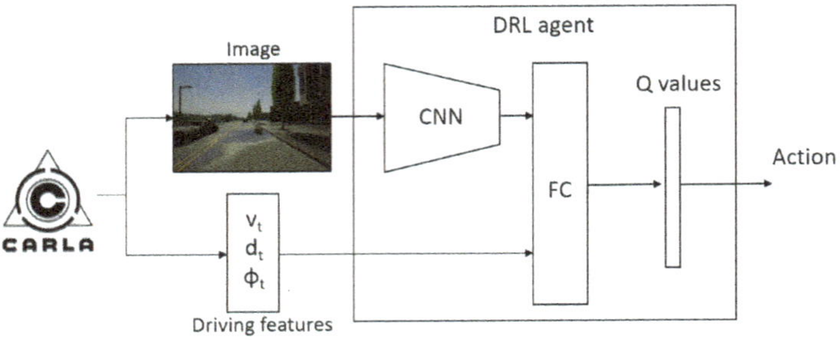

Fig. 5. DQN-CNN agent.

DQN-FC Agent. DQN-Fully-Connected agent differs the previous one in the CNN suppression. In this case, the input of the agent are the waypoints provided by the global planner of CARLA. Once A* algorithm is called, the complete path can be used, so only the next 14 waypoints from actual vehicle position will be introduced into the Fully-connected layers, so as the car moves the waypoints that feed the network will be updated accordingly. Although global planner's waypoints contain 3 position and 3 rotation components, only X and Y values will be used (referenced to the local map of the vehicle). This is due to only variations regarding the lane plane are taken into account. As in the previous agent, this vector will be concatenated with the driving features vector into the Fully-connected layer in order to obtain the action to take (see Fig. 6).

DQN-Flatten-Image Agent. This approach is built as simplification of DQN-CNN agent and is based on converting the RBG image into a resized B/W image, where the navigation path is segmented in a pre-processing stage, so that this image is flattened and converted into visual features vector. The visual vector is concatenated with driving features vector and this data is inserted into a DQN-Fully-Connected agent (see Fig. 7).

DQN-PilotNet Agent. Due to the really hard task of getting a good model training the DQN-CNN agent, we decided to separate the problem into two stages: the first one to obtain waypoints from the images and the second one to train the DQN from the waypoints. This architecture was built with the main

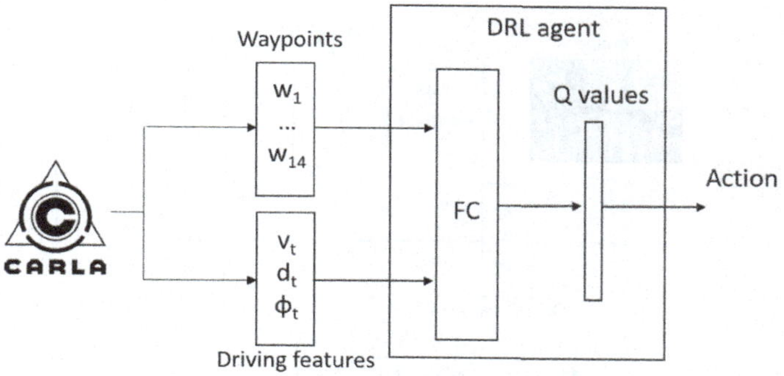

Fig. 6. DQN-fully-connected agent.

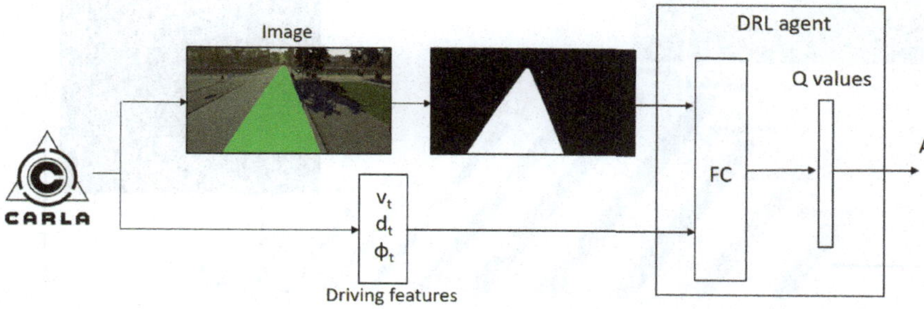

Fig. 7. DQN-flatten-image agent.

objective of obtaining the waypoints through a CNN, but numerous networks can be used for this purpose not being necessary to design an ad-hoc network from scratch. A known network model that works with images is "PilotNet" [27], which was created with the purpose of obtaining the road angle and vehicle speed from the track image. So, assuming a proper PilotNet behaviour, some necessary changes in its architecture are made in Python, and thanks to "Tensorflow" and "Keras" libraries, a well trained model is achieved (see Fig. 8).

A PilotNet architecture as shown in Fig. 9 is able to return the same waypoints that CARLA would provided thought PythonAPI from its planner. This implies having an independence of third party algorithms to obtain the waypoints and in a synchronized way with the DRL Net. Obviously, to work with this network it is necessary a pre-training stage with data recorded from the simulator. However, this is much simpler and shorter than a complete training for the DQN-CNN agent. The remaining agent is identical to the one implemented in DQN-FC agent.

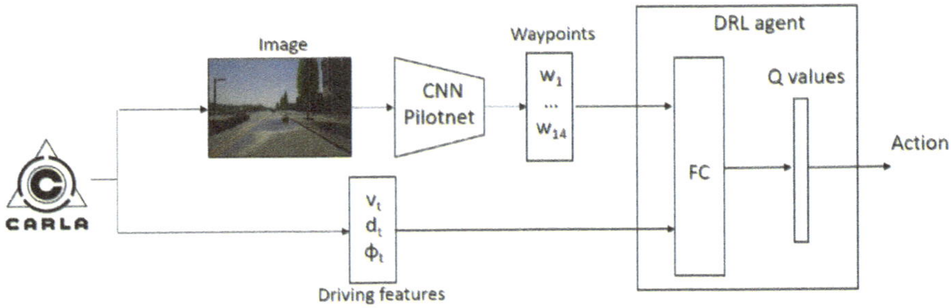

Fig. 8. DQN-PilotNet agent.

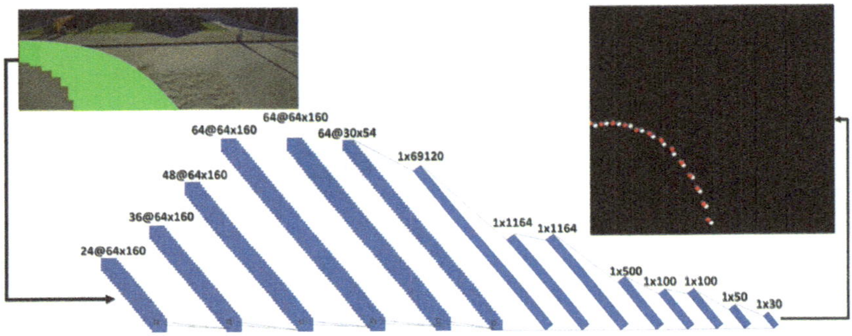

Fig. 9. CNN's architecture based on PilotNet modified to obtain waypoints to feed the Deep Reinforcement Learning training. A 160 × 60 RGB image is provided to the PilotNet to return a (15, 2)-shaped waypoint vector.

6 Results

To test the performance of the implemented DQN Agents, we compare results with other navigation modes such as manual driving (hand-crafted) and autonomous driving using a Classical Waypoint Tracking Controller based on the LQR [28].

The experiment carried out consist on learn how to drive in a trajectory where a origin-destination pair is selected from the "Town01" of CARLA so that the trajectory includes straight sections and curves to the left and right as it is shown in Fig. 10. All the training and testing experiments have been developed using a PC (Intel Core i7-9700, 32 GB) with Nvidia GeForce RTX 2080 and using CUDA-based GPU.

The following steps have been developed in the training process:

1) At the beginning of each episode, the CARLA Python-API generates a path using a global planner (origin, destination and waypoints).
2) Each timestamps (step), the corresponding visual and driving features vector are obtained, which are introduced in the DQN agent.

Fig. 10. Navigation-trajectory on CARLA's Town01 scene obtained by Global Path Planning A* algorithm provided by Carla PythonAPI.

3) In each timestamp of the DQN training process, the DQN weights are updated, the reward is calculated and the Q matrix is updated.
4) The episode ends when the destination is reached or a collision or lane departure occurs.
5) Throughout the process, the accumulated reward is calculated. This value will be used to determine which episode has made the best trajectory (best-reward-episode).
6) Training ends when the maximum number of episodes is reached.

Table 2 shows the results obtained in DQN agents training process. The number of training episodes is 20,000 with the exception of the DQN-CNN where it has been expanded to 120,000. This is so because we take RGB images directly as input instead of waypoints, which supposes a much larger order of data, and consequently, requires a higher number of iterations for the system to converge.

As you can see, all agents reach the final destination in the established episodes, although this does not guarantee that the performed trajectory is adequate. Those DQNs that reached their best-reward-episode earlier (shorter learning time) are DQN-FC and DQN-PILOTNET with 8300 and 13200 episodes respectively. These two models are the ones that use a set of waypoints as visual feature vector as input to the agent's full-connected layer. The other agent that also reaches its best-reward-episode before reaching the maximum number of episodes is the DQN-Flatten. In this case the RGB image information is reduced, converted to b/n and flattened. Finally, the most complex model, the DQN-CNN

Table 2. Training performance

Model	Training episodes	Best episode
DQN-FC	**20000**	**8300**
DQN-FLATTEN-IMAGE	**20000**	**16500**
DQN-PILOTNET	**20000**	**13200**
DQN-CNN	**120000**	**108600**

requires many more episodes to achieve a correct result. In this case it has taken 108,600 episodes. To summarize, those DQN agents with a simpler features learn to drive much faster than those with more complex visual features.

Once the training process of the different DQN agents is completed, the objective is to compare how well the trajectory performed by each one fits to an ideal driving following the center of the lane (ground-truth). To do this, the different DQNs will be compared with each other and also with manual driving (hand-crafted) and an autonomous driving method using by the authors in its real prototype (Smart Elderly Car Project) [29].

Figure 11 shows the trajectories followed by the different DQN agent as well as the manual and classical models. As can be seen, all approaches are close the ideal path and manage to reach the final destination. In general terms, it can be seen that better results are obtained using classical methods. DQN-based approaches show greater oscillations when following the ideal path and present higher errors compared with the classical approach. DQN-CNN and DQN-Flatten agents that use an image as visual feature vector obtain a trajectory with more oscillation and move further away from the ideal path. On the other hand, DQN-FC and DQN-Pilotnet follows a trajectory that is much closer to the ideal and have hardly any oscillations.

To get a more quantitative idea of errors made with respect to the ground-truth (obtained interpolating by means of a spline-line the route points provided by CARLA) we calculate the mean square error (MSE) and the maximum error along the whole path. Numbers are depicted in Table 3.

The best result are obtained by Waypoint Tracking Controller based on LQR Optimal Controller with an MSE of 6 cm, being the maximum distance to the ground-truth 0.74 m in the middle of the first curve.

DQN-FC and DF-Pilonet, which use waypoints as inputs present an small error ranging from MSE = 0.21 m to MSE = 0.33 m. These two agents correctly follow the ideal path in straight sections, but in curves they have maximum errors of 1.32 m and 1.72 m respectively. With these errors and taking into account the lane width is 3.5 m both models model can be applied for self-driving a car.

On the other hand, DQN-Flatten and DQN-CNN present higher errors, ranging from MSE = 0.64 m for DQN-Flatten to MSE = 0.83 m for DQN-CNN. Although the mean error may be acceptable, the maximum error reach 3.15 m and 2.15 m on curves, which are unacceptable because generates road departures.

It must be remarked that the manual driving gets a MSE=0.4 m and a maximum error in curves of 0.74 m. In consequence, if we compare autonomous driving behaviours with the performed by a human errors would be lesser.

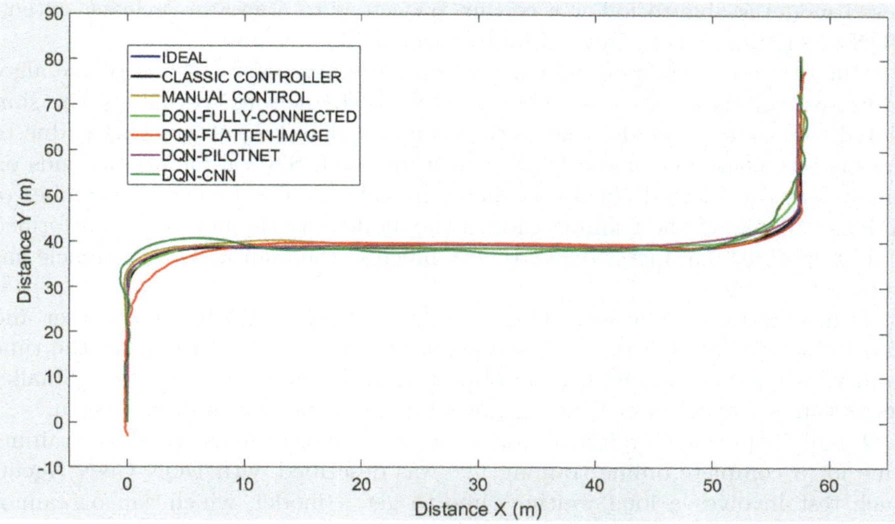

Fig. 11. 2D-trajectories-comparison.

Table 3. Lateral mean square error.

Model	Mean square error	Maximum error
CLASSIC CONTROLLER	**0.06 m**	**0.74 m**
MANUAL CONTROL	**0.40 m**	**1.80 m**
DQN-FC	**0.21 m**	**1.32 m**
DQN-PILOTNET	**0.33 m**	**1.72 m**
DQN-FLATTEN-IMAGE	**0.64 m**	**3.15 m**
DQN-CNN	**0.83 m**	**2.15 m**

Summarising, although all the implemented agents have reached the destination, it is observed that the agents that use directly images as inputs obtain worse results (greater errors following the center of the lane) than those that work directly with waypoints obtained from a external global planner or generated from pre-processing of the images using a CNN. This is due to work with images is more complex (there are more inputs to the neural network) and requires much more agent training time to obtain similar results.

7 Conclusions

In this work, an approach for integrating some Deep Reinforcement Learning algorithms into a navigation system was proposed, more specifically DQN-based algorithms. In spite of the results shown in previously section where each proposed model is able to follow a certain trajectory to a greater or lesser extent, DQN algorithm presents several limitations for this purpose.

One of them is the real difficulty to tune the necessary outputs of the algorithm, as well as its discrete values related to the actions to take by the simulated car, to get a model that performs a complete trajectory. This is due to the discrete character of the DQN algorithm itself. Several factors depends on the amount of selected classes as shown in Table 1. The higher the number of actions, the higher the training time of the model and the better the performed trajectory. This parameter depends very much of the map where the vehicle has to travel.

Other limitation is the training time, controlled by Epsilon parameter and that decays as training progresses. Epsilon decay selection establishes the time from which a valid model can be obtained, and how good it will be. Smaller decay values causes fewer training times but worse models, and vice versa.

Finally, the worst limitation found in DRL algorithms is the huge training time for a complete online training like was described with DQN-CNN Agent. Each test involves a long waiting time to get a model, which can or cannot work. DQN-PilotNet Agent was built in order to solve the large DQN-CNN training times by including a pre-trained neural network, as explained above, that decouple the features extraction in two steps. According to the results achieved by this agent, we can claim that including the modified PilotNet is a good practical solution for our application.

Despite the above difficulties, simulated trajectories have been carried out, unlike many of the articles in the state of the art. These results are promising, since even though DQN is a discrete algorithm, the final objective of the work has been achieved, which was that a car could learn by itself how to follow a trajectory on an urban map in simulation, efficiently and in a safety way.

The reported results leaves the door open to the implementation of a continuous Deep Reinforcement Learning algorithm to solve the limitations observed in the DQN, with the idea of reaching better models with shorter training times. Another future work is to export our best simulated models to our real car prototype to really check the potential of these DRL techniques.

Acknowledgment. This work has been funded in part from the Spanish MICINN/FEDER through the Techs4AgeCar project (RTI2018-099263-B-C21) and from the RoboCity2030-DIH-CM project (P2018/NMT- 4331), funded by Programas de actividades I+D (CAM) and cofunded by EU Structural Funds.

References

1. Chan, C.: Advancements, prospects, and impacts of automated driving systems. Int. J. Transp. Sci. Technol. **6**(3), 208–216 (2017)

2. Urmson, C., Anhalt, J., et al.: Autonomous driving in urban environments boss and the urban challenge. J. Field Robot. **25**(8), 425–466 (2008)
3. Montemerlo, M., Becker, J., et al.: Junior: the stanford entry in the urban challenges. J. Field Robot. **25**(9), 569–597 (2008)
4. Wang, F.: AI and intelligent vehicles future challenge (IVFC) in China: from cognitive intelligence to parallel intelligence. In: ITU Kaleidoscope: Challenges for a Data-Driven Society (ITU K), pp. 1–2 (2017)
5. Zhao, J. et al.: TiEV: the tongji intelligent electric vehicle. In: International Conference on Intelligent Transportation Systems(ITSC), pp. 1303–1309 (2018)
6. Liang, M., Yang, B. Wang, S. Urtasun, R.: Deep continuous fusion for multi-sensor 3D object detection. European Conference on Computer Vision (ECCV), pp. 641–656 (2018)
7. Barea, R., et al.: Vehicle detection and localization using 3D LIDAR point cloud and image semantic segmentation. In: 21st International Conference on Intelligent Transportation Systems (ITSC), pp. 3481–3486 (2018). https://doi.org/10.1109/ITSC.2018.8569962
8. Silver, D., et al.: Mastering chess and shogi by selfplay with a general reinforcement learning algorithm. CoRR, vol. abs/1712.01815 (2017)
9. Mnih, V., et al.: Human-level control through deep reinforcement learning. Nature **518**(7540), 529 (2015)
10. Kendall. A et al.: Learning to drive in a day. In: 2019 International Conference on Robotics and Automation (ICRA), Montreal, QC, Canada, pp. 8248–8254 (2019). https://doi.org/10.1109/ICRA.2019.8793742.
11. Lillicrap, T. et al.: Continuous control with deep reinforcement learning. In: International Conference on Learning Representations (ICLR) (2016)
12. Dosovitskiy, A., Ros, G., Codevilla, F., Lopez, A., Koltun, V.: CARLA: an open urban driving simulator. arXiv preprint arXiv:1711.03938 (2017)
13. Quigley, M., et al.: ROS: an open-source robot operating system. In: ICRA Workshop on Open Source Software (2009). https://www.ros.org/
14. Levinson, J., et al.: Towards fully autonomous driving: systems and algorithms. In: 2011 IEEE Intelligent Vehicles Symposium (IV), pp. 163–168. IEEE (2011)
15. Chen, J., Yuan, B., Tomizuka, M.: Deep imitation learning for autonomous driving in generic urban scenarios with enhanced safety. arXiv preprint arXiv:1903.00640 (2019)
16. Codevilla, F., Miiller, M., Lopez, A., Koltun, V., Dosovitskiy, A.: End-to-end driving via conditional imitation learning. In: 2018 IEEE International Conference on Robotics and Automation (ICRA), pp. 1–9 (2018)
17. Yurtsever, E., Capito, L., Redmill, K., Ozguner, U.: Integrating deep reinforcement learning with model-based path planners for automated driving. ArXiv:2002.00434 (2020)
18. Vitelli, M. Nayebi, A.: CARMA: a deep reinforcement learning approach to autonomous driving. Technical report, Stanford University (2016)
19. Ganesh, A., Charalel, J., Sarma, M.D., Xu, N.: Deep reinforcement learning for simulated autonomous driving. In: Stanford University (2016)
20. Jaritz, M, Charette, R., Toromanoff, M., Perot, E., Nashashibi, F.: End-to-end race driving with deep reinforcement learning. In: ICRA 2018, pp. 2070–2075 (2018). https://doi.org/10.1109/ICRA.2018.8460934
21. Chen, J., Shengbo, E., Tomizuka, M.: Interpretable end-to-end urban autonomous driving with latent deep reinforcement learning. arXiv preprint arXiv:2001.08726 (2019)

22. Fan, J., Wang, Z., Xie, Y., Yang Z.: A theoretical analysis of deep Q-learning. ArXiv:1901.00137 (2020)
23. CARLA: Open-source simulator for autonomous driving research.https://carla.org
24. Sentdex: Carla-RL (2020).https://github.com/Sentdex/Carla-RL
25. Keras. https://keras.io/
26. Tensorflow. https://www.tensorflow.org
27. Bojarski, M., et al.: End to end learning for self-driving cars (2016). arXiv: 1604.07316
28. Gutiérrez, R., López-Guillén, E., Bergasa, L.M., Barea, R., Pérez, Ó., Gómez-Huélamo, C., Arango, F., del Egido, J., López-Fernández, J.: A waypoint tracking controller for autonomous road vehicles using ROS framework. Sensors **20**, 4062 (2020)
29. Smart Elderly Car Project. https://www.robesafe.uah.es/index.php/en/smartelderlycar

Mobile and Social Robots

Concept Drift Detection and Adaptation for Robotics and Mobile Devices in Federated and Continual Settings

Fernando E. Casado[1(✉)], Dylan Lema[1], Roberto Iglesias[1], Carlos V. Regueiro[2], and Senén Barro[1]

[1] CiTIUS (Centro Singular de Investigación en Tecnoloxías Intelixentes), Universidade de Santiago de Compostela, 15782 Santiago de Compostela, Spain
{fernando.estevez.casado,dylan.lema,roberto.iglesias.rodriguez, senen.barro}@usc.es

[2] CITIC, Computer Architecture Group, Universidade da Coruña, 15071, A Coruña, Spain
carlos.vazquez.regueiro@udc.es

Abstract. Service robots and other smart devices, such as smartphones, have access to large amounts of data suitable for learning models, which can greatly improve the customer experience. Federated learning is a popular framework that allows multiple distributed devices to train deep learning models remotely, collaboratively, and preserving data privacy. However, little research has been done regarding the scenario where data distribution is non-identical among the participants and it also changes over time in unforeseen ways, causing what is known as *concept drift*. This situation is, however, very common in real life, and poses new challenges to both federated and continual learning. In this work, we propose an extension of the most widely known federated algorithm, FedAvg, adapting it for continual learning under concept drift. We empirically demonstrate the weaknesses of regular FedAvg and prove that our extended method outperforms the original one in this type of scenario.

Keywords: Federated learning · Continual learning · Nonstationarity · Concept drift · Federated averaging · Catastrophic forgetting · Rehearsal

1 Introduction

Today, we are experiencing an increasing integration of service robots in our homes, companies, etc. These robots have the ability to perform intended tasks based on current state and sensing, without human intervention, ranging from partial autonomy to full autonomy. Nevertheless, it is true that the problems that will arise when robots are taken from laboratories to real environments, or the way people will react to them, are highly unpredictable. Thus, undesirable and inaccurate behaviors will inevitably appear.

L. M. Bergasa et al. (Eds.): WAF 2020, AISC 1285, pp. 79–93, 2021.
https://doi.org/10.1007/978-3-030-62579-5_6

Hopefully, there is also a high chance that these robots will be connected to the cloud, allowing them to transfer data from their environments to the companies that built them. This transfer of data will be justified by the need of detecting and improve robot limitations, or to detect future advances. In addition, learning from the crowds represents a quick way to improve performance in a short period of time, by creating consensus-based models in which each individual provides only a small portion of data. However, the problem is how to achieve a learning process that is robust to bias and noise without putting at risk customer's privacy.

Traditional and offline server-based techniques are the most common approaches to carry out the machine learning process. In the context of distributed devices, such as service robots, this involves collecting data from all the devices, also called *clients*. This data, typically sensor measurements, images, videos, etc., must be later uploaded and processed centrally in a cloud-based server. However, in this kind of distributed scenarios, cloud-centric learning can be highly inefficient or even infeasible, since it involves facing the following challenges:

- **Scalability**, in terms of storage, communication, and computational costs. Central storage is never a scalable solution. Transferring large amounts of data over the network can take a long time, and communication may be a continuous overhead. Similarly, central processing can take much more time than parallel computing.
- **Data privacy and sensitivity**. Central data collection puts user privacy into risk. In recent years, governments have been implementing data privacy legislations to protect the consumer, limiting data collection and storage only to what is consented by the user and absolutely necessary for processing.
- **Nonstationarity**. Most machine learning algorithms assume that data comes from a static distribution. However, it does not hold in many real-world applications, where the underlying distribution of the data streams changes over time in unforeseen ways. This phenomenon is widely known as *concept drift* [1,2]. If a concept drift occurs, the inducted pattern of past data may not be relevant to the new data, leading to poor outcomes. This problem is related to two conflicting objectives: retaining previously learned knowledge that is still relevant and replacing the obsolete one with current information. This is known as the *stability-plasticity dilemma*.

With the aim of addressing one or other of the above-mentioned challenges, several machine learning paradigms and frameworks have been proposed in the literature. In this way, *continual learning* (CL) [2,3], has been positioned as the reference paradigm for addressing the problem of nonstationarity, as well as continuous and adaptive learning. On the other hand, *federated learning* (FL) [4,5] deals with scalability and user privacy. Finally, *glocal learning* (GL) [6,7] has recently been proposed as a way of approaching all these problems as a single one. Glocal learning can be understood as federated learning, but with special emphasis on nonstationarity, and with less attention paid to extreme privacy preservation. Obviously, all these strategies are not exclusive to robotics. On the contrary, they are being applied to a wide variety of problems on other smart devices, such as smartphones and wearables.

In the present work, we propose an improvement of the most widely known FL algorithm, *Federated Averaging* (FedAvg) [4], adapting it for continual learning under concept drift. In particular, we extend the algorithm to be able to deal with continuous single-task problems where the underlying distribution of data might be non-IID (independent and identically distributed) among the different client devices, but that also can change in unforeseen ways over time. Our proposal can be considered an adaptation of the original FedAvg method to the glocal learning framework, something that has not been done before. We perform an experimental validation employing a dataset for human activity recognition using the smartphones of 10 different users. We have chosen smartphones, instead of robots, to test our proposal because it is easier to get a small community of people (and hence devices) working and sharing experiences. Our experimental results show the weaknesses of regular FedAvg in this kind of distributed, non-IID, and continual settings, and also a remarkable improvement in the learning ability of our extended version of the method over the original one. We also show a reduction in storage, communication, and computational costs for each of the participating devices.

The paper is structured as follows: Sect. 2 provides a review of the state of the art. Section 3 gives a formal definition of federated learning and the problem of concept drift. In Sect. 4, our proposal is introduced. Section 5 presents the experimental results. Finally, some conclusions are drawn in Sect. 6.

2 Related Work

Continual learning [2,3] is built on the idea of learning continuously and adaptively about the external world, being able to smoothly update the prediction model to take into account different tasks and data distributions but still being able to re-use and retain useful knowledge during time. It is inspired by the human learning process, as a person learns to perform numerous tasks with large diversity over his/her lifespan, making use of the past knowledge to learn about new concepts and task without forgetting previously learned ones. CL deals with high and realistic time-scales, where data becomes available only during time, previous perception data is not accessible and it is imperative to build on top of previously learned knowledge.

In early works on CL, the most important challenge was on how to perform knowledge transfer from past concepts or tasks to new ones without having to train a new model from scratch. Nevertheless, this problem is relatively straightforward to solve using DNNs, and the popularity of deep learning is on the rise. Thus, recent works focus on how to deal with the aforementioned stability-plasticity dilemma, concept drift detection, and catastrophic forgetting. *Catastrophic forgetting* [2] refers to the phenomenon of a neural network experiencing performance degradation at previously learned concepts when trained sequentially on learning new ones. Since by definition the continual learning setting deals with sequences of concepts, classes and/or tasks, catastrophic forgetting is an important challenge to be tackled.

Federated learning [4] is a distributed learning framework under differential privacy, which consists of a local learning, in the devices, and a global parameter aggregation in a central server in the cloud. There is a single learner, which is a Deep Neural Network (DNN), whose weights are initialized on the server. In each federated learning round, every participating device receives the current weights of the model from the server, performs stochastic gradient descent (SGD) on its own local dataset and sends back the gradients. Then, the contributions of each device are aggregated on the server, thus updating the model. FL has been widely applied to solve complex classification and regression tasks, such as object recognition in images or predictive text in smartphone keyboards [5].

The most popular framework for global parameter aggregation is *Federated Averaging* (FedAvg) [4]. Some authors have evaluated the FedAvg algorithm on non-IID scenarios. McMahan et al. [4] synthesize pathological non-identical user splits from the MNIST dataset, sorting training examples by class labels and partitioning into shards such that each client is assigned with 2 shards. They demonstrate that FedAvg on non-identical clients still converges to 99% accuracy, though taking more rounds than identical clients. In a similar manner, Zhao et al. [8] generate extreme partitions on the CIFAR-10 dataset, forming a population of 10 clients. These settings are somewhat unrealistic, as practical federated learning would involve more complex distributions than simple partitions. Other authors look at more realistic client data distributions. For example, Caldas et al. [9] use Extended MNIST dataset with partitions over writers of the digits, rather than simply partitioning over digit class.

All the above-mentioned works are characterized by analyzing the impact of non-identical client data partitions in FedAvg performance. However, little work has been done studying the effect of non-IID data streams shifting over time. Yoon et al. [10] propose a method for federated continual learning with Adaptive Parameter Communication, Fed-APC, which additively decomposes the network weights into global shared parameters and sparse task-specific parameters. Thus, they are able to perform inter-client knowledge transfer and prevent inter-client interference. They validate their approach on several settings, including multi-incremental-task continual scenarios. When comparing their method with existing federated baselines, they show significantly higher accuracies and also a reduced communication cost. Nevertheless, to the best of our knowledge, there is no work in federated learning that has addressed concept drift detection and adaptation in the underlying distribution of data.

As we will describe later, there are important benefits as a consequence of the explicit detection of concept drifts. In short, we can say that it helps us to answer two key questions: *what* needs to be learned and *when* it should be learned. Thus, in this article we address the impact of these unpredictable changes, focusing on a single-task scenario where different drifts happen over time.

3 Concept Drift in Federated Settings: Problem Definition

In standard federated settings, the learning process is carried out over multiple rounds of local learning and global aggregation. At each round r, each client $c_c \in \{c_1, \ldots, c_C\}$ and the server s perform two different learning stages: *local parameter update* and *global parameter aggregation.*

The model parameters are initialized on the server. At the beginning of each round, a random subset of clients, $\mathcal{C}^{(r)} \subseteq \{c_1, c_2, \ldots, c_C\}$, is selected. The server sends the current global algorithm state to each of these clients, i.e, the current model parameters. In the local parameter update step, each client c_c performs stochastic gradient descent (SGD) on its own local dataset and sends the updated parameters $\theta_c^{(r)}$ to the server. Then, the server aggregates the parameters $\theta_c^{(r)}$ sent from all the clients in $\mathcal{C}^{(r)}$ into a single parameter. The way in which this aggregation is carried out is usually the average, although there are variations depending on the aggregation method used. In the particular case of FedAvg, the local parameters from each client are aggregated applying a weighted average:

$$\theta_G^{(r)} \leftarrow \sum_{c=1}^{|C^{(r)}|} \frac{n_c}{N} \theta_c^{(r)}, \tag{1}$$

where N is the total number of data instances at each round, and n_c is the number of instances per client c_c. Then, the updated model is sent back to the devices, a new random subset of clients is selected, and a new training round starts.

That is the original FedAvg procedure proposed by Google in 2016 [4]. However, as we will show in Sect. 5, naive training of FedAvg on real-world scenarios may result in catastrophic forgetting, because local datasets from the crowd are by nature nonstationary and non-IID. When data distribution is not stationary, a concept drift is observed in the data stream. When there is no external information concerning this drift, the model will have to detect it and adapt to it by itself in order to avoid decreasing its performance.

Concept drift is a continual learning issue [2]. Formally, we can define it as follows: Given a time period $[0, t]$, a set of samples, denoted as $S_{0,t} = \{d_0, \ldots, d_t\}$, where $d_i = (X_i, y_i)$ is one observation (or data instance), X_i is the feature vector, y_i is the label, and $S_{0,t}$ has a certain joint probability distribution of X and y, $P_{0,t}(X, y)$. Therefore, concept drift can be defined as a change in the joint probability at timestamp t, such that $\exists t : P_{0,t}(X, y) \neq P_{t+1}(X, y)$.

We now extend the conventional concept drift definition to the federated learning setting, with multiple clients and a global server. Now the goal is to train a global model in a distributed and parallel way using the local data of the C available clients. Each client will have a different bias because of the conditions of its local environment, and likewise, its data stream may change in different ways over time. Therefore, there may occur concept drifts that affect all clients, some of them, or just a single one. Thus, we can generalize the problem in the following way: Given a time period $[0, t]$, a set of clients $\{c_1, \ldots, c_C\}$, and a set of local samples for each client, denoted as $S^c_{0,t} = \{d^c_0, \ldots, d^c_t\}$, where $d^c_i = (X^c_i, y^c_i)$ is one data instance from client c_c, X^c_i is the feature vector, y^c_i is the label. Each local dataset $S^c_{0,t}$ has a certain joint probability distribution $P^c_{0,t}(X, y)$. A *local concept drift* occurs at timestamp t for client c_c if $\exists t, c : P^c_{0,t}(X, y) \neq P^c_{t+1}(X, y)$.

Nevertheless, note that a local concept drift does not necessarily have a direct impact on the global federated model. It may be the case that a local drift on device c_c will result in a change in the distribution of c_c, but not in the joint distribution of all clients, $P^G_{0,t}(X, y)$. In that case, the federated model will not be affected by the local change and this one can be ignored. Thus, we can define a *global concept drift* as a distribution change at time t in one or more clients $\{c_i, \ldots, c_j\} \subseteq \{c_1, \ldots, c_C\}$ such that $\exists t : P^G_{0,t}(X, y) \neq P^G_{t+1}(X, y)$.

If a global concept drift happens the model might lower its performance. Thus, as we will explain in Sect. 4, we need to extend the FedAvg algorithm giving it the ability to detect these global changes and adapt to them itself.

4 Concept Drift Detection and Adaptation for FedAvg

Research on learning under concept drift is typically classified into three components [1]: (1) drift detection (whether or not drift occurs), (2) drift understanding (when, how, where it occurs) and (3) drift adaptation (reaction to the existence of drift). In this work, we have focused on global drift detection and adaptation.

4.1 Drift Detection

Drift detection refers to the techniques and mechanisms that characterize and quantify concept drift via identifying change points or change intervals in the underlying distribution of data. In our proposal, if a concept drift is identified it means that the federated model is no longer a good abstraction of the knowledge of all the devices, so it must be adapted in some way. Drift detection algorithms are typically classified into three categories [1]: (1) error rate-based methods, (2) data distribution-based methods and (3) multiple hypothesis test methods. The first class of algorithms focuses on tracking changes in the online error rate of the model. The second type uses a distance function or metric to quantify the dissimilarity between the distribution of historical data and the new data. The third group basically combines techniques from the two previous categories in different ways. In this work we decided to use a data distribution-based algorithm. The main reason for this decision is that no data labels are needed.

We propose an adapted version of the change detection technique (CDT) originally developed in [11], which is a CUSUM-type CDT on *beta* distribution [12]. Algorithm 1 outlines the method. It is invoked on each of the devices $c_c \in \{c_1, \ldots, c_C\}$ as soon as a new data instance $d^c_i = (X^c_i, y^c_i)$ is available. As this method does not require instance labels to detect the drift, we actually only need to get the new feature vector X^c_i. For each new X^c_i, we will estimate a metric that will help us to quantify the dissimilarity between the old and the new data distributions. Any metric of those already proposed in the literature can be used [1]. In our case, since our experimental evaluation is done on a classification task (Sect. 5), we propose to use the confidence of the classifier as metric. We define the *confidence* of the model on a sample $d = (X, y)$ as the classifier conditional posterior probability, i.e., the probability $P(\zeta_i|X)$ of a class ζ_i from one of the M possible classes $\{\zeta_1, \zeta_2, \ldots, \zeta_M\}$ to be the correct label y. Thus, we allow the detection of global concept drifts, since the confidence of the federated model in each of the local datasets is evaluated.

Algorithm 1: Drift detection algorithm.

1 **Input:** W, X_i, T_h, α, Δ, N_{max}
2 **Output:** W
3 $[\hat{y}_i, w_i] \leftarrow \text{predict}(X_i)$ // Classify the pattern
4 $W \leftarrow W \cup w_i$ // Add the confidence into W
5 **if** $|W| >= N_{max}$ **then**
6     $w_0 \leftarrow \varnothing$ // Remove the oldest element in W
7 **end**
8 $N \leftarrow |W|$
9 $s_f \leftarrow 0$
10 $r \leftarrow \text{random}(0, 1)$ // Generate random number in the interval [0,1]
11 **if** $e^{-2w_i} \geq r$ **then**
12   **for** $k \leftarrow \Delta$ **to** $N - \Delta$ **do**
13         $m_b \leftarrow \text{mean}(w_1 : w_k \in W)$
14         $m_a \leftarrow \text{mean}(w_{k+1} : w_N \in W)$
15   **if** $m_a \leq (1 - \alpha) \cdot m_b$ **then**
16             $s_k \leftarrow 0$
17             $[\hat{\alpha}_b, \hat{\beta}_b] \leftarrow \text{estimateParams}(w_1 : w_k)$
18             $[\hat{\alpha}_a, \hat{\beta}_a] \leftarrow \text{estimateParams}(w_{k+1} : w_N)$
19   **for** $i \leftarrow k + 1$ **to** N **do**
20                 $s_k \leftarrow s_k + \log\left(\frac{f(w_i \mid \hat{\alpha}_a, \hat{\beta}_a)}{f(w_i \mid \hat{\alpha}_b, \hat{\beta}_b)}\right)$
21   **end**
22             $s_f \leftarrow \max(s_f, s_k)$
23   **end**
24   **end**
25   **if** $s_f > T_h$ **then**
26         $W \leftarrow \{\varnothing\}$ // Reinitialize the sliding window
27   driftAdaptation() // Perform adaptation to drift (Section 4.2)
28   **end**
29 **end**

Algorithm 1 works as follows. First of all, the confidence w_i^c of the current model on the new data instance X_i^c is estimated (line 3 in the pseudocode). This confidence is stored in a sliding window W of length $|W|$ (line 4). We will try to detect changes in the distribution of the confidences stored in W. In the original method proposed in [11], authors do not use a sliding window, but a dynamic window, which is reinitialized every time a drift is identified. However, they do not establish any limits on the size of this dynamic window, which is not very realistic, since its size could grow to infinity if no drift is detected. Therefore, we use a sliding window and we set a maximum size N_{max}. The value of N_{max} will be strongly dependent on the task to be solved. In any case, in our experiments we use $N_{max} = 1000$. Once this maximum size is reach, adding a new element to W implies deleting the oldest one (lines 5 to 7).

The core of this CDT algorithm (lines 12 to 28) can be a bottleneck in our system if we have to execute it after inserting each confidence value in W. Therefore, we restrict the number of executions, so that the core of Algorithm 1 will be executed with a probability of e^{-2w_i}, for any confidence value w_i (line 11 in the pseudocode). Hence, the higher the confidence, the lower the probability of executing CDT, and *vice versa*.

In the core of the algorithm, W is divided into two sub-windows for every pattern k between Δ and $N - \Delta$, where N is the length of W (lines 12 to 14 in Algorithm 1). Let W_a and W_b be the two sub-windows, where W_a contains the most recent confidences. Each sub-window is required to contain at least Δ elements to preserve statistical properties of a distribution. When a concept drift occurs, confidence scores are expected to decrease. Therefore, only changes in the negative direction are required to be detected. In other words, if m_a and m_b are the mean values of the confidences in W_a and W_b respectively, a change point is searched only if $m_a \leq (1 - \alpha) \times m_b$, where α is the sensitivity to change (line 15). Same as in [11], we use $\alpha = 0.05$ and $\Delta = 100$ in our experiments, which are also widely used in the literature.

We can model the confidence values in each sub-window, W_a and W_b, as two different *beta* distributions. However, the actual parameters for each one are unknown. The proposed algorithm estimates these parameters at lines 17 and 18. Next, the sum of the log likelihood ratios s_k is calculated in the inner loop between lines 19 and 21, where $f\left(w_i \mid \hat{\alpha}, \hat{\beta}\right)$ is the probability density function (PDF) of the *beta* distribution, having estimated parameters $\left(\hat{\alpha}, \hat{\beta}\right)$, applied on the confidence $w_i \in W$. This PDF describes the relative likelihood for a random variable, in this case w_i, to take on a given value, and it is defined as:

$$f\left(w_i \mid \alpha, \beta\right) = \begin{cases} \frac{w_i^{\alpha-1}(1-w_i)^{\beta-1}}{B(\alpha,\beta)}, & \text{if } 0 < w_i < 1 \\ 0, & \text{otherwise,} \end{cases} \tag{2}$$

where

$$B\left(\alpha, \beta\right) = \int_0^1 w_i^{\alpha-1}(1 - w_i)^{\beta-1} dx. \tag{3}$$

The variable s_k is a dissimilarity score for each iteration k of the outer loop between lines 12 and 24. The larger the difference between the PDFs in W_a and W_b, the higher the value of s_k (line 20). Let k_{max} is the value of k for which the algorithm calculated the maximum s_k value where $\Delta \leq k \leq N - \Delta$. Finally, a change is detected at point k_{max} if $s_{k_{max}} \equiv s_f$ is greater than a prefixed threshold T_h (line 25). As in the original work, we use $T_h = -\log(\alpha)$. In case a drift is detected, the sliding window W is reinitialized and a drift adaptation strategy is applied (lines 26 and 27). We will discuss this strategy now.

4.2 Drift Adaptation

Once a drift is detected, the model should continue to be trained according to that drift. Otherwise, catastrophic forgetting may happen. When it comes to neural networks, as it is the case, the most common methods to avoid that are (1) regularization methods, (2) rehearsal approaches, and (3) generative replay [2]. Regularization in continual learning consist in modifying the update of weights when learning from new concepts in order to keep memory of previous knowledge. Rehearsal or replay approaches gather all methods that save raw samples as memory of past tasks. Generative replay approaches train generative models on the data distribution, being able to afterwards sample data from past experience when learning on new one.

Rehearsal methods have been proven to achieve the best results [13], since they ensure that the memories are not degraded through time. In a cloud-centric setting, their main drawback is the need of a separate memory of raw and unprocessed data which is a vanilla way of saving knowledge that does not respect data privacy and may involve high storage costs. However, in the continual federated setting this is not an issue since small pieces of old memories can be kept by each of the local clients. Thus, we propose to use a simple rehearsal method.

In our extension of FedAvg, each time a new global concept drift κ_i^c is detected by client c_c (using the detection algorithm from Sect. 4.1), certain amount of data observations S_i^c will be kept in representation of the new concept. For that, we propose that each client has two different data storages: a *short-term memory* and a *long-term memory*. The short-term memory is required to store the data instances the device has acquired in the last time interval. This recent data is kept for a limited amount of time and is processed to check whether a drift occurs. Samples of past concepts will be stored in the long-term memory, where they will be kept for a long time. Formally, the long-term memory of a client c_c at time t contains a dataset $S_{0,t}^c = \{S_0^c \cup S_i^c \cup ... \cup S_l^c\}$ that gathers representative data from each of the concepts $\{\kappa_0^c, \kappa_i^c, \ldots, \kappa_l^c\}$ detected by c_c so far, where $0, i, \ldots, l$ are the timestamps where the concepts were detected, being $0 < i < \cdots < l \leq t$.

In classification tasks, data from each class can be acquired in any order. Thus, to ensure a minimally balanced rehearsal dataset with representation of all classes, we establish a heuristic rule to handle the data which is stored in the long-term memory. We define a minimum amount of data, L, so that there must be at least $\frac{L}{2M}$ examples from each class in the long-term memory representing each concept, where M is the number of possible classes. In our experiments, we have used $L = 1400$, which is quite a significant amount of data, but it ensures that the memories do not degrade over time. We also assume that we can store data of unlimited concepts.

The first concept κ_0^c, since it is the first one, is not detected using the detection method from Sect. 4.1. Instead, at the beginning of the training process, the first data acquired by c_c will automatically belong to concept κ_0^c. Once enough data ($\frac{L}{2M}$) has been acquired for the initial concept, a limited number of training rounds R is performed. In our experiments we have used $R = 5$, although this value will depend on the problem. After completing those R first rounds, each client continues to acquire new data, saving it in the short-term memory and processing it. Only when a new concept drift is detected by any of the clients, new data related to the new concept will be stored in the long-term memory and further R training rounds will be carried out.

4.3 Benefits of Explicit Drift Detection: Saving Resources on Devices

Our extended FedAvg proposal not only allows a continual learning that avoids catastrophic forgetting, it also answer two important questions often overlooked in the literature on federated learning: *what* needs to be learned and *when* it should be learned.

In a naive implementation of FedAvg, there is no given criterion on what data to store and what data to discard. By default, unless there are task-dependent or implementation restrictions, devices will save all the data they capture. In the same way, FedAvg does not set any limits on the number of training rounds to be performed, they could be infinite. It is true that only a random subset of devices is selected in each round, so they are not training all the time. In addition, you can limit training even further based on conditions such as whether the device has access to WiFi or is connected to a power supply (this is more intended for smartphones). However, these criteria do not take into account whether there is a real need for further training. Finally, training constantly and without proper criteria also implies an increase in the required bandwidth. In fact, in many works in the literature of federated learning authors state that the main bottleneck is the communication cost [4,10].

Following our approach, it is possible to determine which data is relevant and which is not anymore thanks to the integration of the concept drift detector and a long-term memory. Besides, it can be assumed that it is only necessary to continue training when a new drift is detected, so that the model can be adapted to it. Thus, it is possible to significantly reduce the number of communications between clients and server, as well as the computational burden on clients.

5 Experimental Results

In order to evaluate the performance of our extended version of FedAvg, we have selected a real-world complex task: human activity recognition on smartphones. In particular, we have used the dataset created by Shoaib et al. [14], which contains data of seven physical activities: walking, sitting, standing, jogging, biking, walking upstairs and walking downstairs. This dataset is characterized by the fact that, for each activity, data was recorded with the smartphone placed in 5 different locations (Fig. 1): (1) on the belt position using a belt clip, (2) in the left jeans pocket, (3) in the right jeans pocket, (4) on the right upper arm using a sports armband, and (5) on the right wrist. This last aspect is crucial since it will allow us to analyze the performance of the concept drift detection and our extended version of FedAvg.

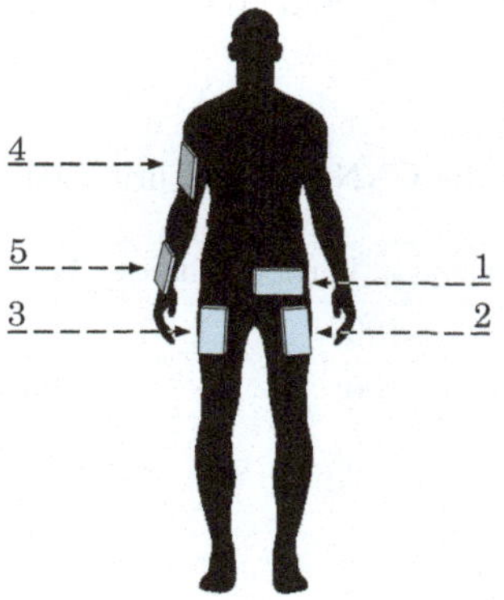

Fig. 1. Overview of the phone positions on a participant.

There were ten participants involved in data collection, who performed each of the activities for about 20 min (4 min for each phone position). All ten participants were male, between the ages of 25 and 30. The experiments were carried out indoors, except biking. The orientation of the smartphones was portrait for the upper arm, wrist and two pockets and landscape for the belt position. Data was collected from the inertial sensors of the smartphone, including the tri-axis accelerometer, gyroscope, and magnetometer. Data acquisition rate was limited to 50 samples per second, which is enough to recognize human physical activities.

We pose the problem as a multiclass classification task, in which our goal is to correctly predict which of the 7 activities is being performed by the user. The information regarding the position of the smartphone is very valuable to us. We will use it later to emulate concept drifts. In order to create a federated learning setting, we choose 7 of the 10 participants as clients, in which both FedAvg and our improved version will be running. Data from the 3 remaining users will be used for testing. Since there are relatively few devices, at each federated round all 7 clients participate in the local training, instead of selecting a random subset.

We divided the original raw data into time windows of 124 samples, which is equivalent to about 2.5 s. We decided to use only the information from the accelerometer and the gyroscope. Thus, the input shape is a 6-dimensional time window composed of 124 values from each of the axis of the accelerometer (a_x, a_y, a_z) and the gyroscope (ω_x, ω_y, ω_z). Each user has a total of 5000 time windows or instances, 1000 for each phone location. In total, there are 35000 samples for training (distributed among the 7 clients) and other 15000 for testing.

For the model, we used a Convolutional Neural Network (CNN). Our architecture has 6 input channels and consists of two 1D convolutional layers, one max-pooling layer, one flattening layer, two dropout layers and two fully connected layers. The total number of parameters is 764399. Table 1 shows the details of the architecture. A dropout rate of 0.2 was used in both dropout layers. We use a batch size of 100 instances. On each training round, 10 epochs are performed by each client, and then the global parameter aggregation is done in the cloud. We assume all clients begin to acquire data at the same time and at the same frequency.

Table 1. Details of the CNN architecture used in the experiments.

Layer name	Kernel size	# kernels	Stride	Feature map.	# params
conv1	1 × 10	100	1	115 × 100	6100
conv2	1 × 10	100	1	106 × 100	100100
max_pool	1 × 2	–	1	53 × 100	0
dropout1	–	–	–	53 × 100	0
flattening	–	–	–	1 × 5300	0
fully_con1	–	–	–	1 × 124	657324
dropout2	–	–	–	1 × 124	0
fully_con2	–	–	–	1 × 7	875

In order to have a baseline, we first conducted federated learning under the unrealistic assumption that data is acquired in an identically distributed manner over time and for all users. Thus, we simply shuffle all the data of each user (5000 samples) in a random way without taking into account the position of the phone, thus forcing an IID scenario. We applied regular FedAvg algorithm during a total of 25 rounds of local training and server aggregation. At the end of the process, we achieve over 80% accuracy on test. Figure 2 shows the evolution over time. The thick black line is the overall accuracy, whilst each of the thin coloured lines represent how well the model performs when evaluated on the test data specific to each phone position.

Results from Fig. 2 are quite good but, in real life, data is usually non-IID and evolves over time. Therefore, now, for each user, we will sort the data in the data stream by the phone position, thus forcing a non-stationary distribution that will change a total of 4 times. Each change in the device location will involve a change in the data underlying distribution, that is, a concept drift. Data is sorted according to the phone position in the same way for all users: 1st belt, 2nd left pocket, 3rd right pocket, 4th upper arm, 5th wrist. For each location, the respective 1000 data observations are randomly sorted. This is not a totally realistic scenario since in real life each user would acquire data in a totally different manner, but it helps us to constrain the changes and check their impact during training. Again, we applied regular FedAvg algorithm with

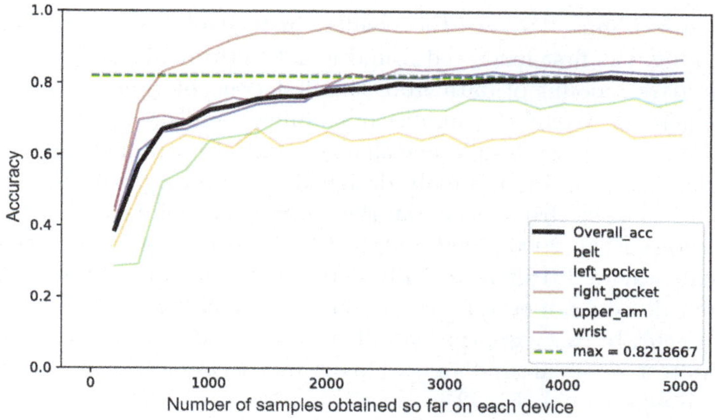

Fig. 2. Test results for standard FedAvg in an IID setting. The thick black line is the overall test accuracy.

exactly the same configuration as before, training a total of 25 rounds. Figure 3 shows the results. The vertical dashed lines indicate where a distribution change occurs, which is every 1000 patterns for all the users. It can be seen that, although the general tendency of the model is to improve, it forgets concepts as it learns others. For example, from iteration 3000 it begins to learn about the upper arm position, which causes a drop in accuracy for data related to belt and left pocket. The final accuracy of the model on the test set is around 60%.

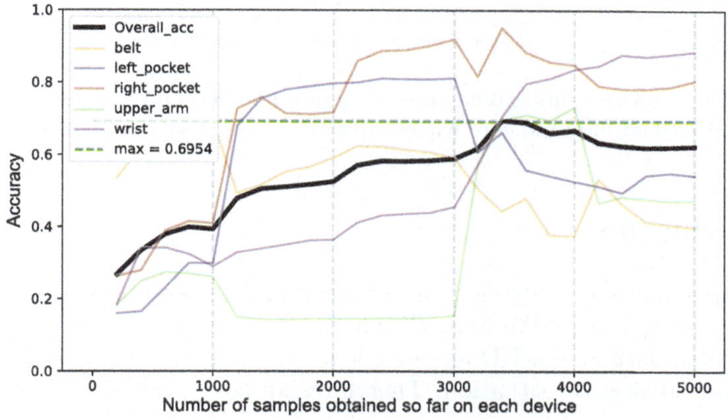

Fig. 3. Test results for standard FedAvg in a non-IID and nonstationary setting. The vertical dashed lines indicate where a distribution change occurs.

Finally, we repeat the last experiment but this time using our extended version of FedAvg. This time, the training process is aware of the existence of the concept drifts. Figure 4 shows the results. Instead of starting training at the very beginning, the first federated round is not performed until all 7 users have a representative amount of data from the first concept (belt position) in their long-term memory. We fix the number of training rounds per concept to 5. Thus, in this case, as there are 5 different changes, model is trained for a total of 25 rounds. This has been intentionally designed to be on a par with the previous setting. The concept drifts occur at the same time points as in the previous case (iterations 1000, 2000, 3000 and 4000). However, now the vertical dashed lines indicate where a drift is actually detected by one or several of the clients applying our detection method. We can notice that all drifts are detected shortly after they occur. In this case, the overall accuracy at the end of the learning process on the test set is around 78%. This is much closer to the result obtained with the baseline model (Fig. 2).

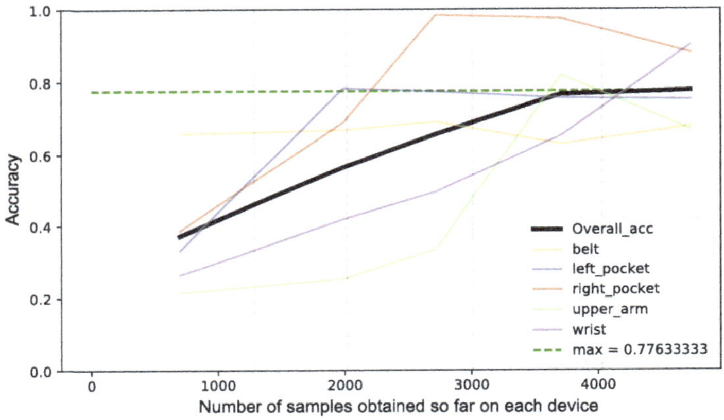

Fig. 4. Test results for our extended FedAvg in a non-IID and nonstationary setting. The vertical dashed lines indicate when a drift is detected by our algorithm.

6 Conclusions

In this paper we have addressed the problem of federated and continual learning under concept drift. We have shown the shortcomings of regular federated learning when data is non-IID among clients and nonstationary over time. This is a common real-world situation. Therefore, we have proposed a new method, which is an extension of the original Federated Averaging (FedAvg), but capable of detecting concept drifts and adapting to them. In order to detect the drifts, we use a distribution-based algorithm, which uses a confidence metric to quantify the dissimilarity between the distribution of historical and new data. We define a short-term and a long-term memory for each of the clients. When a drift happens, we adapt the federated model to it by applying rehearsal using the data in the long-term memory. Thus, we avoid catastrophic forgetting. Moreover, we provide an answer to the two key questions of *what* needs to be learned and *when* it should be learned. This allows us to save storage, communication and computational resources. We have tested our proposal in a real multiclass classification problem, human activity recognition, and we have also shown that our proposal outperforms regular FedAvg in this kind of scenario.

Acknowledgments. This research has received financial support from AEI/FEDER (EU) grant number TIN2017-90135-R, as well as the *Consellería de Cultura, Educación e Ordenación Universitaria* of Galicia (accreditation 2016–2019, ED431G/01 and ED431G/08, and reference competitive group ED431C2018/29), and the European Regional Development Fund (ERDF). It has also been supported by the *Ministerio de Universidades* of Spain in the FPU 2017 program (FPU17/04154).

References

1. Lu, J., Liu, A., Dong, F., Gu, F., Gama, J., Zhang, G.: Learning under conceptdrift: a review. IEEE Trans. Knowl. Data Eng. **31**, 2346–2363 (2018)
2. Lesort, T., Lomonaco, V., Stoian, A., Maltoni, D., Filliat, D., Díaz-Rodríguez, N.: Continual learning for robotics: definition, framework, learning strategies, opportunities and challenges. Inf. Fusion **58**, 52–68 (2020)
3. Parisi, G.I., Kemker, R., Part, J.L., Kanan, C., Wermter, S.: Continuallifelong learning with neural networks: a review. Neural Networks (2019)
4. McMahan, H.B., Moore, E., Ramage, D., Aguera-Arcas, B.: Federated learning of deep networks using model averaging. arXiv preprint arXiv:1602.05629v1 (2016)
5. Li, Q., Wen, Z., He, B.: Federated learning systems: vision, hype and reality for data privacy and protection. arXiv preprint arXiv:1907.09693 (2019)
6. Casado, F.E., Lema, D., Iglesias, R., Regueiro, C.V., Barro, S.: Learning from the individuals and the crowd in robotics and mobile devices. In: Iberian Robotics Conference, pp. 632–643. Springer (2019)
7. Casado, F.E., Lema, D., Iglesias, R., Regueiro, C.V., Barro, S.: Collaborative and continual learning for classification tasks in a society of devices. arXiv preprint arXiv:2006.07129 (2020)
8. Zhao, Y., Li, M., Lai, L., Suda, N., Civin, D., Chandra, V.: Federated learning with non-iid data. arXiv preprint arXiv:1806.00582 (2018)
9. Caldas, S., Wu, P., Li, T., Konečný, J., McMahan, H.B., Smith, V., Talwalkar, A.: Leaf: A benchmark for federated settings. arXiv preprint arXiv:1812.01097 (2018)
10. Yoon, J., Jeong, W., Lee, G., Yang, E., Hwang, S.J.: Federated continual learning with adaptive parameter communication. arXiv preprint arXiv:2003.03196 (2020)
11. Haque, A., Khan, L., Baron, M.: Sand: semi-supervised adaptive novel class detection and classification over data stream. In: Thirtieth AAAI Conference on Artificial Intelligence, pp. 1652–1658 (2016)
12. Baron, M.: Convergence rates of change-point estimators and tail probabilities of the first-passage-time process. Can. J. Stat. **27**(1), 183–197 (1999)
13. van de Ven, G.M., Tolias, A.S.: Three scenarios for continual learning. arXiv preprint arXiv:1904.07734 (2019)
14. Shoaib, M., Bosch, S., Incel, O.D., Scholten, H., Havinga, P.J.: Fusion of smartphone motion sensors for physical activity recognition. Sensors **14**(6), 10146–10176 (2014)

QoS Metrics-in-the-Loop for Better Robot Navigation

Renan Salles De Freitas[1], Adrián Romero-Garcés[2], Rebeca Marfil[2], Cristina Vicente-Chicote[3], Jesús Martínez-Cruz[2], Juan F. Inglés-Romero[4], and Antonio Bandera[2(✉)]

[1] Blue Ocean Robotics, Odense, Denmark
renan028@gmail.com
https://www.blue-ocean-robotics.com/
[2] Universidad de Málaga, Málaga, Spain
ajbandera@uma.es
[3] Universidad de Extremadura, Cáceres, Spain
[4] Biometric Vox S.L., Murcia, Spain

Abstract. Navigation is an essential capability in most robotic solutions, being basic in scenarios such as inspection and maintenance, or factory intralogistics. Moreover, with the advance of service robotics and its increasing use in all kinds of real-world applications, nowadays, robots are expected to operate well in unstructured and dynamic environments, where for instance they could work close to human counterparts. In this context, traditional navigation approaches are no longer valid due to their constrained flexibility. Dealing appropriately with the intrinsic variability of open-ended environments requires robots to adapt themselves according to the situation they perceive in order to achieve the required quality of service. This paper describes a model-based framework for dealing with adaptive robot navigation. Our proposal relies on the systematic use of models for dynamically reconfiguring the robot navigation behavior, defined in terms of Behavior Trees, according to the runtime prediction and estimation of quality of service metrics based on system-level non-functional properties.

Keywords: Runtime self-adaptation · Quality of Service metrics · Behaviour Trees

1 Introduction

Robotics technology is currently spreading from the highly structured environments with well-defined use cases of large-scale mass manufacturing to a wider range of market domains. In contrast to the structured scenarios for industrial robotics, these new application areas demand robots to interact with different unstructured and dynamic environments, often sharing space and tasks with people. It is required that robots will be endowed with the particular functionalities required to accomplish the very specific use cases that each application

L. M. Bergasa et al. (Eds.): WAF 2020, AISC 1285, pp. 94–108, 2021.
https://doi.org/10.1007/978-3-030-62579-5_7

area requires, fulfilling non-functional properties such as safety or performance. For being deployed in these challenging scenarios, robots must be autonomous, but this autonomous capability must be revisited to have an effective impact on the adoption of robots. Robots must perform their missions with *dependable autonomy* [2]. Among the core technologies involved in providing adequate levels of dependability in robotic systems, one relevant topic is self-adaptation. If an autonomous robot is the one able to operate in a real world setting without any form of external control for extended periods of time, we can assume that, in the previously described scenario, it is mandatory that an autonomous robot will be able to monitor its context, internal and external, for self-adapting its behaviour when any relevant change is perceived.

Considering that navigation is an essential capability in most robotic solutions, this paper focuses on describing our effort for endowing the self-adaptation ability to the navigation stack of a robot system. Robot navigation is a complex task, involving a collection of sub-problems that can range from the high-level decision of how to organize the trajectory (i.e. determining the ordered sequence of relevant places to visit) to the low-level reactive obstacle avoidance. Besides, navigation is closely linked to Mapping and Localization and tied to hardware and application-specific components (e.g. person-related perceptors for social navigation). With the advance of service robotics and its increasing use in all kinds of real-world applications, nowadays, robots are expected to operate well in unstructured and dynamic environments, where for instance they could work close to human counterparts. Dealing with variability in these open-ended environments requires robots to adapt themselves according to the current situation in order to achieve the required Quality of Service (QoS). QoS will be highly dependent on the specific nature of each application. For instance, a robotic vacuum cleaner or a weed-killing robot in the agri-food domain should cover the working space in as much as possible. A healthcare robot assisting elderly people should put special emphasis on safety and social awareness when moving around in a retirement home or a geriatric day-care hospital. Performance and punctuality are basic features for a robot working in intralogistics inside a factory. All these requirements (coverage, safety, social-awareness, punctuality...) express non-functional properties (NFPs), which define how a system performs rather than what it does. Being able to measure the degree of fulfilment of NFPs during the operation of the robot, is essential to identify when an adaptation is needed and how much the system improves after its application [1].

This paper describes a framework for endowing navigation with the desirable runtime self-adaptation. At design-time, the framework is intended to support the modeling of (i) variation points (which determine the decision space of the adaptation process, i.e., the answer to what can be adjusted). Variation points will be linked to elements in the specification of the robotic behavior (e.g., tasks and skills); (ii) contexts, mainly expressed in terms of QoS metrics; and (iii) adaptation policies, explicating how to configure the variation points depending on the current context in order to optimize relevant non-functional properties (NFP), such as safety or performance. All these topics are covered in Sect. 2.

The specific deployment of the proposal into a robotics architecture is described in Sect. 3, which also describes how modules exchange information, with the aim of providing a final, context-aware course of actions. Section 4 describes our experimental setting and provides some examples that allow to validate the proposal. In this example, the proposed approach is built on the open source Flexible Navigation stack, from the Hochschule Ulm (Ulm University of Applied Sciences, HSU) [3]. Experiments show how the framework is able to monitor the context and then adapt the robot's course of action to external and internal events. Finally, conclusions and future work are drawn in Sect. 5.

2 Adaptive Robot Navigation at Design-Time

The proposed framework gathers a comprehensive set of model-based tools (in the form of Eclipse plug-ins). Both modeling and code generation tools are provided, enabling the creation of composable systems with adaptive navigation capabilities. At design time, our framework allows the modeling of (1) variation points, which specify what can be configured at runtime and, thus, determine the decision space for the adaptation process; (2) contexts, mainly expressed in terms of QoS metrics; and (3) adaptation rules, describing how, depending on the current situation, the affected variation points should be configured in order to optimize relevant NFPs. Our navigation use cases are specified in this paper using Behaviour Trees [4,5]. It is worth noting that other alternative methods for decision making could also be considered, as long as they provide ways to switch the current course of action based on the recommendations received from the adaptation logic.

2.1 Nominal Behaviour Tree

In order to illustrate our approach, we have set a simple use case: the robot will be deployed in a retirement home to search for people and remind them that it is time to take their medicines. As the retirement home has several rooms, the robot will have to go through them to detect and recognize people, and, if she is the person it is looking for, then approaches her for giving the message. As aforementioned, this task is described using a nominal Behaviour Tree (BT). Figure 1 shows the map of the retirement home and this nominal BT. The scenario has been built using Webots [7]. To design the BT, we have used the Groot integrated development environment [8].

The nominal BT in the figure is composed of a main tree and several subtrees. The one depicted at the top of Fig. 1 is the main tree, whose aim is to activate the `peoplerecognition` ability, and then move the robot from room to room, looking for patient `Thomas` (note that this example is limited to two rooms for the sake of clarity). If `Thomas` is not in these rooms, the mission ends and the robot returns to `Room8` in order to recharge its batteries. The BT depicted at the bottom of the figure implements the `SearchThomasRoomX` subtree. Briefly, it moves the robot to the door of the room and, from there, to a second goal inside

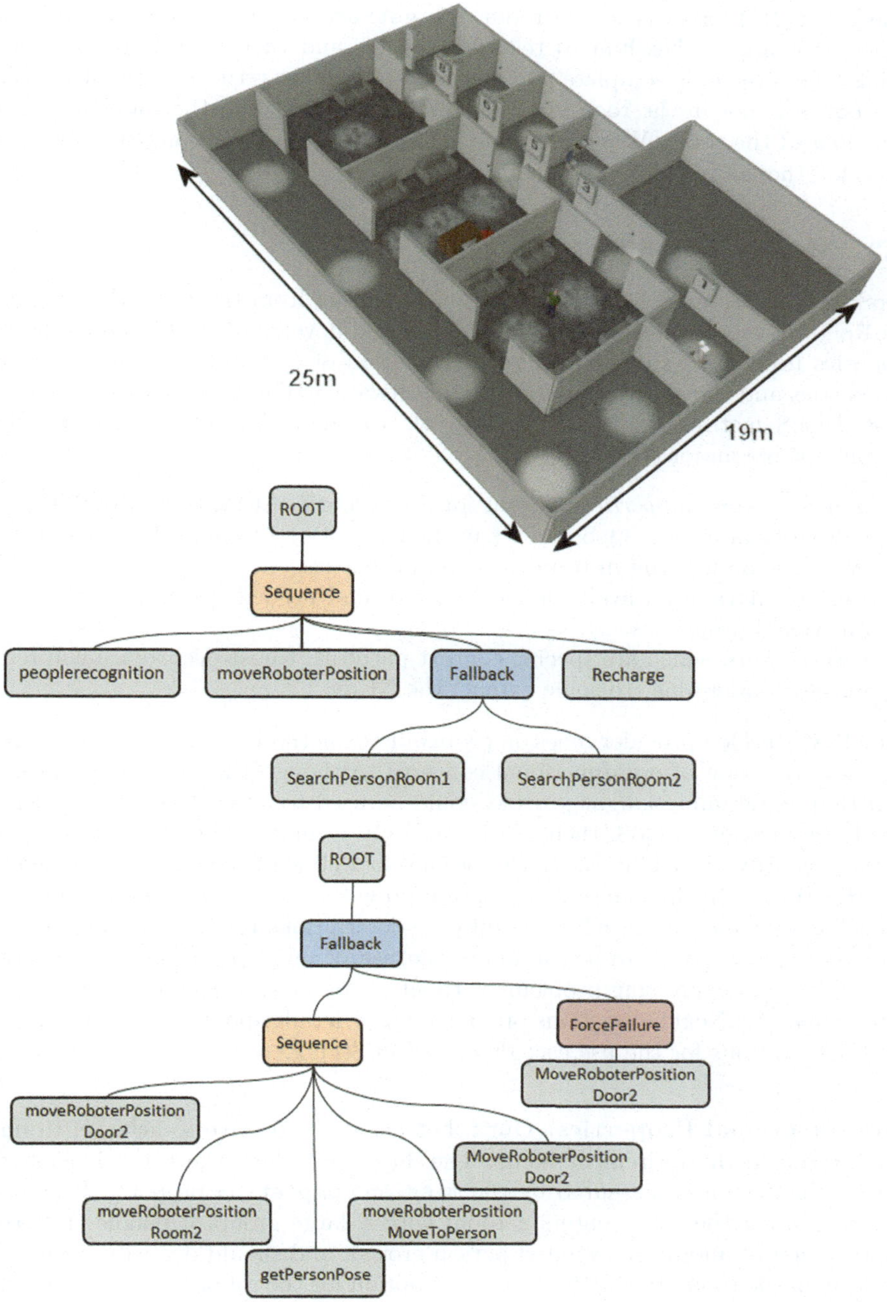

Fig. 1. Map of the retirement home (powered by Webots) and nominal BT. There are two BTs. The upper one is the main one and the one below extends the `SearchThomasRoomX` node (see text for details)

the room. If `Thomas` is in the room, the `getPersonPose` returns his position. The robot approaches him to tell the message and then returns to the door. Thus, the mission is complete and the course of action returns to the main BT. If `Thomas` is not in the room, the `getPersonPose` fails and the robot moves to the door of the room. When the main BT resumes the execution, the robot will look for `Thomas` in the second room.

2.2 Modelling QoS Metrics

QoS metrics are modelled using the tools resulting from the RoQME Project[1]. In RoQME, our team worked on the design and development of a model-driven robotics framework supporting the specification of system-level non-functional properties and the generation of runtime software artifacts to estimate the associated QoS metrics [1]. In this framework, the modelling of the QoS metrics involves three major concepts:

- *non-functional properties* relevant for the scenario. At runtime, RoQME provides estimates (i.e., QoS metrics in the range [0, 1]) to indicate how well the system is performing in terms of these properties.
- *context* information available to the robot and needed to compute the non-functional properties.
- *observations*, which are specific context patterns, whose occurrence will reinforce or undermine (to some extent) the NFPs.

The RoQME QoS provider relies on probabilistic methods to address uncertainty in the estimation of non-functional concerns. Although there are many probabilistic programming languages that could be used to apply Bayesian inference to the estimation of QoS, their use is generally complex. They require to accurately quantify probabilities, which does not fit well with the natural way people express things. In this sense, our proposal provides a simple modelling language to define QoS metrics, in which qualitative descriptions predominate over quantitative ones [1]. As a result, it limits the use of numbers, promoting a more natural way of expressing problems without introducing ambiguity in its execution semantics. Next subsections provide further details about the NFPs, context and observations for the use case described in Sect. 2.1.

Non-functional Properties. Our robot has one objective to achieve: to find one person in the retirement home. The degree to which the robot is close to locate the person is measured by the `mission_completion` property. It should increase when the robot enters a room with a large group of people (i.e., the expectation of finding the wanted person grows), and should decrease if there is nobody in the room. In this case, the best option for the robot is to move quickly to a new room to continue the search. Other non-functional properties considered in our example are listed in Listing 1.1. Apart from `mission_completion`, we

[1] https://robmosys.eu/wiki/baseline:environment_tools:roqme-plugins.

will take into account `safety` and `power_autonomy`. The first one decreases when there are people or objects in the robot surroundings. The second one is directly linked to the battery level. According to the language syntax, it is necessary to state a reference value between 0 and 1, which is the default value of the QoS metric. Note that properties use different reference values. Thus, for the `mission_completion` one, the reference value is equal to 0,5. This means that, when the mission is launched, it is far away to be closed (but neither can it be said that it will not be completed). `Safety` and `power_autonomy` are initially set to 1, since the situation starts out as'safe' and the battery is full.

Listing 1.1. Non-functional properties

```
property safety reference 1
property power_autonomy reference 1
property mission_completion reference 0,5
```

Context and Observations. The runtime estimation of NFPs lies on the monitoring of the whole scenario (e.g., is there people in the room?; is it difficult to navigate due to the large density of furniture/people...?) and the information coming from the robot sensors (e.g. battery level). Listing 1.2 enumerates the contexts which are available for our use case. Basically, the robot is equipped with a number of software components that inform about the presence of people, the battery level, the presence of very close objects, etc. The modules in charge of providing this information will be introduced in Sect. 4.

Listing 1.2. Contexts definition

```
context PeopleInRoom : boolean
context WantedPersonFound : boolean
context BatteryLevel : number
context BatteryCharging : boolean
context RobotBump : boolean
context PersonBump : boolean
context ObjectIdentified : boolean
context SearchingPerson : boolean
```

Observations describe how contexts impact the NFPs. They are based on specified context patterns that define whether the non-functional property will be improved (reinforced) or worsened (undermined) and how much (very high, high. low and very low). Listing 1.3 shows the observations specified for the `safety` property. Basically, the QoS metric associated with `safety` decreases when there are very close objects or people, or when the robot detects too many people on a room. The opposite situations increase the metric. We have also defined other observations for the remaining NFPs of the use case.

Listing 1.3. Observations for the `safety` property

```
o8 : eventWhen(ObjectIdentified) reinforces safety
o9 : eventWhen(!PeopleInRoom) reinforces safety
o10 : (RobotBump -> !RobotBump) or
```

```
         (PersonBump -> !PersonBump) reinforces safety low
o11 : eventWhen(PeopleInRoom) undermines safety low
o12 : eventWhen(PersonBump) or eventWhen(RobotBump)
         undermines safety high
```

2.3 Modelling Robot Reconfigurations

Variation points are the elements that can be reconfigured in the robot to improve its performance, in terms of the NFPs that we previously defined with RoQME. We propose a new modelling language, called MIRoN, that brings together RoQME and BTs to allow the specification of variation points. MIRoN considers two different types of variation points, namely, parameters and optional/alternative behaviours. In any case, each variation point will be linked to a node in the nominal BT to indicate the target that will be adapted at runtime. Therefore, a variation point can be associated with a parameter defined in a node, which will leave its update to the adaptation process; or a variation point can add optional/alternative behaviours (i.e., variants expressed as behaviour subtrees) to the nominal BT. The upper part of Fig. 2 shows two variation points in MIRoN and how they connect targets in the nominal BT and variants.

From a MIRoN model, we will be able to generate an extended BT including the variability expressed by the variation points. At runtime, the robot will execute this new BT together with the adaptation engine. The latter will be responsible for closing the open variability by assigning concrete variants and values to the variation points, taking into account the RoQME QoS metrics. Regarding the extension of the nominal BT, since parameters do not modify the course of action, the nominal BT does not change as a result of a parameter change. That is, parameter updates are not executed through the BT but by accessing specific skills[2] in the robot system. Conversely, when a variation point applies an optional/alternative branch to the nominal BT, the corresponding variants are woven into the nominal BT. Figure 3 shows the extended BT associated to the main one in Fig. 1. The `Sequence` branch runs in parallel to a new branch (`Repeat`) where a `variant` value can be recommended by the adaptation engine (`VariantAction`), and annotated in the internal blackboard of the BT[3]. The robot will perform the `Recharge` action if variant "one" is selected. In the default case, the robot will look for Thomas in the rooms, as described in Sect. 2.1.

Thus, taken into consideration the framework described in previous Sections, each proposed model is linked: (a) to a RoQME model, since the contexts are linked to RoQME contexts and QoS metrics; (b) to a nominal BT model, describing the basic (non-adaptive) behaviour of the robot; and, optionally (c) any number of additional BT models, describing each behaviour variant (see Fig. 2.[Top]).

---

[2] In this paper, a task describe how a robot does something in an abstract and independent manner, and a skill provides access to functionalities realized by components for the usage within tasks.

[3] See https://www.behaviortree.dev/tutorial_02_basic_ports/ for details about the blackboard.

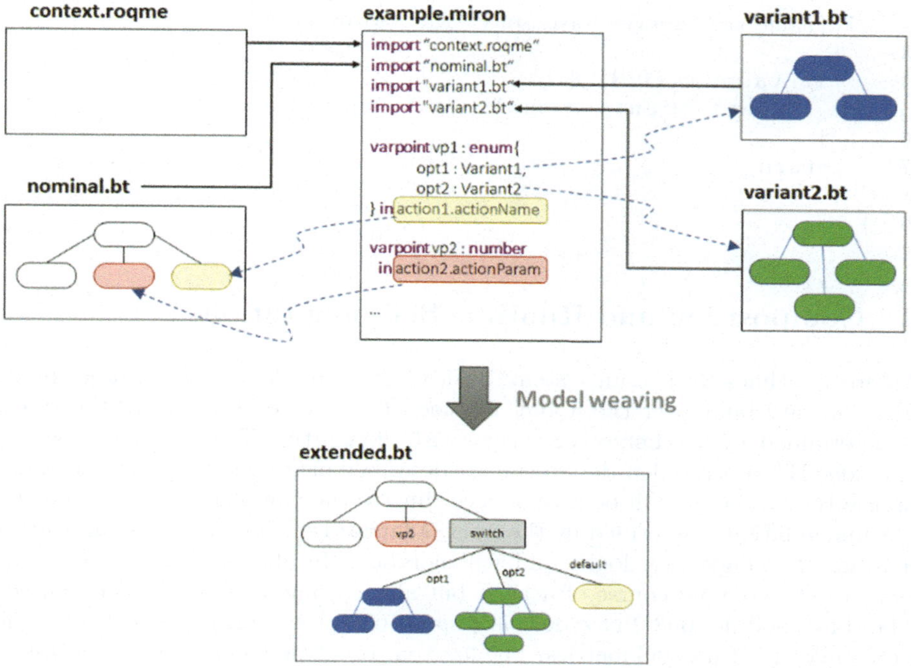

Fig. 2. [Top] Models involved in the process: a RoQME and a MIRoN model and several BT models (nominal and variants). [Bottom] Extended BT model generated by weaving the BT variants in the nominal BT model, according to the MIRoN specification.

Besides, both the RoQME and BT models are linked, in turn, to other elements of the robot system (perceptive and actuation components).

Finally, the runtime self-adaptation is driven by the application of a set of rules to the streams of QoS metrics. The adaptation logic must tune the variation points for maximizing the NFPs. In our case, Listing 1.4 provides the two simple rules employed for change the maximum velocity or stop the normal execution encoded on the nominal BT for sending the robot to charge their batteries.

Listing 1.4. Adaptation logic

```
void EstimateListener::dataAvailable(...){

 double value = data.value();
 const std::string name = data.name();

 if(name == "safety" && value != -1)
  {
   double sq_value = sqrt(value);
   changeVelocity(1000*sq_value/(sq_value+sqrt(1-value)));
   return;
  }
```

```
  if(name == "power_autonomy" && value != -1)
  {
    if(value <= 0.3) {
     variant_client_->sendVariant("one");
    }
  return;
  }
}
```

3 Composition and Runtime Self-adaptation

Figure 4 outlines the overall system. Basically, we can identify three major modules that are built over the `Robot System`. The course of action of the robot is determined by the `Behaviour Tree (BT) Executor`. This module runs the extended BT generated at design-time as described in Sect. 2. The different leaves on the BT (variants) will be chosen according to the nominal execution and the recommendations prescribed by the `Self-Adaptation Engine`. This engine will manage the adaptation logic, and will not be only able to suggest modifications to the nominal course of action, but also to tune how a skill is executed. The third module (`QoS Provider`) is responsible for capturing context information and estimating QoS metrics. The `Navigation Stack` endowed in the Robot System allows us to update parameters at runtime. It is also necessary that the Robot System will be able to capture the required internal and external context (`Perceptive Modules`).

Briefly, at runtime, (1) the QoS Metrics Provider continuously estimates system-level NFPs based on the robot (internal and external) contextual information; (2) these QoS metrics are used by the adaptation process to evaluate the degree of fulfilment of the required NFPs. At one point, if the robot navigation shows a poor performance on any of the measured metrics, the adaptation logic will change the parameters of the skill under execution (e.g. the maximum velocity of the robot as the safety shows low values) or will annotate information about the variant on the extended BT to choose. The way for communicating this decision to the BT is by updating a variant value in the internal blackboard of the BT. This is the mission of the `VariantAction` in the extended BT shown in Fig. 3. This value is managed by the `Switch2` node included in this BT. Thus, this `Switch2` node chooses variants that were stated at design-time, but according to measures captured at runtime. In any case, the BT Executor is always the module in charge of determining the course of action, sending the Robot the stream of skills to be executed. As stated before, QoS metrics are always evaluated at runtime, allowing the adaptation engine to make decisions on the variation points defined at design time to improve the NFPs.

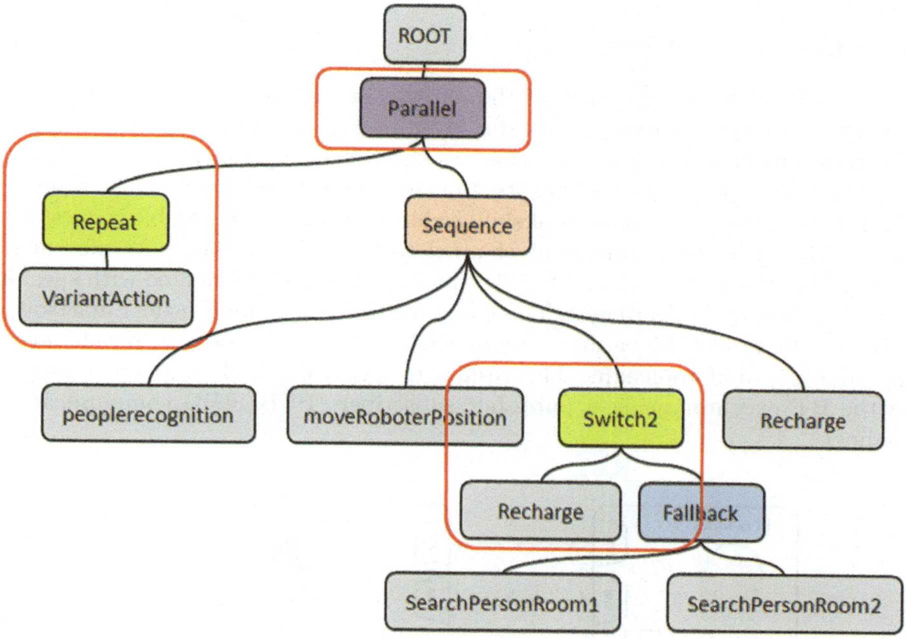

Fig. 3. Extended BT model generated from the nominal BT defined at Fig. 1.

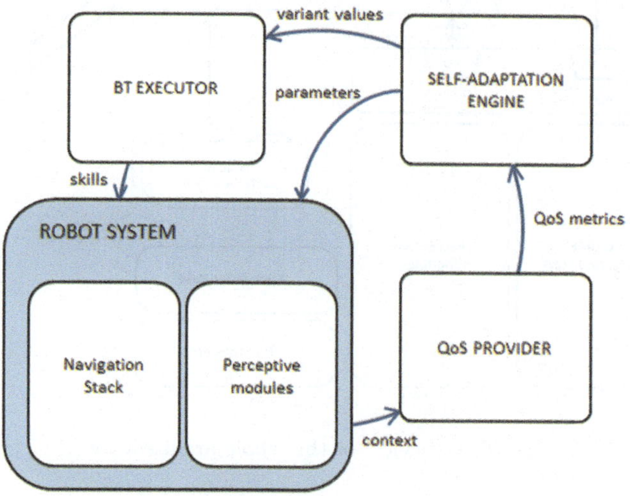

Fig. 4. Main modules and their interaction

4 Experimental Evaluation

4.1 Experimental Setting

Figure 5 provides an overview of the current instantiation of our proposal. It includes the four modules already displayed in Fig. 4, together with some additional details. The BT Executor employs the BehaviorTree.CPP library, developed by Davide Faconti[4] in the context of the MOOD2Be Project[5]. The result of the BT execution is a sequence of skills, which are sent via a reliable request/reply (REQ/REP) communication channel to the robot system. As aforementioned, this system includes a Flexible Navigation Stack, augmented with specific modules for context-awareness. These modules have been built using Smartsoft[6], a systematic model-driven and component-based robotics software development methodology and toolchain. The results of executing each skill are reported to the BT executor using a publisher/subscriber (PUB/SUB) communication channel.

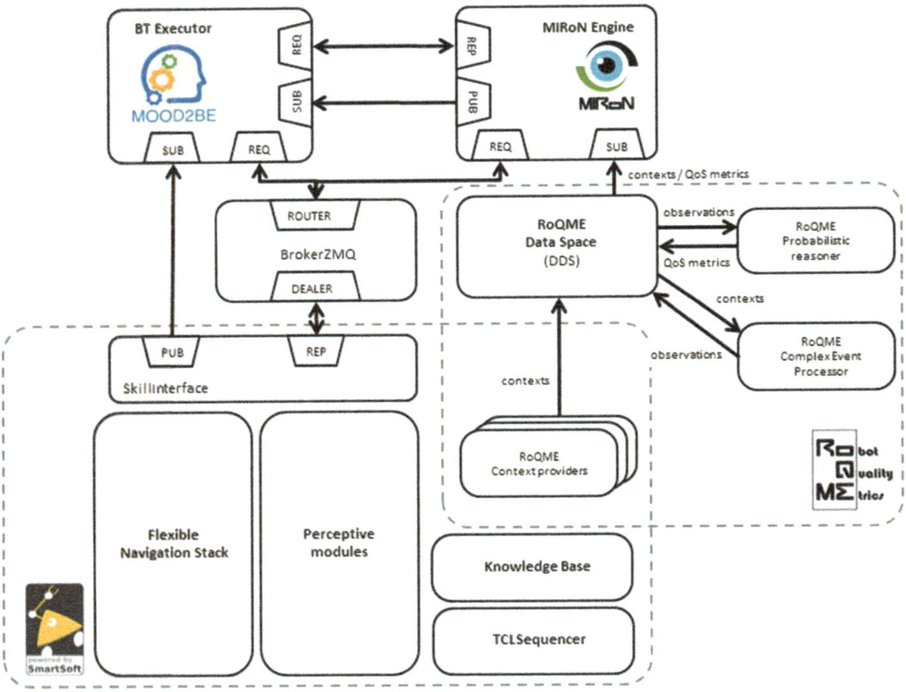

Fig. 5. Overview of the whole architecture.

[4] https://github.com/BehaviorTree/BehaviorTree.CPP.

[5] https://robmosys.eu/wiki/community:behavior-tree-demo:start.

[6] https://robmosys.eu/wiki/baseline:environment_tools:smartsoft:start.

Context is captured in the Smartsoft architecture and sent using publisher/-subscriber communication channels to the RoQME module in charge of providing the QoS metrics. For generating the stream of metrics, the RoQME module internally uses a Complex Event Processor (CEP) and a Probabilistic Reasoner [1,6]. The QoS metrics are the inputs of the Self-adaptation engine, which will generate two types of messages:

- Parameter updates (directly sent to the navigation stack). These parameters do not imply changes on the course of action, but fine tuning the execution of a skill.
- Variant values (captured as a pair variant - value). Selected variants are updated on the internal blackboard of the BT Executor and aim at modifying the course of action.

The stream of parameter changes shares the input port to the Smartsoft architecture with the stream of skills coming from the BT Executor. This forces us to use a broker for managing both communication streams.

The whole Smartsoft-based system is detailed next, in Fig. 6. This figure identifies several navigation-related and context-awareness modules, including those responsible for managing the skills. This instantiation is running over the simulation environment generated using Webots. Thus, the perceptive components implement sensors running on this simulated world.

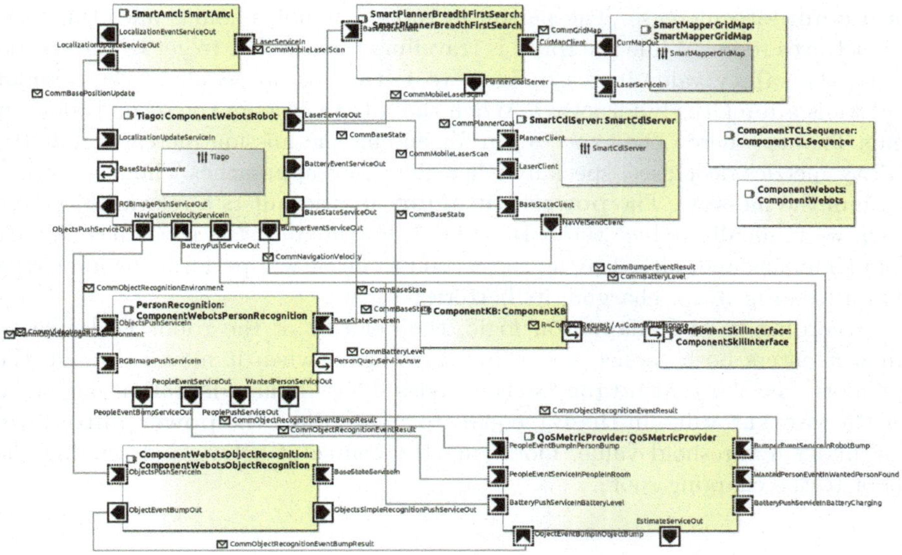

Fig. 6. Architecture of the Smartsoft-based robot system

The Flexible Navigation Stack includes software components for path-planning (`SmartPlannerBreadthFirstSearch`), reactive navigation (`SmartCdlServer`), mapping (`SmartMapperGridMap`) and localization (`SmartAmcl`). The skills are receiving in the `SkillInterface` and managed using the Smartsoft `TCLSequencer` and `Knowledge Base`). These last components implement the skill execution using the Domain-Specific Language SmartTCL (Task Coordination Language)[7]. The rest of components in Fig. 6 allows our system to work on the simulated world: `Webots` and `Tiago:WebotsRobot` provide the world and Tiago robot, respectively; and `WebotsPersonRecognition` and `WebotsObjectRecognition` implement the person and object recognition modules. The `QoSMetricProvider` captures the context that is needed for updating our properties and sends the QoS metrics to the adaptation engine. Topics not useful for the use case are not considered. For instance, the `ObjectsSimpleRecognitionPushServiceOut` from `WebotsObjectRecognition` is not connected to the `QoSMetricProvider`, as we are not using this key information.

4.2 Runtime Monitoring and Actuation

Figure 7, shows how the QoS metrics change when the robot is moving through the environment depicted in Fig. 1 for searching Thomas. When the robot enters the first room, the **safety** value decreases (the entrance is narrow, so obstacles are close to the robot (`RobotBump`), but also there is a person in this first room (`PeopleInRoom`). The presence of people in the room makes the **mission_complete** increase. The person in room 1 is not Thomas and, thus, the robot leaves room. While the robot is travelling the corridor to get to the second room, the **safety** value increases again to 1 (there is no people in the corridor and walls are relatively far away). When the robot enters in the second room, it finds Thomas there. The expectation for ending the mission increases, but the **safety** metrics decreases, specially when the robot approaches Thomas for giving him the message. The **power_autonomy** metric suffers a relevant decrease when we manually reduce the battery level. Fortunately, the robot has already found Thomas and, after leaving the second room, it will perform the `Recharge` action (see Fig. 3) for charging its batteries.

According to the adaptation logic, the velocity of the robot is decreased when it enters both rooms, but it increases again when it moves through the corridor. The `VariantAction-Switch` nodes allowed the adaptation engine to set the `variant` value in the BT equals to "one" when the **power_autonomy** was under a threshold value, modifying the course of action and sending the robot to the charging room.

[7] https://wiki.servicerobotik-ulm.de/about-smartsoft:robotic-behavior:smarttcl.

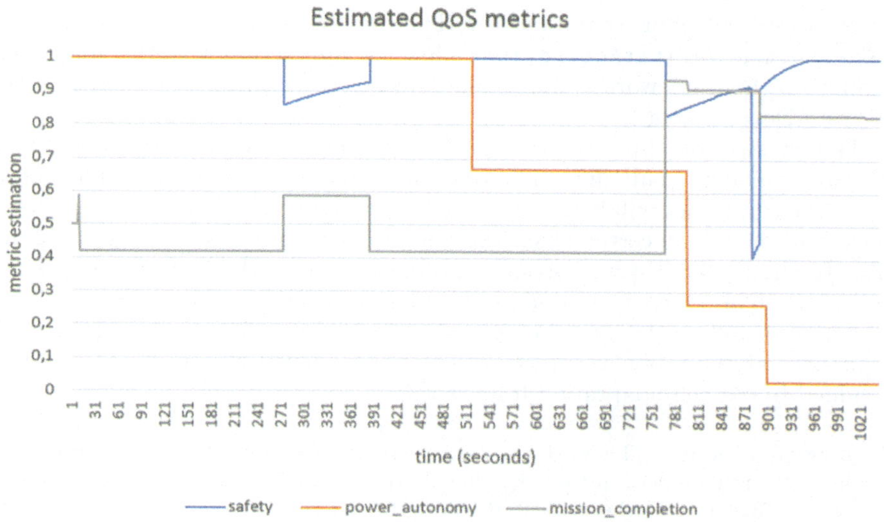

Fig. 7. Evolution of the QoS metrics

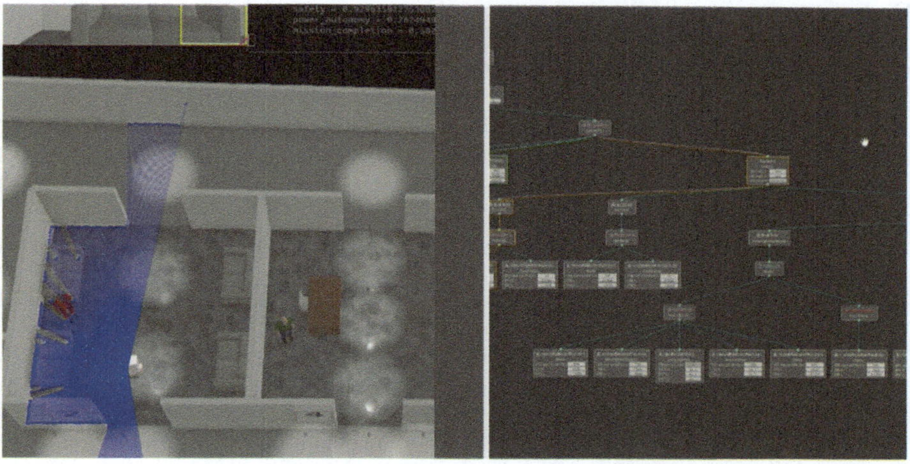

Fig. 8. Snapshot showing the robot moving in the simulated environment and the evolution on the extended BT

5 Conclusions and Future Work

This paper shows a complete framework for self-adapting the navigation behaviour of a robot. The framework provides a continuous monitoring of the internal and external context and its encoding using QoS metrics, and is able to generate recommendations for changing parameters or modifying the robot's course of action at runtime. Regarding the decision making process, our proposal integrates a Behaviour Tree Executor. This Behaviour Tree can deploy specific nodes (`Switch`, `VariantAction`) that allow the interaction with the adaptation engine. All the framework is implemented using the principles of Model-Driven Software Development.

The example on this paper is a simple one, but the described framework is valid for managing many different situations as it is not bound to the navigation domain. Future work will focus on evaluating complex missions in real scenarios. These will include uncertainty on the measurement processes, giving an added value to the probabilistic reasoner used for estimating the QoS metrics. These scenarios will also force us to update the adaptation engine and its internal logic. We will also analyse the inclusion of rewiring techniques for robot modules (that is, changes at the robot architectural/component level) as a new possibility to be added to our self-adaptation framework.

Acknowledgements. This work has been partially funded by the MIRoN Integrated Technical Project funded, in turn, by the EU H2020 RobMoSys Project under Grant Agreement 732410, and the project RTI2018-099522-B-C4X, funded by the Spanish Ministerio de Ciencia, Innovación y Universidades and FEDER funds.

References

1. Vicente-Chicote, C., Inglés-Romero, J.F., Martínez, J., Stampfer, D., Lotz, A., Lutz, M., Schlegel, C.: A component-based and model-driven approach to deal with non-functional properties through global QoS metrics. MODELS Workshops (2018)
2. Hernández, C., Bermejo-Alonso, J., Sanz, R.: A self-adaptation framework based on functional knowledge for augmented autonomy in robots. Integr. Comput.-Aided Eng. **25**, 157–172 (2018)
3. Flexible Navigation Stack [RobMoSys Wiki]. https://robmosys.eu/wiki/domain_models:navigation-stack:start
4. Dromey, R. G.: From requirements to design: formalizing the key steps. SEFM (2003).https://doi.org/10.1109/SEFM.2003.1236202
5. Colledanchise, M., Ögren, P.: Behavior trees in robotics and AI: an introduction, CoRR abs/1709.00084 (2017). http://arxiv.org/abs/1709.00084
6. Vicente-Chicote, C., García-Pérez, D., García-Ojeda, P., Inglés-Romero, J.F., Romero-Garcés, A., Martínez, J.: Modeling and estimation of non-functional properties: leveraging the power of QoS metrics. In: Ferrández Vicente, J., Álvarez-Sánchez, J., de la Paz López, F., Toledo Moreo, J., Adeli, H. (eds.) IWINAC 2019. LNCS, vol. 11487, pp. 380-388. Springer, Cham (2019). https://doi.org/10.1007/978-3-030-19651-6_37
7. Cyberbotics Ltd. Webots: open source robot simulator. https://www.cyberbotics.com/
8. Faconti, D.: Groot, an integrated development environment for the design and execution of behaviour trees. https://github.com/BehaviorTree/Groot

DSR_d: A Proposal for a Low-Latency, Distributed Working Memory for CORTEX

Pablo Bustos[1(✉)], Juan C. García[1], Ramón Cintas[1], Esteban Martirena[1], Pilar Bachiller[1], Pedro Núñez[1,2], and Antonio Bandera[2]

[1] RoboLab, School of Technology, University of Extremadura, Cáceres, Spain
pbustos@unex.es

[2] ISIS, School of Telecommunication Engineer, University of Málaga, Málaga, Spain

Abstract. Robotics Cognitive Architectures (RCA) are becoming a key element in the design of robots that need to be aware of its surrounding space and of their role in it. This is especially important for robots that interact with people in household, eldercare or industrial collaborative scenarios. We have proposed in earlier works an RCA called CORTEX designed for social robots operating in HRI environments. One of CORTEX's main elements is a working memory designed as a graph-like data structure that is accessed by all the computational modules in charge of some relevant function in the system. Our current implementation is based on the concept of a real-time database, where one of the modules stores, receives and publishes changes to all modules. In this paper, we propose a new design of this element based on the Conflict-free Distributed Replicated Data Types (CRDT) theory of distributed data types. The new working memory presents important advantages over existing designs that are demonstrated with several experiments.

Keywords: Social robotics · Cognitive architectures · CRDT · Working memory

1 Introduction

Cognitive robotics is concerned with endowing robots with the capacity to plan solutions for complex goals and to enact those plans while being reactive to unexpected changes in their environment. To pursue this goal, cognitive architectures for robotics attempt to provide a reasonable structure where all the functionalities of a working cognitive robot can be fit. If the final goal is to endow these architectures within future service robots, the new design should provide an adequate response to the demanding requirements imposed by the human-robot interaction scenario.

L. M. Bergasa et al. (Eds.): WAF 2020, AISC 1285, pp. 109–122, 2021.
https://doi.org/10.1007/978-3-030-62579-5_8

Despite the large efforts for making cognitive and social robots a potential counterpart of human users, it is currently not easy to find successful stories, neither in the industrial segment, nor in the academic world. In the exhaustive review by Kotseruba et al. [1], they assert that only some architectures implement multiple skills for complex scenarios. One of the cited architectures is CORTEX [2,3], a proposal emerged from previous National and European projects granted to our consortium. CORTEX is a long term effort to build a series of architectural designs around the simple idea of a group of agents that share a distributed, dynamic representation acting as a working memory. This data structure is called Deep State Representation (DSR) due to the hybrid nature of the managed elements, geometric and symbolic, and concrete (laser data) and abstract (logical predicates) [4]). Figure 1 shows the main elements of CORTEX in its current state.

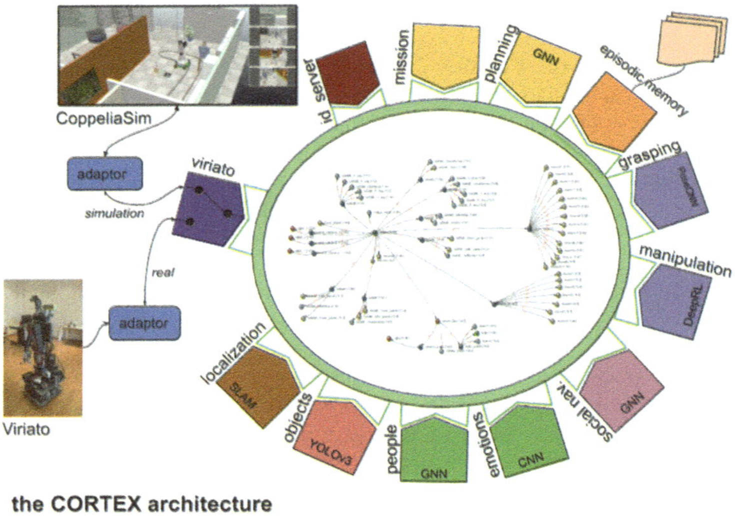

Fig. 1. The illustration shows a possible instance of the CORTEX architecture. The central part of the ring contains the DSR graph that is shared by all agents, from whom a reference implementation is presented here. Coloured boxes represent agents providing different functionalities to the whole. The purple box is an agent that can connect to the real robot or to a realistic simulation of it, providing the basic infrastructure to explore prediction and anticipation capabilities.

The CORTEX architecture exposes four important dimensions of the RCA design space: the trade-off between decoupling and sharing; the trade-off between top-down/bottom-down control; the functional content of the agents; and the granularity of the functional decomposition. Decoupling is the main engineering asset to tackle complexity. Accordingly, agents are defined as software encapsulated groups of components that provide some specific, limited functionality[1]. In this sense, encapsulation provides decoupling but isolates each agent from the valuable contextual information created by the others. To combine their functionalities and show some degree of intelligence, agents must be able to share information[2]. Each of them must know something about the others, otherwise their goals will remain local, and complex sequential tasks will be outside the reach of their individual abilities. The second dimension mentioned above is the top-down/bottom-up control trade off, by which the system has to allow for the existence of simultaneous and opposite streams of control. Top-down control in CORTEX is generated by, at least, one deliberative agent with the capacity to reason about high level tasks and compute efficient plans to drive the behavior of the robot. Bottom-up reactions to unforeseen situations are locally handled by agents while trying to fulfil plan steps. It should be noted that the planning agent must react itself if the execution of the plan fails. Finally, these reactions might trigger a cascade of changes in other agents through the shared memory, giving raise to large behaviour modifications. The third design axis deals with how the functional space, that is to be managed by the RCA, is partitioned in local domains. In CORTEX, this is the problem of defining the role of each agent in the overall problem space. Finally, the fourth dimension raises the issue of how large the functional domain of these agents can be in order to fulfill two requirements: being simple enough to facilitate the design of complex architectures; and being computationally realizable in terms of CPU usage, communication delays and software maintainability.

Since the conception of CORTEX, different design choices have been studied to integrate these features. As briefly noted below and further described in the references, several implementations of the CORTEX architecture have been deployed in different types of robots working in real world scenarios during these last years. All these use cases have a deep human-robot interaction component and include situations such as attracting potential consumers to a stand in public spaces [2], conducting geriatric tests to elderly people [6], conducting physical therapies with children [7], searching and bringing objects to humans or as a companion to the householder of an adapted apartment [8], or human-aware navigation in different types of environments with people and objects [9]. The implementations of these experimental setups have grown to a considerable software complexity, reaching up to 45 interconnected components.

On such large configurations, where all agents contribute from their functional domains to a common working memory, the access to data can emerge as a relevant bottleneck. The solution presented in this work is based on two key technologies that have the potential to reduce bandwidth consumption by the set of agents and maintain local immediate responsiveness to operation on the

[1] The term *agent* is used here as a synonym of module and, as such, no specific features such as goal-seeking or autonomy are imposed.

[2] We do not claim here that a shared representation is the only way to share information among agents. The *dynamicist* approach is a well-known alternative [5].

graph. The first one is a high-performance pub/sub middleware[3] implementing RTPS (Real-time Publish-Subscribe Protocol) and configured to use UDP reliable multicast. The data sent by an agent is broadcasted to all the others at once, minimizing the required bandwidth compared to point-to-point connections. The second one is a theoretical development named Conflict-free Replicated Data Types (CRDT) that provides data structures that can be safely and asynchronously edited by a distributed set of processes without a central server. By combining both technologies we have been able to create a new, highly efficient, distributed graph (the DSR_d). Contrary to our previous design, each local DSR_d copy can be now asynchronously edited by all the agents. Their changes are sent as incremental state modifications through the network using minimal resources and, when these updates are received, the local graph in each agent is updated concurrently with the user's operation, using a thread-safe interface to the graph. This new server-less version will only exist as the set copies maintained by the agents, that will eventually converge to the same final state after all editing is stopped. To our knowledge there are no other published CRDT graphs with the functionality required here.

The rest of the paper is organised as follows: Sects. 2 and 3 describe the internal structure of the graphical representation of the DSR and the relevance of the eventual consistency in our context. The implementation of the CRDT graph is presented in Sect. 4, where we also provide preliminary experimental validation. Finally, conclusions and future work are drawn in Sect. 6.

2 The Deep State Representation

Conceptually, the DSR represents a network of entities and relations among them. Relations can be unary or binary predicates, while the entities may have complex numeric properties such as pose transformation matrices that represent the kinematic relations of objects in the world and the robot's parts. Mathematically, the DSR is internalized as a directed graph with attributed edges. As a hybrid representation that stores information at both geometric and symbolic levels, the nodes of the DSR store concepts that can be symbolic, geometric or a combination of both. Metric concepts describe numeric quantities of objects in the world, which can be structures such as a three-dimensional mesh, scalars such as the mass of a link, or lists such as revision dates. Edges represent relationships between nodes. Two nodes may have several kinds of relationships but only one of them can be geometric. The geometric relationship is expressed with a fixed label called *RT*. This label stores the transformation matrix (expressed as a rotation-translation) between them (Fig. 2).

The DSR can be described as the union of two *quivers*: one associated with the symbolic part of the representation, $\Gamma_s = (V_s, E_s, s_s, r_s)$, and the other related to the geometric part, $\Gamma_g = (V_g, E_g, s_g, r_g)$. A quiver is a quadruple, consisting of a set V of nodes, a set E of edges, and two maps $s, r : E \rightarrow V$.

[3] https://www.eprosima.com/index.php/resources-all/performance/40-eprosima-fast-rtps-performance.

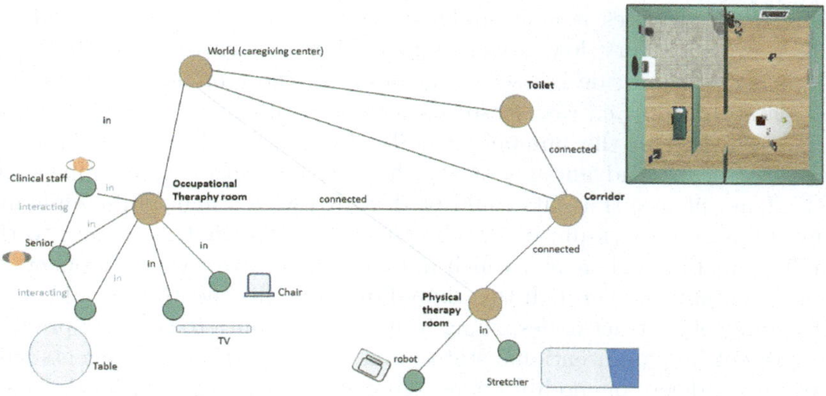

Fig. 2. Unified representation as a multi-labelled directed graph. For instance, the edge labelled as connected denotes a logic predicate between nodes and it belongs to Γ_s. On the other hand, the edge starting at the **Occupational therapy room** and ending at **the chair** is geometric and encodes a rigid transformation (RT' and RT respectively) between them. Geometric transformations can be chained or inverted to compute changes in coordinate systems.

These maps associate each edge $e \in E$ with its starting node $\mathbf{u} = s(e)$ and ending node $\mathbf{v} = r(e)$. Sometimes we denote an edge by $e = \mathbf{uv} : \mathbf{u} \rightarrow \mathbf{v}$ with $\mathbf{u} = s(e)$ and $\mathbf{v} = r(e)$. Within the DSR, both quivers are finite, as both sets of nodes and edges are finite sets. A *path* of length m is a finite sequence $\{e_1, ...e_m\}$ of edges such that $r(e_k) = s(e_{k+1})$ for $k = 1...m - 1$. A path of length $m \geq 1$ is called a *cycle* if $s(e_1)$ and $r(e_m)$ are identical.

This embedding of geometric and symbolic structures within the graph is exploited in the implementation described below through specific access APIs, that provide specialized methods to compute long-range kinematic transformation among distant nodes, in the first case, and PDDL conversion or some forms of reasoning, in the second.

3 Eventual Consistency and CRDTs

One problem with the standard server-oriented design that is currently being used in CORTEX is that an increase in the graph's size or in the data density will quickly increase latency and reduce throughput across the communication network. Once latency is too high, the context provided by the DSR becomes very difficult to use, since events in the world will happen faster than the reaction time required to compute a response. In the worst case, an agent could be always working on its local copy but being unable to publish its result. To address this problem we need to find a solution that provides both a minimum band-width use and a minimum amount of data exchanges among agents. Both requirements must hold while a certain level of coherence among the local copies is guaranteed.

This work proposes a new implementation of the DSR that is fully distributed, provides very low-latency and a high and sustainable throughput. Additionally, this new design scales up nicely in the number of agents, in the size of data objects (*i.e.*, raw sensor data from a camera or a LIDAR) injected at high rates and in the number of nodes (objects) and edges (relationships or predicates). The advantages of this high-performance working memory in CORTEX are plenty: i) agents could be designed with a finer grain while maintaining full access to graph; ii) the designer could exploit this feature to think of smaller functional elements, which implies simpler software components that still can communicate through this shared medium; iii) the graph could contain a wider range of abstract nodes, going from raw laser data to complex predicates that represent human intentional states; and iv) the participating agents will be able to inject, delete or modify nodes and edges as the result of local decisions extended with the global graph data. We aim at latencies below 50 ms for a wide range of configurations and payloads. With this value and assuming that most real-world interaction loops, at a mid-abstraction level, run 10 Hz, the agents will view the internal representation is always updated.

It is well-known that low-latency or high-availability in a distributed network of partitioned nodes is subject to the CAP theorem [10]. One cannot have both low-latency and consistency (*i.e.*, having an up-to-date copy of the data) at the same time if the data is stored in several nodes that can loose or degrade their connection at any time. Consistency requires synchronous updating of the state. If one wants consistency, it may happen that after receiving a write operation two nodes cannot communicate, their updates cannot be synchronously transmitted to the rest and a blockage will occur, thereby forfeiting low-latency. If one chooses low-latency, you need to allow that at least one node updates its state to provide a quick response, making the other nodes inconsistent and thus forfeiting consistency. It should be noted that even if all agents run on the same board or in boards connected by a network switch, they are partitioned nodes that can enter or leave the group at any time, and each one could have their own group of clients -e.g. local threads- requesting information.

In this proposal we choose low-latency and give up strong consistency in favour of a form of weak consistency named *eventual consistency* [11,12]. Under this form of consistency, changes made to one copy of the shared object are asynchronously propagated to the rest, but if all update activity stops, after a period of time all replicas of the data object will converge to the same state. This means that at some point in the future, all data in the nodes become indistinguishable from one another. In general, to ensure convergence, nodes must exchange information with another about what writes they have seen. While strong consistency requires synchronous updates to keep the semantics of a single-system image (SSI), eventual consistency uses asynchronous updates. When one node receives an update, it is executed locally, without synchronization. Then, it is sent to other nodes. All updates eventually reach all nodes, asynchronously and possibly in different orders.

CRDTs are distributed data types that are always available and eventually converge when all operations are processed at all nodes [13]. CRDTs come in two flavors: *operation-based* and *state-based.* In operation-based designs a representation of the local operation is created in the local node and shipped then to all other nodes. Once received, the representation of the operation is applied to the local copies of the receiving nodes. Alternatively, in a state-based design an operation is applied only on the local node's copy. Periodically, the node propagates its local changes to other nodes through shipping its entire copy. The other nodes incorporate the received copy with their local one via a *merge* function that deterministically reconciles both copies[4]. Although operation-based CRDTs allow for simple implementations and require smaller band-width, they require exactly-once causal broadcast and a new operation has to be defined for each kind of modification to the graph. State-based systems do not have this limitation but they may incur in high communication overheads when shipping large size local copies. Recently, a new improvement on state-based systems has been proposed that mitigates this problem. Delta-state RDT are based on propagating only a *representation of the effect* of the update operation on the copy [14,15]. Instead of shipping the complete copy, this new design allows the nodes to send only δ-states (i.e. incremental changes) of their copies.

4 Design and Implementation

The design of the distributed graph (DSR_d) is organized in three loosely coupled layers, as shown in Fig. 3. The deepest one contains the RTPS logic, as provided by eProsima in its FastDDS open source library. The next layer contains the CRDT logic, based on the C++ code kindly provided by C. Baquero[5]. We have used a CRDT implementation of a map data type as the core structure to wrap inside a graph with the characteristics described in Sect. 2. The third layer is the one exposed to the users of the tool and holds a view of the graph accessible through a thread-safe API and specialized sub-APIs. This design allows the replacement of each layer in case a better technology becomes available.

One of the main goals in the creation of this tool has been is to simplify robotics code development when the system is composed of many distributed components. There are several features that we believe are mandatory to achieve this goal:

1. *A very simple deployment procedure with a minimum set of configuration parameters.* To start an instance of CORTEX, the user only needs to get the robot or the simulator ready and run first a special agent named *idserver*. This agent provides a unique identifier creation service required by all agents to add new nodes to the graph, being this one of the requirements of the underlying theory. When *idserver* starts, it can read a previously stored graph on disk,

[4] Convergence is guaranteed by defining merge as a least-upper bound over a join-semilattice.

[5] https://github.com/CBaquero/delta-enabled-crdts.

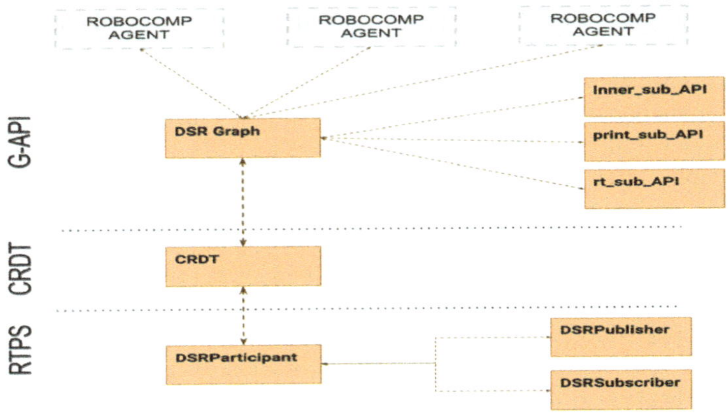

Fig. 3. Layered design of the working memory. From bottom-up, each layer is conceived to be replaceable and provides additional functionality to the system.

as a JSON file. At this point, any other entering agent can ask for an updated copy of the graph, using a topic reserved for this purpose. Once it gets the copy it can safely start its operation.

2. *Automatic code generation of all parts non dependant on the user code.* All these new agents are still software components created within the RoboComp development framework [16]. Thus, all generic code for these agents is automatically generated from a very simple DSL specification where only the name is mandatory. Since no topics, ports, interfaces or IPs are needed any more, a coder knowing the framework can create and start a new agent connected to an existing CORTEX instance in less than thirty seconds.
3. *A simple, intuitive and thread-safe user-level API to access the graph.* Concurrency originated between the user access to the graph and the merge operations triggered by incoming data from other agents, or by the user accessing the graph from several threads, is handled by the core methods using the new `std::shared_mutex` primitive and the associated `std::shared_mutex` operation provided in C++17. To access a node, the user requests a copy of it and the operation is protected with a read only `"shared_lock"` operation, that allows read access to several concurrent threads. When the user finishes editing the node, it is copied back to the graph using, this time, a `"unique_lock"`, that waits until all reading threads finish. Inserting new nodes or edges is not problematic and deleting nodes removes also all incoming and outgoing edges using the `"unique_lock"` mechanism.
4. *A dictionary of node and edge types, and of all possible names and types of both, node and edge attributes.* One common source of errors when building sets of components that share a structured representation like a graph is the misnaming of nodes, edges and attributes among coders. This situations usually take a long time to debug. The goal with DSR_d is to completely eliminate the source of these errors. The current solution uses a common

repository of types and names that all the team must use. However, we are working in a more radical solution using the new *constexpr* functionality provided in C++17 to keep in an include file all names and types of nodes, edges and attributes. This will result in compile-time errors when the user attempts to create or edit an element using the wrong name or data type.

5. *A graphical and textual display of the current state of the graph integrated in each agent.* Integrated in the generated code, each agent can optionally include a graphical interface, implemented with Qt5, that provides several, selectable, simultaneous, real-time views of the graph. These representations are interactive allowing the inspection of graph elements, and are intended as a powerful high-level debugging tool. Four views are provided: a standard planar drawing of the graph; a 2D representation of the geometric nodes embedded in the graph using Qt 2D drawing framework; a 3D view of the same content using OpenSceneGraph[6]; and a textual tree view which can be used to hide or show groups of nodes and edges in large graphs. Additional views can be easily added since the core methods of G-API emit Qt signals whenever a change is made. Another important feature that facilitates the debugging of complex CORTEX configurations, is the possibility to draw inside this graphical representations without perturbing the graph. An example could be the drawing of landmarks, targets, bounding boxes or robot paths used or produced by an agent.

All the code that managed the graph is compiled as a dynamic library and linked to each agent. The user *sees* the graph G as a local data structure that is accessed through its G-API.

5 Examples and Experiments

Since DSR_d is still a prototype under heavy testing, optimization and debugging, we will show here a series of examples of increasing complexity, each one with a corresponding video. We still don't have metrics that could give an idea of the final performance of the tool, except for the throughput and latency figures provided by eProsima on FastDDS, which are among the best of their class. The synchronization layer using CRDTs adds some overload that will depend on the impact of the *merge* operation on each node update. Also, it is difficult to provide quantitative comparisons here since we don't know of similar tools designed with the same purpose.

The first example[7] shows the initiation procedure of a CORTEX instance running against the CoppeliaSim robotics simulator. The example shows a two-step process. First, CoppeliaSim is started using a non-agent RoboComp component named *ViriatoPyRep*. This is Python program that imports PyRep [17] to run

[6] http://www.openscenegraph.org/.

[7] Video: *first-example* available in the list https://www.youtube.com/playlist?list=PLDkfV8Ufc2i1p-viGTV3QKupFD2ute535.

CoppeliaSim in the same loop and loads a model of RoboLab's adapted apartment, ALab. As a component, it provides RoboComp interfaces for all elements of the robot and for all external virtual devices installed in the apartment. In the second step, the agent *idserver* is started and it reads an initial graph from a JSON file. This stored graph contains the same scene that was loaded in CoppeliaSim but without the person. Once the agent has started, it opens a window with the views described before. Figure 4 shows a composition of the simulated scene and the views provided by the agent.

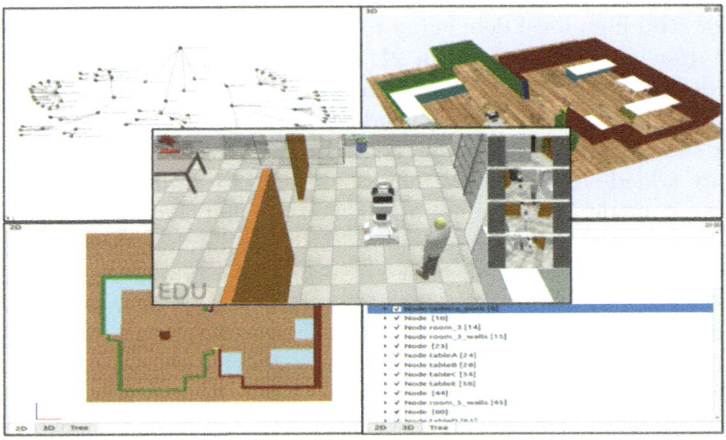

Fig. 4. Composition for Example 1 showing in the first plane the simulated scene of the ALab with the robot, a person and three additional cameras placed on the walls. Each of the four background captures shows a different representation of the graph made available to the user as part of the generated code. Top-left, is the standard graph representation; top-right is the OSG 3D view automatically computed from the graph; bottom-left is Qt 2D view also computed from the graph; and bottom-right is a textual tree view.

The second example[8] shows the deployment of a second agent, *ViriatoDSR*, that acts as a bridge between the simulated robot and the graph. On startup, this agent broadcasts, on a special topic, a request for a fresh copy of the graph. The request is served by the agent *idserver*, already running. *ViriatoDSR* is connected to *ViriatoPyrep* using RoboComp's interfaces and the Ice middleware, and to the graph using FastDDS. All low-level sensor data from the robot is injected into the graph, including the robot's RGBD camera streams. The use of reliable multicast reduces the bandwidth to the net load of one transmission plus some control information. The first running agent, *idserver*, receives the updates on the attributes of the nodes corresponding to the camera, the LIDAR and the

[8] Video: *second-example* available in the list https://www.youtube.com/playlist?list=PLDkfV8Ufc2i1p-viGTV3QKupFD2ute535.

position of the robot in the scene. The video shows how this data can be accessed from the UI in real-time. It also shows nodes and edges flashing when they are being updated by incoming data. See Fig. 5.

Fig. 5. Composition for Example 2 in text. A second agent has been started that acts as a bridge between the Python adapter, ViriatoPyrep and the graph. All received data is injected as it arrives and propagated to the other agent. The left part of the figure shows the first agent's UI, with the views for the LIDAR (an attribute of node *laser* that hangs from node *omnirobot*) and the camera. The right part shows the same data for the second agent. The central image corresponds to the scene being simulated. The capture in the lower side of the image corresponds to an instantaneous measure of the band-width measured with the Linux utility *iftop*. The 35 Mb/s figure corresponds to one RGB stream and shows that bandwidth remains constant with the number of agents.

In the third example[9], a third agent is introduced that access the number and pose of the avatars included in the scene. The agent used here is a simplified version of the real one that uses OpenPifPaf[10] and other DNNs to detect complete bodies, track them and estimate their orientation [18]. When this agent detects a person, it inserts it in the graph creating a new node and connects it to the world coordinate system using an RT edge. The avatar is shown in the 2D and 3D views and the new node is propagated to the other agent, where it is also depicted. The edge's rotation and translation attributes are hereinafter updated whenever a change is read from the simulator. See Fig. 6.

[9] Video: *third-example* available in the list https://www.youtube.com/playlist?list=PLDkfV8Ufc2i1p-viGTV3QKupFD2ute535.

[10] https://github.com/vita-epfl/openpifpaf.

The last example[11] shows how a fourth agent is started. This new participant takes the RGB image from the node camera's in its updated copy of the graph

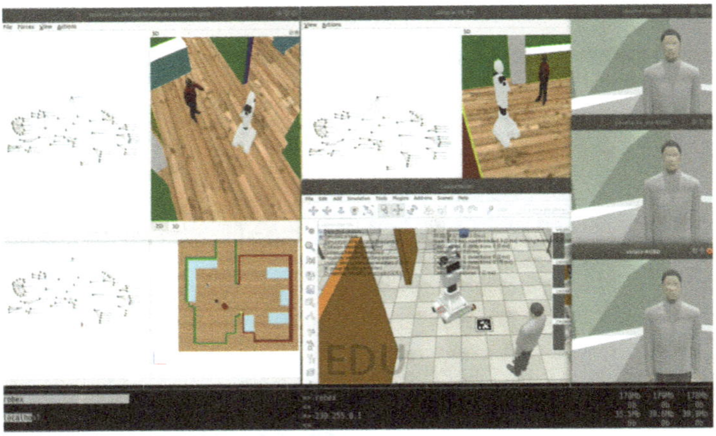

Fig. 6. Composition for Example 3 in text. The figure shows three agents running and sharing the graph. When the third agent initiates, it injects the detected person into the graph and the new nodes are propagated to the other agents. The three similar UIs show the reconstructed scene from different points of view. On the right side, the flow of images arriving to each agent is shown. Again, the total bandwidth in the multicast address is kept constant.

Fig. 7. Composition for Example 4 in text. A fourth agent is added that integrates the DNN YOLOv3 to detect object in the virtual scene, The left part of the composition shows the UIs of two agents, one of them displaying the LIDAR data in its graph. The right part shows the current scene in CoppeliaSim and the window in the center shows the detected objects over the processed image.

[11] Video: *fourth-example* available in the list https://www.youtube.com/playlist?list=PLDkfV8Ufc2i1p-viGTV3QKupFD2ute535.

and processes it with the DNN YoloV3[12]. The algorithm detects three objects in the virtual scene that are created as new nodes in the graph and made available to the other agents. See Fig. 7.

6 Conclusions and Future Work

The results presented in this work are a first step towards a fully operational distributed DSR_d. We have proposed a design based on a new combination of δ-CRDTs and an high-performance pub/sub middleware. The initial experiments have shown that several agents can share a non-trivial graph and edit its nodes for long periods of time without compromising its integrity. At the end of the trials, all copies have been checked to have the same contents. Also, the inspection of the traffic published by the middleware shows that only data belonging to the modified nodes or edges are published and, still, eventual consistency is achieved every time. Additionally, the window of consistency in these tests using only one host is very short as expected. All these results are good indicators of the potential that a distributed, low-latency, working-memory can have in future RCAs, where software complexity will be a major problem for developers and roboticists. As aforementioned, dealing with the software complexity that comes with an increasing number of agents, is one of our more urgent priorities and more tools will be needed to assist in debugging these large ensembles of processes.

Acknowledgments. This work has been partially funded by the EU RobMoSys project (H20202-732410), the project RTI2018-099522-B-C4X, funded by the Spanish Ministerio de Ciencia, Innovación y Universidades and FEDER funds, the EU INTERREG-POCTEC project 0043-EURAGE-4-E, and the Extremaduran Goverment projects GR15120 and IP IB16090.

References

1. Kotseruba, I., Gonzalez, O., Tsotsos, J.: A review of 40 years of cognitive architecture research: focus on perception, attention, learning and applications. Technical report (2016). http://arxiv.org/abs/1610.08602
2. Romero-Garcés, A., Calderita, L.V., Martínez, J., Bandera, J.P., Marfil, R., Manso, L.J., Bandera, A., Bustos, P.: Testing a fully autonomous robotic salesman in real scenarios. In: ICARSC, pp. 1–7 (2015)
3. Bustos, P., Manso, L., Bandera, J., Romero-Garcés, A., Calderita, L., Marfil, R., Bandera, A.: A unified internal representation of the outer world for social robotics. ROBOT **2**, 733–744 (2015)
4. Calderita, L., Bustos, P., Mejías, C.S., Fernández, F., Viciana, R., Bandera, A.: Asistente Robótico Socialmente Interactivo para Terapias de Rehabilitación Motriz con Pacientes de Pediatría. Revista Iberoamericana de Automática e Informática Industrial RIAI **12**(1), 99–110 (2015). https://doi.org/10.1016/j.riai.2014.09.007

[12] https://pjreddie.com/darknet/yolo/.

5. Beer, R.: Dynamical approaches to cognitive science. Trends Cogn. Sci. **4**(3), 91–99 (2000)
6. Voilmy, D., Suarez, C., Romero-Garcés, A., Reuther, C., Pulido, J., Marfil, R., Manso, L., Lan, K., Iglesias, A., González, J., Garcia, J., Garcia-Olaya, A., Fuentetaja, R., Fernández, F., Duenas, A., Calderita, L., Bustos, P., Barile, T., Bandera, J., Bandera, A.: CLARC: a cognitive robot for helping geriatric doctors in real scenarios. ROBOT **1**, 403–414 (2017)
7. Pulido, J., González, J., Suarez-Mejias, C., Bandera, A., Bustos, P., Fernández, F.: Evaluating the child-robot interaction of the naotherapist platform in pediatric rehabilitation. Int. J. Soc. Robot. 16– (2017)
8. Bandera, A., Bandera, J.P., Bustos, P., Férnandez, F., García-Olaya, A., García-Polo, J., García-Varea, I., Manso, L.J., Marfil, R., Martínez-Gómez, J., Núñez, P., Perez-Lorenzo, J.M., Reche-Lopez, P., Romero-González, C., Viciana-Abad, R.: LifeBots I: building the software infrastructure for supporting lifelong technologies. Adv. Intell. Syst. Comput. **693**, 391–402 (2018). https://doi.org/10.1007/978-3-319-70833-1_32
9. Vega-Magro, A., Manso, L., Bustos, P., Núñez, P., Macharet, D.: Socially acceptable robot navigation over groups of people. In: RO-MAN 2017 - 26th IEEE International Symposium on Robot and Human Interactive Communication, vol. 2017-January 2017. https://doi.org/10.1109/ROMAN.2017.8172454
10. Brewer, E.: CAP twelve years later: how the "rules" have changed. Computer **45**(2), 23–29 (2012). https://doi.org/10.1109/mc.2012.37
11. Bailis, P., Ghodsi, A., Hellerstein, J.M., Stoica, I.: Bolt-on causal consistency. In: Proceedings of the ACM SIGMOD International Conference on Management of Data, pp. 761–772 (2013). https://doi.org/10.1145/2463676.2465279
12. Terry, D.: Replicated data consistency explained through baseball. Commun. ACM **56**, 82–89 (2013). https://doi.org/10.1145/2500500
13. Baquero, C., Almeida, P.S., Shoker, A.: Pure Operation-Based Replicated Data Types (609551), 1–30 (2017). http://arxiv.org/abs/1710.04469
14. Almeida, P.S., Shoker, A., Baquero, C.: Delta state replicated data types. J. Parallel Distrib. Comput. **111**, 162–173 (2018). https://doi.org/10.1016/j.jpdc.2017.08.003
15. Enes, V., Almeida, P.S., Baquero, C., Leitao, J.: Efficient synchronization of state-based CRDTs. Proceedings - International Conference on Data Engineering 2019-April, pp. 148–159 (2019). https://doi.org/10.1109/ICDE.2019.00022
16. Manso, L., Bachiller, P., Bustos, P., Calderita, L.: RoboComp: a Tool-based Robotics Framework. In: Ando, N., Balakirsky, S., Hemker, T., Reggiani, M., von Stryk, O. (eds.) Simulation, Modeling and Programming in Autonomous RobotsRobots, vol. 6472, chap. LNCS 6472, pp. 251–262. Springer, Darmstadt, Germany (2010)
17. James, S., Freese, M., Davison, A.J.: Pyrep: Bringing v-rep to deep robot learning. arXiv preprint arXiv:1906.11176 (2019)
18. Rodríguez, D., Bachiller, P., Bustos, P., Vogiatzis, F., Manso, L.J.: Multi-camera torso pose estimation using graph neural networks. Accepted in RO-MAN 2020

Multi-agent System Model of Taxi Fleets

Lluís Ribas-Xirgo(✉)

Universitat Autònoma de Barcelona, 08193 Bellaterra, Barcelona, Spain
Lluis.Ribas@uab.cat

Abstract. Multi-agent models successfully cope with complexity of transportation systems. When used in combination with real systems, they take data from reality to keep the digital twins update and have the capability to control them. The design of the corresponding models is quite a challenge. In this work, we show how to achieve this with simple tools through an example with educational purposes. We also show how the model handles much more sophisticated, actual transportation systems.

Keywords: BDI agents · Cooperating robots · MAS · Mobile robots · Traffic simulation

1 Motivation

Many tools can be used to model complex systems with agent-oriented programs. The main challenge, however, is to make its programming easy. Graphical interfaces are helpful but typically fall short when going down in the hierarchy of the models, and conversely for textual programming.

Mimicking the work of Sakellariou, Kefalas and Stamatopoulou [1], we have created a Lua based environment (Sect. 2) to develop agent-based models with the added advantage of using a graphical interface for programming agent controllers (Sect. 3). This way, a hierarchy of state machines represent the behavior of an agent, from the low abstraction level to the highest. The latter are a kind of state machines we call extended finite-state stack machines or EFS^{2}M [2], which have the capability for representing belief-desire-intention (BDI) agents with the added advantage of increased verifiability.

Our version of the original cooperative taxi model includes a configurable map and taxis moving in accordance to right-hand traffic (RHT) rules, which makes it easy simulating actual, realistic environments for taxi systems and the like. Experimentation with simulator allows analyzing multiple scenarios and take decisions about systems' quality of service, performance and efficiency (Sect. 4). Results show that the development environment supports rapid design of taxi fleet models and that their simulators help tuning systems for their best operation (concluding section).

The main contributions of this work include the definition of EFS^{2}M, a framework to simulate agent-based systems in Lua, and an example applied to taxi fleets, which can be extended to model traffic on e.g. urban areas or inside industrial facilities.

L. M. Bergasa et al. (Eds.): WAF 2020, AISC 1285, pp. 123–134, 2021.
https://doi.org/10.1007/978-3-030-62579-5_9

2 Cooperative Taxis Case

The "cooperative taxis" case [1] consists of carrying passengers from their location, anywhere in a city, to their central airport. The city plan has a Manhattan layout and traffic is right-handed so to give taxis a way to solve conflicts among them. This case is indeed realistic, which is worth modeling for the impact taxi systems have on urban mobility and economics [3, 4]. What's more, it easily extends to other application domains, particularly of mobile robots for indoor logistics.

2.1 Environment

The environment consists of a city area surrounding an airport (Fig. 1, center) which is a 29 × 29 array of patches that represent 5 × 5 city blocks with roads in between. Patches are objects of the environment, each with (x, y, z) properties, where (x, y) are its position in the map and z, an index for the traffic direction over it.

```
###################...
#V<<<<V<<<<V<<<<V<<...
#>A>>>>A>>>>A>>>>A>...
#VA###VA###VA###VA#...
#VA###VA###VA###VA#...
#VA###VA###VA###VA#...
#V<<<<V<<<<V<<<<V<<...
#>A>>>>A>>>>A>>>>A>...
#VA###VA###VA###VA#...
#VA###VA###VA###VA#...
#VA###VA###VA###VA#...
#V<<<<V<<<<V<<<<V<<...
#>A>>>>A>>>>A>>>>A>...
#VA###VA###VA#@#VA#...
#VA###VA###VA@#@VA#...
...
```

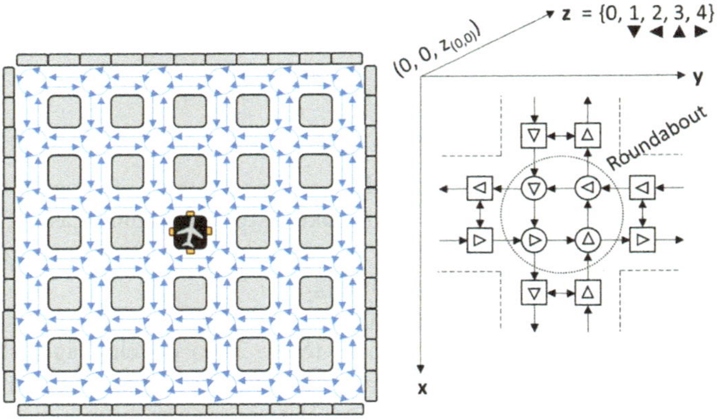

Fig. 1. A section of the text patch map (left), the city plan (center), and the coordinate system, with z being and index for the traffic direction (right).

The text file (Fig. 1, left) that defines the patch map starts from position (0, 0) and increase y each column and x each new line (Fig. 1, right). The "direction" characters 'V', '<', 'A', and '>' define the value for z. Any other valid character sets z to 0, i.e. taxis cannot circulate on it. Patches have attributes to determine whether they are a port or a gate, whether they belong to a roundabout, and to store people standing on them or taxis passing by.

A graph on top of the patches (Fig. 1, right) holds the information about the traffic network. Each patch $P = (x, y, z)|\{z > 0\}$, contains a triplet $P_{\text{out}} = \langle \mathcal{F}, \mathcal{L}, \mathcal{R} \rangle$ telling the direction of front, left and right patches, which are non-zero if a taxi can move from that patch to the corresponding direction, i.e. if this direction does not oppose to that of the destination patch or leads to a patch with $z = 0$.

2.2 The Lua Framework

Simulation is a key element in the development of systems and, particularly, of MAS. It gives clues about the correctness of the design of the behavior of agents. One of the typical applications of MAS is for the transportation.

In this work, we go, literally, three steps beyond the original one [1]. The first is to change the original Netlogo development environment to Lua; the second is to change to a rule-based traffic style, and the last, and the most significant, is to use state machines, including a type of our own, for the design of the agents. Anyway, we shall use the problem of taking passengers to their closest airport gate to illustrate the three.

The development environment in Lua, like the Netlogo one, includes libraries for the ACL functions and for BDI agents. The study case builds on top of these through defining two BDI derivative classes for the passengers and the taxis. A specific module ("ScenarioMap.lua") makes up for the lack of the Netlogo IDE environment and simulates the scenario as well as provides methods for the agents to interact with it. Additionally, we have developed a module for reading XML files from state machines drawn with http://app.diagrams.net/ and synthesizing code.

The main module, the *simulation motor*, takes care of controlling the simulation of the whole and capturing data from it by repeatedly activating agents' behaviors and calls for environment updates.

Differently from the previous works where they could go to any free patch around, in this work, taxis move around in an RHT fashion to solve conflicts and have a more fluent traffic.

Before moving, taxis reserve patches and, in cases where they see another taxi reserving the same patch, they decide whether they should move on or give way. Taxis giving way reserve the patches where they stay, which can cause other taxis giving way, and so on until there are no conflicts at all. Deadlocks happen when all taxis in a conflict-solving situation give way, and they look for alternatives until some move.

Agent behaviors described by state machines make their design straightforward and, more important than that, more verifiable. Although you can program them in Lua using the provided library, it is highly recommended to use a graphical input to generate the behavior-related code.

3 Belief-Desire-Intention Agents

The fundamental architecture of these agents is that of a controller that builds beliefs out of its own knowledge and inputs from sensors and incoming messages, and devises plans of actuation intended to satisfy its own desires, which would be like the reference set points of a lower level controller.

Agents can change plans on the way by manipulating the stack of the so-called extended finite state stack machines (EFS^{2}M), which we have proposed as an alternative to represent BDI behaviors [2].

The proposed EFS^{2}M model is completely compatible with other state machine models and thus, all the controllers of the controllers' stack of an agent can be modeled and synthesized in the same way.

3.1 State Machine Stack Structure

A hierarchical concurrent state machine models the overall controller architecture of an agent, where each level of the hierarchy is a layer of its controller stack.

For the study case, only two levels are required. The lowest is a reactive controller that directly interacts with the environment and only communicates with the highest. The high level is a deliberative controller able to communicate with other agents.

Communication between agents follows the Contract Net Protocol, while communication between layers of the same controller is specific (Fig. 2). In our case, the latter follow a handshake protocol to prevent from event misses. Passengers *desire* to go the airport and, for that, they make calls for proposals to choose the least cost transportation, which initiate a specific Contract Net Protocol that goes on by waiting for some time until eventually they accept one proposal. In this case, passengers do not send reject messages. Upon acceptation, taxis might eventually send a failure message if they cannot reach the passenger. To minimize inter-agent communication, they do not send informative messages.

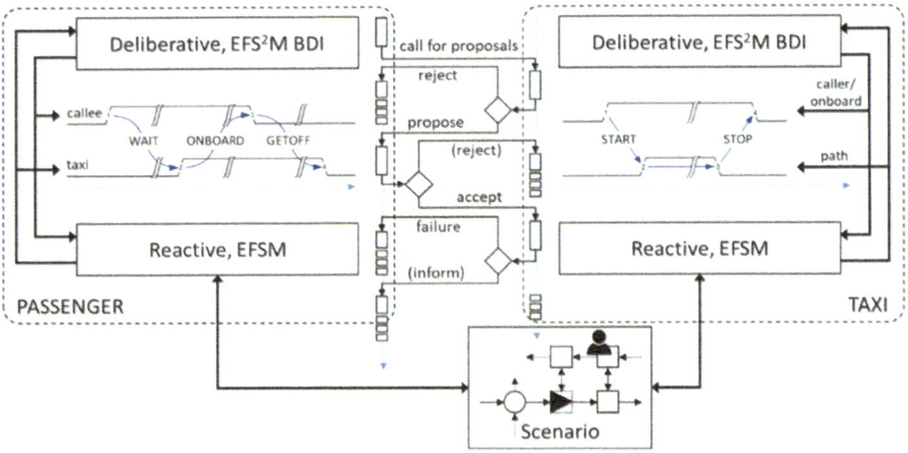

Fig. 2. Architecture of agents organizes their behavior into one deliberative layer on top of a reactive one, which communicate internally (middle left and right). Inter-agent communication follows a Contract Net Protocol (center, top down read).

While communications amongst agents depend on timeout values to force termination of missing acknowledgements, internal communications use explicit synchronization protocols, which are faster. They work under the assumption that no messages are missing.

Transmission of internal messages works as follows: senders write values on shared variables and receivers get the values by reading them. For instance (see left chronogram in Fig. 2), deliberative parts of passengers change the value for the winner taxis or "callees" from nil to their references. These changes are messages for the reactive parts to look for the corresponding taxis to reach the passengers' patches. When it happens, they change the value of taxi from nil to the callee reference, so to inform the deliberative part they are onboard. At that moment, deliberative parts acknowledge the receipt of the message by making callee variables return to the initial nil values. Similarly, the variables *taxi* will be set back to nil when passengers go out of the taxis at the gate patches. Hence, communication follows a 4-phase handshake, each phase starting with signals changing from/to nil.

Communication between deliberative and reactive parts of taxi agents follow a handshake protocol of two phases. In this case, the output variables of the deliberative part are the caller reference, *C*, and whether *C* is in the passengers' onboard set. In the chronogram of Fig. 2 (right side), they share the same waveform, with low-level value meaning nil or false, and high-level meaning *C* or true (*C* is in the corresponding set). Reactive parts have one output (path) which can be empty or not. In the implementation, empty paths because of having reached the target patch distinguish from empty paths due to non-existent paths to destination. For the sake of simplicity, we have not included the latter in the chronogram.

Deliberative parts initiate first communication by asking reactive parts to follow a path (by changing caller/onboard from nil/false to *C*/true), and the latter (second phase) acknowledge receipt of the message by setting variable *path* to a non-null value. Reactive parts initiate second transmission by returning path to nil (to indicate either a failure or a success getting to pick-up/drop-off patch), which is acknowledged by deliberative parts when caller is set to nil or *C* is taken out from the onboard set.

3.2 Extended Finite-State Machines

Figure 3 shows the EFSM of the reactive part of taxis. An EFSM is a FSM with an extended state, which, on top of a state variable, has other variables, which also hold values from cycle to cycle [5]. Assignments to variables of the model can be immediate or delayed [6]. *Immediate assignments* take effect at the beginning of the current cycle. *Delayed assignments* are represented by appending a plus sign to the variable name (e.g. `step+=2`) and take effect at the end of the current cycle, like the update for the next state. Conditions at arcs evaluate *after* immediate assignments, and, of course, *before* setting variables to their delayed values.

The EFSM for the reactive behavior of taxis starts at the WANDER state. In this state, it calls `move_random()` to decide which patch to occupy next and reserve it. Decisions at patches with multiple outputs, excluding those that lead to U-turns, are taken randomly. The *making a reservation* simplifies how other vehicles can detect its movement vector. Cases where two or more vehicles reserve the same patch compare to actual traffic situations where they have colliding movements. These conflicts are solved assuming that all vehicles will behave fairly and according to RHT rules.

The taxi will move randomly, i.e. stay at WANDER, until input variable caller gets a non-nil value. At that instant, it computes a path to the caller position (not shown in the figure), sets the variable step to 2 (because the first element of the path is the taxis' current patch), and moves to ON-WAY state. Paths can include U-turns.

At the ON-WAY state, it simply follows the path and increases the `step` value if it is less than the length of the path, `#path`.

Notice that there is a variable `inc` that is assigned immediately to the return value of `follow_path()` so that it can be used at the same instant in other computations for delayed assignments. Variable `inc` becomes 1 when the taxi moves one step or 0 if it stays at place whether for giving way or reaching the end of the path. In the latter case, `path` equals `#path` and it moves to PICK-UP. In the former case, it is because it has given way to other taxis.

In fact, a series of calls to `giving_way()` follow both `move_random()` and `follow_path()`. This sequence ends when the taxi reserves its current patch (i.e. stays at place) or wins all bids on the wanted new patch. When all taxis involved in a conflict-solving situation give way, a deadlock occurs. In the simulation framework, we have opted by a detect-and-solve approach rather than others (e.g. [7–9]) because systems can be adapted to make deadlocks infrequent.

The path following state terminates when reaching the last step of the path. Then, the state machine moves to PICK-UP and waits for the caller to leave the road and get in, i.e. for the corresponding variable be set to nil again, and the set of passengers onboard be increased by the caller's reference.

Once onboard, it passes to the CARRY state, which works much like ON-WAY by calling the same `follow_path()` procedure, including managing conflicts by appropriately giving way and detecting eventual deadlocks.

The machine stays at ON-WAY until the taxi reaches the last step of the path. At this moment, it passes to DROP-OFF where it waits for the passenger to get out. Once the passenger is no longer onboard, i.e. that the number of passengers q has decreased, it returns to the initial WANDER state.

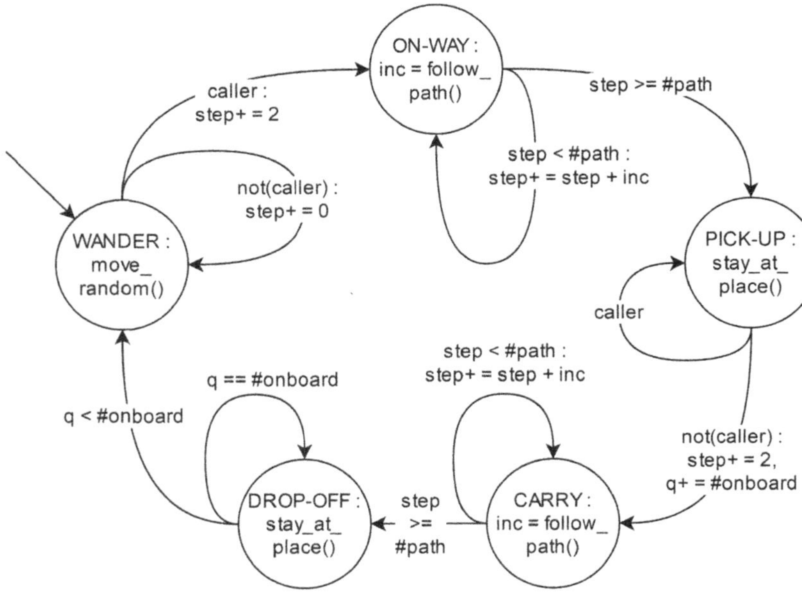

Fig. 3. EFSM of the taxi for the reactive behavior.

Passengers' EFSM interfaces environment with deliberative part to help detecting when the called taxi has arrived and when it has reached the destination. Starting state is CALL, from which passes to WAIT when a taxi acknowledged the acceptation of its proposal, then from WAIT to EN-ROUTE and, finally, from EN-ROUTE to ARRIVED, which is a sink state, i.e. they do nothing else.

Apart from interfacing environment with high-level modules of passengers, state monitoring of these EFSM provides a simpler way for measuring quality of service, as we shall see in Sect. 4.

3.3 Extended Finite-State Stack Machines

An EFS^{2}M is an EFSM with the state variable being a stack. Graphical representations of this type of state machines follow the same syntax that that of EFSM with additional symbols for stack manipulation.

Figure 4 shows the EFS2M of the deliberative part of taxis. Empty bubbles at the beginning of arrows mean state stack pops and bubbles at the end of arrows mean state stack pushes and, consequently, they must contain the name of the state to push. Blank push bubbles mean to push the current state. Hexagons at the end of an arrow mean to pop one state from the stack and make it the next one. (Although not used, hexagons at the beginning of an arrow mean to remove all states from the state stack.)

The initial state of the deliberative EFS^{2}M is *look_for_passenger*, which waits for an acceptance of a previous proposal. At this state, it replies to any call for proposals or *CfP* with an estimation of the cost of taking the proponent to the airport. Rejections are ignored.

When a proposal is accepted, it makes a plan to render the passenger the required transportation. To do so, it stacks several states before getting to the last one. In the graph, the plan is read in reverse order: *check_trip*, *go_to_pick_up*, *carry_passenger_to_dest*, *drop_passenger*, and *look_for_passenger* again (recall that empty bubbles at the tip of arrows mean pushing the current state into the stack.)

The *check_trip* state checks whether there is a path to pick up the passenger. If such path does not exist, it sends a failure message to the caller and pops out all the stacked states but the first one, which is *look_for_passenger*, and then, it goes to it, because it is the only one left in the stack (and the arrow points to a hexagon, i.e. to the top of the stack). On the contrary, going to the top stack state implies changing state to *go_to_pick_up*.

Go_to_pick_up ends when it runs into the same patch than the caller and moves on to *carry_passenger_to_dest*. At this state, it waits until it finds itself at the destination patch (in the example, one of the airport gates).

The last part of the plan is to *drop_passenger*, where passenger is removed from the taxi onboard list by function `getoff()`. The use of a list makes it easy to extend the application to model taxis capable of pooling.

Passengers have the goal of being idle, i.e. of reaching the state *idle* at the destination patches. Their intentions include the following plan, which is self-explanatory: *send_CfP_to_taxis*, *collect_proposals* and *evaluate_proposals*.

After evaluating proposals, they wait for taxis to arrive and, once aboard, for taxis to reach the airport gate. At that moment, they get out and stay idle for the rest of the simulation.

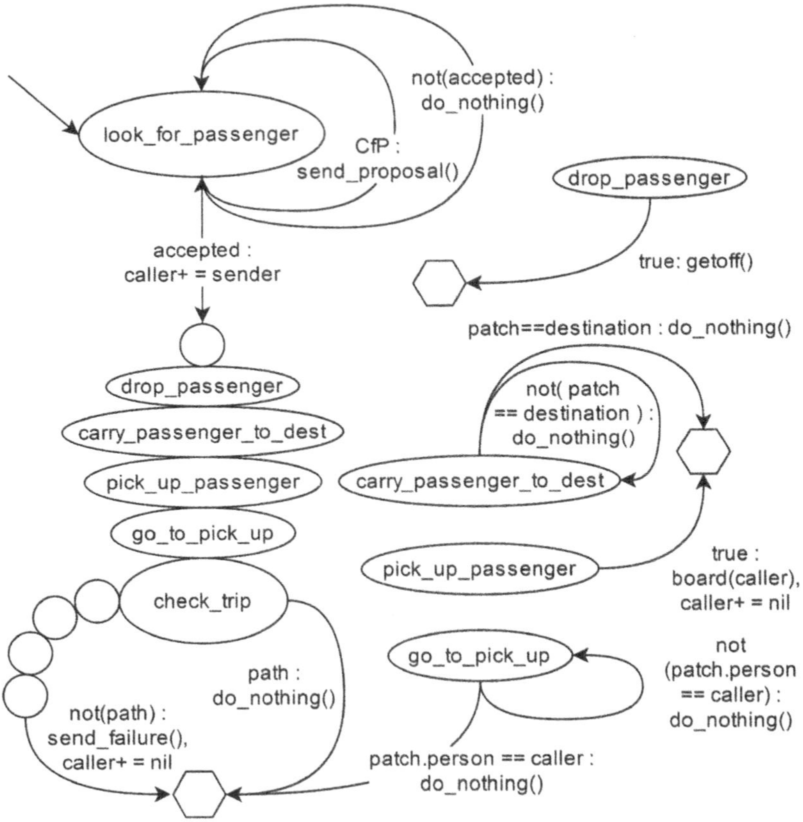

Fig. 4. EFS^{2}M of the taxi for the deliberative behavior.

4 Experimentation

In order to validate the Lua framework and to verify agents' behavior, we have developed the following experiment. In fact, agents' behavior has been validated in previous works [10, 11], but not with the RHT mode and not within this framework.

We shall analyze the effects of traffic saturation and supply/demand ratio on the quality of service [12]. Traffic saturation is the percentage of road room taken up by taxis with respect to total road length. In the test scenario, this value is obtained from the quotient between the number of taxis and its maximum value, which is 500, as we do not allow gate patches to be initial positions for taxis. This leaves 4 free patches for taxis to move, much like in a number slide puzzle fashion, but they have a minimal room for solving conflicts in movement, at least.

In this paper, supply/demand ratio stands for how many available taxis are per client call [13], though it relates directly with the same term in economics. Values less than one should increase the probability for clients to choose a closer taxi than values over one, where taxis accept client proposals in first-come, first-served basis.

When running the experiment, taxis are randomly placed at the beginning of it and passengers are created up to keep the supply/demand ratio at the given value.

Experiments are run until a fixed number of passengers arrive at the airport, also with random initial locations for them.

The quality of service is measured with respect to the minimum average time of a passenger to get to the closest airport gate for any supply/demand ratio. The time starts with the client's first call for proposals and stops when getting off the taxi.

We expect that high supply/demand ratios with low traffic densities to have better quality of services than the other way around. However, in a real case, system efficiency should also be considered. Anyway, this experiment is focused on showing how the framework can be used to analyze this kind of parameters and how changes on the agents' behavior can affect the outcome.

The first experiment consists of simulating the environment with one active passenger at a time and from 1 to 500 taxis on roads. Each case is repeated until 1000 passengers arrive at the airport or up to 100000 ticks to cut off excessive simulation times. (In a regular laptop, the 500000 cases were simulated in about 23 h, i.e. 0,17 s/case.) Data given are in number of ticks, i.e. of simulation steps, as the scenario is difficult to fit in a real scenario because of its simplicity. Yet, at a top speed of 40 km/h per taxi, with roads 20 m wide and blocks of 60 × 60 m, each tick would correspond to 1.8 s in real life for what would be a small neighborhood area of a city.

If we plot the service time with respect to the number of taxis (Fig. 5), we can see that it gets slightly lower as the number of available vehicles increase (section *a*) up to a point where traffic density starts hindering free movement (beginning of section *d*). From that point, service time worsens.

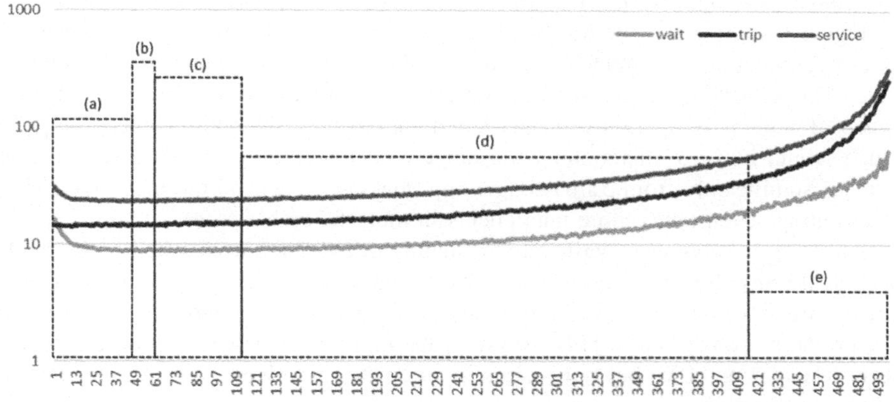

Fig. 5. Service time (ticks) as a function of number of taxis.

The quality of service has a sweet region (*b*), from 40 to 50 taxis (traffic density < 10%) as random movement of taxis minimize waits. Up to a traffic density of 0.2, values are quite close to minimum average service time (23.2 ticks). Anyway, in this scenario, values of region *d* would also be acceptable, as the quality of service would be in accordance with traffic density perception. Anyway, at the end of the region, for densities above 0.8, the increase of services times become exponential.

This observation adds to common sense and literature stating that high traffic densities must be avoided: they end up in poor quality of service and system efficiency and cannot be compensated with variable speed limitations. (Note that simulated case would not be different with vehicles moving smoothly instead of making starts and stops.)

Decomposing service time into waiting and trip time reveals that passengers' waits take longer than their ways to airport gates. The main reason for that is because, on average, taxis and passengers are 27/2 = 13,5 patches away for large, homogeneous random samples for taxis, where 27 is the length of roads, while average distance between passengers and airport gates is half the side of the square environment, i.e. 13,5/2 = 6,75 patches.

In previous works, we have also introduced a repetitive procedure in the Contract Net Protocol to make series of auctions so new bidders can enter and existent ones, modify, bids for proposals [10, 11]. This would reduce the effect of making bad guesses in response to CfPs. Other techniques can also be applied. Anyway, if the system at hand does not operate in extreme situations where quality of service is noticeable affected, it might be better not to apply them so to keep system efficiency.

The second experiment consisted on obtaining the average service time for a range of traffic densities and supply/demand ratios. For the sake of simplicity, we shall use number of taxis and of active clients instead, respectively.

Number of runs for every case is 320 but is progressively diminished to 10 for the largest number of taxis and active passengers. Taxis are limited to 256, slightly above 50% of road capacity, to prevent from simulations of very slow and stalled traffic, where the service times are too high. The number of active passengers per tick, i.e. those willing to arrive at the airport at the same instant ranges from 1 to 384. Any time a passenger gets out a taxi, a new one becomes active and calls for proposals.

Figure 6 summarizes the results of this experiment. Each point is the average of a series of simulation runs and tells the number of ticks servicing clients, in a logarithmic scale. The solid line on the surface indicates the minimum service time, which is projected in the passengers-taxi plane (dotted line) to show the optimal operation point, which would be of 32 taxis along the range of demand. Anyway, the service times increase significantly from 32 passengers up, making a client in the most demanding case waiting seven times more than clients in more favorable cases.

Although the former experiment was mainly oriented to check whether the model can handle any situation, cases with lots of circulating taxis help simulating high density situations, which are much more likely to occur near e.g. airports or accesses to transportation hubs in cities [14]. Anyway, the high demand scenarios would be less likely to occur on cities but not so less in automated warehouses were every item to pick is a passenger.

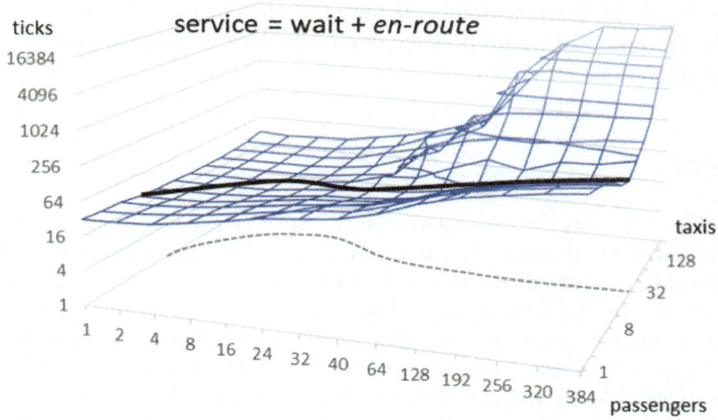

Fig. 6. Service time (ticks) vs. supply/demand (active passengers per tick) and traffic density (number of taxis).

This experiment also helps to locate the optimal operation point of the system under analysis. In this case, using 32 taxis reveals as the best choice for most of the cases. However, note that for demands under 32 clients per tick, the best choice is 40. In fact, from 32 clients per tick upwards, the service time increases exponentially (the vertical axis is logarithmic). This decline on the ideal number of taxis from this point comes from the fact they concentrate around the airport gates and must wait more often from 32 and up.

The quality of service for the best cases, goes from 23.76 ticks (single passenger with 16 taxis moving around randomly) to 143.43 (384 active passengers at any time). For practical applications, however, the limit would be at 16 active passengers and 40 taxis (25.3 ticks).

5 Conclusion

Making a model for taxi fleets enables analyzing a variety of scenarios, ranging from actual taxi fleets in cities to any transportation system based on vehicles.

We have developed a framework to develop a multi-agent model for any system resembling fleets of taxis. The framework includes a library of Lua models and two examples of agents, one for passengers and the other for taxis.

Agents' behavior is graphically described with state machines, including one we have specifically created for BDI agents (EFS^2M).

We have tested the framework with a rather simple example [15], although changing map configuration adapts to a wide range of scenarios.

With that example, we have shown how to simulate systems to analyze quality of service and efficiency, and to determine their operation range.

Results highlight the potentiality of the environment for studying real cases and assist in the definition of traffic policies, including lane directions, priorities at crossings, speed limitations, et cetera.

References

1. Sakellariou, I., Kefalas, P., Stamatopoulou, I.: MAS coursework design in NetLogo. In: Beer, M., Fasli, M., Richards, D. (eds.) Proceedings of the International Workshop on the Educational Uses of Multi-Agent Systems (EDUMAS 2009), pp. 47–54 (2009)
2. Rivas, D., Das, P., Saiz-Alcaine, J., Ribas-Xirgo, Ll.: Synthesis of controllers from finite state stack machine diagrams. In: IEEE 23rd International Conference on Emerging Technologies and Factory Automation (ETFA), pp. 1179–1182, Turin (2018)
3. Salanova-Grau, J.M., Estrada, M., Aifadopoulou, G., Mitsakis, E.: A review of the modeling of taxi services. Procedia – Soc. Behav. Sci. **20**, 150–161 (2011). https://doi.org/10.1016/j.sbspro.2011.08.020
4. Hörl, S.: Agent-based simulation of autonomous taxi services with dynamic demand responses. In: 6th International Workshop on Agent-based Mobility, Traffic and Transportation Models, Methodologies and Applications (ABMTrans), Procedia Computer Science 109C, Elsevier, pp. 899–904 (2017)
5. Ptolemaeus, C. (ed.): System Design, Modeling, and Simulation Using Ptolemy II, Ptolemy.org (2014)
6. Berry, G.: The Esterel v5 Language Primer Version v5 91. Collège de France (2004)
7. Kim, K.H., Jeon, S.M., Ryu, K.R.: Deadlock prevention for automated guided vehicles in automated container terminals. In: Kim, K.H., Günther, H.O. (eds.): Container Terminals and Cargo Systems. Springer, Heidelberg (2011)
8. Li, Z.W., Zhou, M.C.: Deadlock resolution in automated manufacturing systems – a novel Petri net approach. Int. J. Prod. Res. **48**, 5541–5542 (2009)
9. Perronnet, F., Buisson, J., Lombard, A., Abbas-Turki, A., Ahmane, M., Moudni, A.: Deadlock prevention of self-driving vehicles in a network of intersections. IEEE Trans. Intell. Transp. Syst. (2018). https://doi.org/10.1109/tits.2018.2886247
10. Rivas-Alonso, D., Jiménez-Jané, J., Ribas-Xirgo, Ll.: Auction model for transport order assignment in AGV systems. In: Proceedings of the 19th International Workshop of Physical Agents (WAF), Nov. 2018, Madrid, Spain (2018). https://doi.org/10.1007/978-3-319-99885-5
11. Rivas, D., Ribas-Xirgo, Ll.: Agent-based model for transport order assignment in AGV systems. In: IEEE 24th International Conference on Emerging Technologies and Factory Automation (ETFA), pp. 947–954 (2019). https://doi.org/10.1109/etfa.2019.8869302
12. Kim, H., Oh, J.-S., Jayakrishnan, R.: Effect of taxi information system on efficiency and quality of taxi services. Transp. Res. Rec. **1903**, 96–104. https://doi.org/10.3141/1903-11
13. Zhang, R., Ghanem, R.: Demand, supply, and performance of street-hail taxi. IEEE Trans. Intelligent Transp. Syst. (2019)
14. Qin, G., Li, T., Yu, B., Wang, Y., Huang, Z., Sun, J.: Mining factors affecting taxi drivers' incomes using GPS trajectories. Transp. Res. Part C Emer. Technol. **79**, 103–118 (2017). https://doi.org/10.1016/j.trc.2017.03.013
15. Ribas-Xirgo, Ll.: MASLua. https://sourceforge.net/projects/maslua. Accessed 03 Sept 2020

Human-Robot Interaction

Can a Social Robot Learn to Gesticulate Just by Observing Humans?

Unai Zabala, Igor Rodriguez(✉), José María Martínez-Otzeta, and Elena Lazkano

Computer Science and Artificial Intelligence, Faculty of Informatics, UPV/EHU, Manuel Lardizabal 1, 20018 Donostia, Spain
igor.rodriguez@ehu.eus
http://www.sc.ehu.es/ccwrobot

Abstract. The goal of the system presented in this paper is to develop a natural talking gesture generation behavior for a humanoid robot. With that aim, human talking gestures are recorded by a human pose detector and the motion data captured is afterwards used to feed a Generative Adversarial Network (GAN). The motion capture system is capable of properly estimating the limbs/joints involved in human expressive talking behavior without any kind of wearable. Tested in a Pepper robot, the developed system is able to generate natural gestures without becoming repetitive in large talking periods. The approach is compared with a previous work, in order to evaluate the improvements introduced by a computationally more demanding approach. This comparison is made by calculating the end effectors' trajectories in terms of jerk and path lengths. Results show that the described system is able to learn natural gestures just by observation.

Keywords: Social robots · Motion capturing and imitation · Generative Adversarial Networks · Talking movements

1 Introduction

Social robotics [4] aims to provide robots with artificial social intelligence to improve human-machine interaction and to introduce them in complex human contexts. The demand for sophisticated robot behaviors requires to model and implement human-like capabilities to sense, to process, and to act/interact naturally by taking into account emotions, intentions, motivations, and other related cognitive functions.

Talking involves spontaneous gesticulation; postures and movements are relevant for social interactions even if they are subjective and culture dependent. Aiming at building trust and making people feel confident when interacting with them, socially acting humanoid robots should show human-like talking gesticulation. Therefore, they need a mechanism that generates movements that resemble humans' in terms of naturalness.

L. M. Bergasa et al. (Eds.): WAF 2020, AISC 1285, pp. 137–150, 2021.
https://doi.org/10.1007/978-3-030-62579-5_10

Gestures (head, arms and hands movements) are used both to reinforce the meaning of the words and to express feelings through non-verbal signs. The scope of the present work is limited to beats, i.e. conversational movements of body parts synchronised with the flow of speech but not associated with particular meaning [18].

In [23] made use of gestures selected from a set of handmade animations to produce human-like arm gestures in a humanoid robot. Those gestures were randomly concatenated and reproduced according to the duration of the speech. That approach was prone to produce repetitive movements and resulted in unnatural jerky expression specially noticeable when the robot had to give a large speech. That drawback was somehow overcome by using a generative approach for motion generation as shown in [24]. There, Generative Adversarial Networks (GANs) were confronted to the task of motion generation with other well known generative models.

However, the set of built in animations used in the experiments did not contain the variety of gestures that we humans make and that should be exhibited by talking social robots.

Thus, the present approach intends to make one step forward by using a markerless motion capturing and imitation system using OpenPose [6], a real-time multi-person 2D pose estimation that uses a non-parametric representation to learn to associate body parts with individuals in the image. This is a computationally demanding choice and the motivation for choosing it over other less complex approaches is to take advantage of the rich hand model it extracts. Hand motion conveys a lot of valuable expression during natural speech, and thus the aforementioned model should help to better extract hands motion characteristics. The approach is tested in a Pepper robot and compared along with an OpenNI based approach, in order to evaluate the improvements introduced by the presented choice.

The rest of the paper is structured as follows: Sects. 2 and 3 review the human motion capturing problem and generative models respectively, the two main subsystems used in the present work. Next, in Sect. 4, the developed approach is described step by step. Section 5 details the experimental setup, i.e. the hardware and software involved and the configuration of the learning subsystem. Obtained results are discussed in Sect. 6 together with the motion analysis of the robot made with the aim of numerically compare the improvements introduced by the motion capture system. The paper concludes with Sect. 7 tracing the further research work.

2 Motion Capturing Systems

According to Beck et al. [3], there are three main robot motion generation approaches: manually creating motion, motion capturing, and motion planning; for manual creation, it is required to set each joint position of the humanoid robot for each key frame (time step); the motion capture-based approach tries to mimic human gestures, recording human movements and mapping these data to a humanoid robot [20]; and the motion planning approach relies on kinematics and/or dynamics equations to solve a geometric task. Beck et al. found that the motion capturing approach produces the most realistic results, because the robot reproduces previously captured human movements.

Motion capturing and imitation is a challenge because humans and robots have different kinematic and dynamic structures. Motion capture (MoCap) is the process of recording motion data through any type of sensor. Applications of such systems range from animation, bio-mechanics, medicine to sports science, entertainment, robotics [32] of even study of animal behavior [25]. MoCap systems rely on optical technologies, and can be marker-based (e.g. Vicon[1]) or markerless like RGB-D cameras. While the former ones provide more accurate results, the latter ones are less prone to produce gaps (missing values) that need to be estimated [19,28]. Besides, markerless approaches are welcome by users that don't need to wear devices or clothes that can be cumbersome.

Trying to capture human poses as natural as possible, markerless technique seems to be more appropriate due to the freedom of movement of the speaker. Microsoft's Kinect depth sensor has been very popular as markerless skeleton tracker device due to its availability [1,10]. The Kinect uses a learning based method for human body pose recovery. Mainly, it infers body parts from depth images using a randomized decision forest learned from over a million training examples [26].

The advent of deep learning systems motivated the development of more powerful model-based approaches for human pose estimation. OpenPose [7] is a model-based real-time multi-person 2D human pose estimator that provides body, hand, facial and foot keypoints. As most deep learning based techniques, it requires a powerful GPU and needs specific software dependencies in order to run efficiently. However, the face and hand information that provides turns it into a promising alternative for the task we are involved in.

3 Generative Models and Their Applications

Generative models are probabilistic models capable of generating all the values for a phenomenon. Unlike discriminative models, they are able to generate not only the target variables but also the observable ones [27]. They are used in machine learning to (implicitly or explicitly) acquire the distribution of the data for generating new samples. There are many types of generative models. For instance, Bayesian Networks (BNs) [8], Gaussian Mixture Models (GMMs) [9] and Hidden Markov Models (HMMs) [21] are well known probability density estimators.

In a nutshell, a BN is a model representation for reasoning under uncertainty [8]. Formally, its representation is a directed acyclic graph where each node represents a random variable and the edges represent dependence relations between them; GMMs attempt to find a mixture of multi-dimensional Gaussian

[1] https://www.vicon.com/.

probability distributions that best model any input dataset [9]; and HMMs can be considered a generalisation of mixture models where the hidden variables (or latent variables), which control the mixture component to be selected for each observation, are related through a Markov process rather than independent of each other [21].

Focusing on generative models used for motion generation, in [15] the authors propose the combination of Principal Component Analysis (PCA) [29] and HMMs for encoding different movement primitives to generate humanoid motion. Tanwani [27] uses HSMM (Hidden Semi-Markov Models) for learning robot manipulation skills from humans. Regarding on social robotics, some generative approaches are being applied with different objectives. In [17] Manfrè et al. use HMMs for dance creation and in a later work they try variational auto-encoders again for the same purpose [2].

Deep learning techniques have also been applied to generative models, giving rise to deep generative models. A taxonomy of such models can be found in [11]. In particular Generative Adversarial Networks (GANs) [12] are semi-supervised emerging models that basically learn how to generate synthetic data from the given training data. GANs are deep generative models capable to implicitly acquire the probability density function in the training data, being able to automatically discover the internal structure of datasets by learning multiple levels of abstraction [16]. Gupta et al. [13] extend the use of GANs to generate socially acceptable motion trajectories in crowded scenes in the scope of self-driving cars. In [24] GANs showed to overcome other generative approaches such as HMM and GMM when confronted to the task of motion generation.

4 OpenPose Based Motion Mapping

Our proposed method to generate human motion in a humanoid robot includes arms, head and hands motion mapping (upper-body part) since legs are not involved in talking beats. A 3D sensor is used as MoCap device, because both RGB data and depth information are needed. OpenPose will use RGB data to extract human's 2D skeleton information, and the 3D skeleton will be built from depth data. No matter what motion capture system being used, there is a need to transfer human motion to the robot joints (see [20] for a good overview of every aspect of the motion imitation task). This can be done by inverse kinematics, calculating the necessary joint positions given a desired end effector's pose [1]. This process is effective but its complexity requires high computational load. Only the information of the end effectors is considered and employed to estimate appropriate positions of every joint. Alternatively, direct kinematics adapts captured angles to the robot [22,31]. This kind of method is more straightforward and is the approach taken in the system presented here.

Figure 1 depicts the process to transfer human motion to the robot. The observation step consists of the process of extracting the necessary human motion information, in this case the 3D skeleton according to the OpenPose BODY25 model and hand information. These data are low pass filtered before converting into robot motion, in order to stabilize them and avoid trembling.

During the mapping stage, arms, head and hands positions are mapped to the robot joint space. The reproduction step allows to visually verify that the mapping is appropriate. The Pepper humanoid robot must be able to reproduce the upper-body motions according to the mapping results.

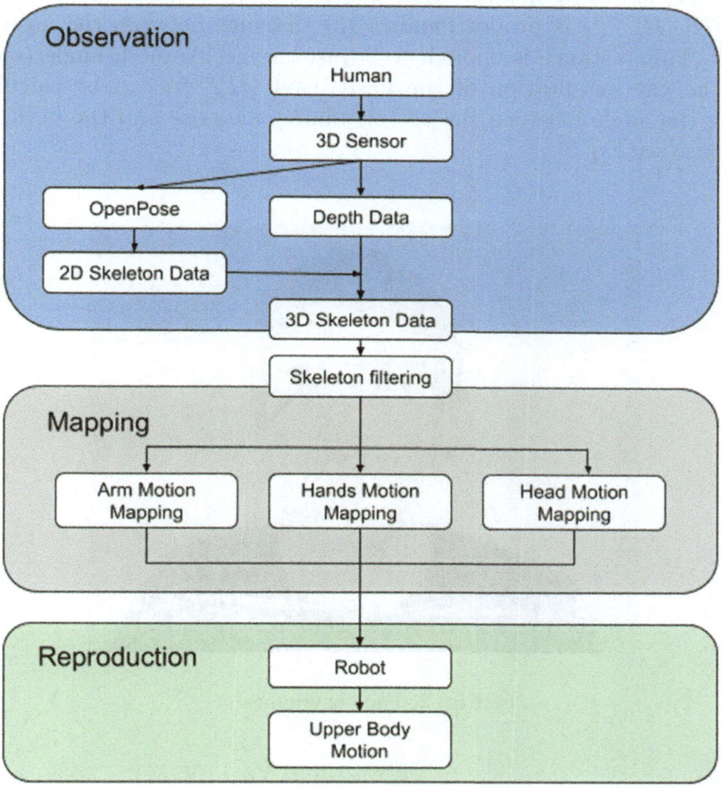

Fig. 1. Human motion imitation process

Sections 4.1, 4.2 and 4.3 describe more in detail the mapping process.

4.1 Arms Motion

The literature reveals different approaches to calculate the robot arm joint positions [14,31]. This mapping process depends upon the robot's degrees of freedom and joints range. For the Pepper robot arms we are dealing with, some upper-body link vectors are built through the skeleton points in the human skeleton model, and joint angles are afterwards extracted from the calculation of the angles between those vectors. For the sake of simplicity, the formulae involved in that process are not going to be reproduced here (see [30] for more detailed information).

4.2 Head Motion

OpenPose detects basic face features such as the nose, the eyes and the ears (see Fig. 2). To map humans head position into the robot, we use the nose position as reference. Nose's position shifts horizontally when moving the head left to right and gets closer to or away from the neck when looking up or down. Thus, head's pitch (H_{ϕ}^{robot}) is proportional to the distance between the nose and the neck joint. This distance is enough to adjust the neck's pitch angle (see Eq. 1). Instead, the yaw orientation of the head itself (H_{ψ}^{robot}) can be calculated by measuring the angle between the vector joining the nose and the neck, and the vertical axis (see Eq. 2).

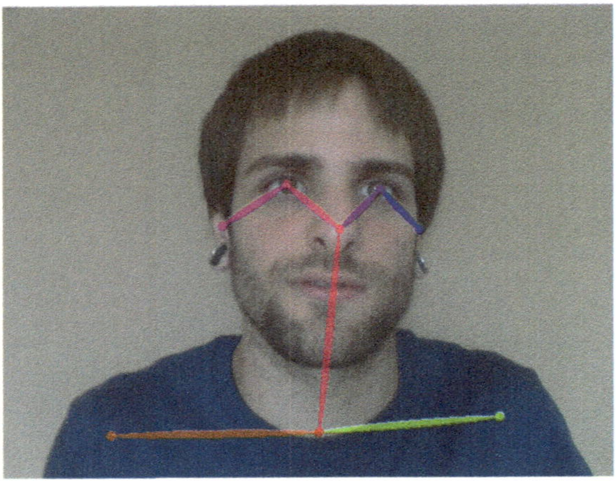

Fig. 2. Face keypoints

$$\begin{aligned} \overline{NN} &= dist(Nose, Neck) \\ H_{\phi}^{robot} &= rangeConv(\overline{NN}, robotRange) \end{aligned} \tag{1}$$

$$H_{\psi}^{robot} = rangeConv(-\arcsin(\overline{NN}_x), robotRange) \tag{2}$$

Note that using the nose information instead of the more general head central pose allows for a more detailed realistic capture of the head motion.

4.3 Hands Motion

OpenPose differentiates left and right sides without any calibration and gives 21 keypoints per hand, four per finger plus wrist.

The mapping of the hand movements is processed as follows:

1. First, it is determined whether a hand is showing the palm or the back. This requires to calculate the angle between the horizontal line and the line joining the thumb and the pinky fingertips (keypoints 4 and 20). This is represented in Eq. 3), where FT stands for fingertip and OFT represents the new origin of a fingertip. Afterwards, the fingers' points are rotated in such a way that the pinky lies at the right of the thumb and the number of fingers over the $Y = 0$ line is calculated (Eq. 4). For the right hand, at least two fingers should lie over that line to consider the palm is being showed as shown if Eq. 5 (the opposite condition for the left hand).

$$\begin{aligned} \forall i FT^i \quad OFT^i_{x,y} &= FT^i_{x,y} - Thumb_{x,y} \\ \alpha &= \arctan(OFT^{pinky}_y, OFT^{pinky}_x) \end{aligned} \tag{3}$$

$$\begin{aligned} \forall i \;\; FT'^i_x &= OFT^i_x * \cos(-\alpha) - OFT^i_y * \sin(-\alpha) \\ \forall i \;\; FT'^i_y &= OFT^i_x * \sin(-\alpha) - OFT^i_y * \cos(-\alpha) \end{aligned} \tag{4}$$

$$HandSide = \begin{cases} Back & ((\sum_{i=1}^{3} FT'^i_y) > 0) \geq 2 \\ Palm & otherwise \end{cases} \tag{5}$$

2. Afterwards, each hand's yaw angle (H_ψ) is set correspondingly by measuring the distance between the thumb and the pinky fingertips (Eq. 6). The minimum and maximum values are adjusted according to the wrist's height so to avoid collisions with the touch screen on the chest of the robot.

$$\begin{aligned} \overline{TP} &= dist(FT'^{Thumb}, FT'^{Pinky}) \\ H_\psi &= rangeConv(\overline{TP}, robotRange) \end{aligned} \tag{6}$$

3. Hand opening/closing is also measured as a function of the distance between wrist (keypoint 0) and middle fingertip (keypoint 12) (see Eq. 7).

$$\begin{aligned} \overline{MW} &= dist(FT^{Middle}, Wrist) \\ ClosedOpen &= rangeConv(\overline{MW}, [0.0 - 1.0]) \end{aligned} \tag{7}$$

5 Experimental Setup

5.1 Robot and Computer Platforms

The robotic platform employed in the performed experiments is a Pepper robot developed by Softbank Robotics [2]. Currently, our robot is controlled using the *naoqi_driver*[3] package that wraps the needed parts of the NAOqi [4] API and makes them available in ROS[5].

[2] https://www.ald.softbankrobotics.com/en/robots/pepper.
[3] http://wiki.ros.org/naoqi_driver.
[4] http://doc.aldebaran.com/2-5/naoqi/index.html.
[5] http://www.ros.org.

During the data recording process, a Intel Core i7-7700 CPU (8 × 3.60 GHz) running Ubuntu 16.04 with 16 GB RAM was used, provided with a Nvidia Titan Xp. The MoCap consisted of a Realsense Depth Camera D435 and OpenPose 1.5 developed by CMU Perceptual Computing lab. Standard OpenPose network configuration was used.

As mentioned before, OpenPose is a 2D pose estimator and depth information is needed to properly perform the human to robot mapping procedure. Therefore, `openpose_ros`[6] is used as ros wrapper for OpenPose, and `skeleton_extract_3d`[7] to get depth information. This package has been adapted to include detailed finger information offered by OpenPose.

5.2 Generative Model

GAN networks are composed by two different interconnected networks. The *Generator* (G) network generates possible candidates so that they are as similar as possible to the training set. The second network, known as *Discriminator* (D), judges the output of the first network to discriminate whether its input data are "real", namely equal to the input data set, or if they are "fake", that is, generated to trick with false data.

The training dataset given to the D network contained 2018 unit of movements (UM), being each UM a sequence of 4 consecutive poses, and each pose 14 float numbers corresponding to joint values of head, arms, wrists (yaw angle) and hands (finger opening value). These samples were recorded by registering 4 different persons talking, about 9 min overall.

The D network is thus trained using that data to learn its distribution space; its input dimension is 56. On the other hand, the G network is seeded through a random input with a uniform distribution in the range $[-1, 1]$ and with a dimension of 100. The G intends to produce as output gestures that belong to the real data distribution and that the D network would not be able to correctly pick out as generated. Figure 3 depicts the architecture of the generator and discriminator networks.

6 Results and Motion Analysis

The robot gestures seem natural and motion looks appropriate, head and arms are coordinated and the robot does not produce weird movements. In order to confirm this subjective impression, we decided to compare the above described system with a OpenNI based approach previously developed [30].

[6] https://github.com/firephinx/openpose_ros.
[7] https://github.com/stevenjj/openpose_ros/tree/master/skeleton_extract_3d.

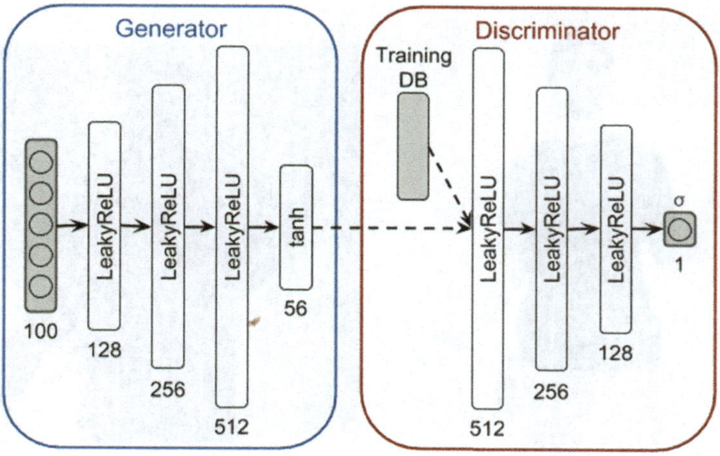

Fig. 3. GAN setup for talking gesture generation.

Figure 4 summarizes the skeleton capturing characteristics of the OpenNI based system. The performance of the robot with both approaches can be appreciated in this video[8]. As it has been previously stated, only beats are being considered. At a glance, the new approach overcomes the three main drawbacks that are detected by observing the robot behavior. On the one hand, head information provided by the skeleton tracking package was not enough for preserving head movements and thus, the resulting motion was poor. On the other hand, the tracker only offered wrist positions and as a consequence, a vision based alternative was developed by segmenting red/green colors of the gloves wore by the speaker for tracking palms and backs of both hands. The opening/closing of the fingers was made at random for each generated movement. Lastly, the robot elbows tended to be too separated from the body and raised up.

The comparison should let us to measure any progress introduced in the robot behavior and to confirm any improvement achieved by the new approach. For quantitatively measuring the quality of the generated movements two measures are used:

- Norm of Jerk: as mentioned in the introduction, the goal is to generate spontaneous smooth movements. The norm of Jerk is a smoothness measure based on root mean square (RMS) jerk quantification [5]. It is calculated according to Eq. 8, where *accel* stands for the acceleration at time *t*.

$$jerk = \frac{1}{T}\sum_{t=1}^{T} ||\dot{accel_t}|| \quad (8)$$

[8] https://www.youtube.com/watch?v=h9wpMEH8JQc.

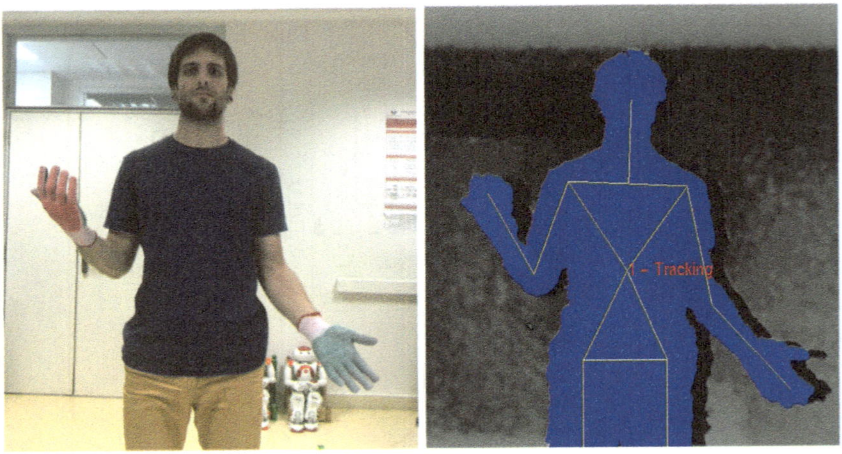

Fig. 4. Wearables needed in the OpenNI based approach

- Length of the generated paths: the length of the path ($lpath$) described by the positions of the hands during time ($\overline{x}_t$) is also another interesting measure. Lower $jerk$ values would lead to lower $lpath$ values, as the movements would be smoother. The measure ($lpath$) is computed as Eq. 9.

$$lpath = \sum_{t=2}^{T} ||\overline{x}_t - \overline{x}_{t-1}|| \tag{9}$$

Two different GANs trained in the same conditions have been used to generate robot motion for about 300 s and during the execution of the movements, the 3D coordinates of the end effectors (i.e. left and right hands) and elbows with respect to the pelvis have been recorded while talking. Head pitch and yaw values were also saved. Table 1 shows the obtained values for each approach.

Overall, OpenPose based motion generation obtains lower (and thus, more desirable) values for every measure. Hands and elbows show smoother motion. Figure 5 compares the elbows' motion paths during those 300 s. Strange elbow poses noticeable in the above referred video are confirmed by the different measures and the plots.

Head jerk is the exception though. There is an explanation for this outlier. In order to be able to capture head pose using the OpenNI based approach, the talking human needed to exaggerate the head movements so that the mapped pose was perceptible on the robot. This drawback can be appreciated during the replication of the movements of the second actress in the simulated robot in this video[9] (timestamp 0:36 to 0:44 s). Small human head movements cannot be captured with fidelity using the OpenNI based approach. As a result, the generated head motion is barely noticeable and hence, produces smaller jerk values.

[9] https://www.youtube.com/watch?v=iW1566ozbdg.

Table 1. Mean values for each measure (ϕ: pitch, ψ: yaw)

		OpenPose based	OpenNI based
Lhand	E_{jerk}	0.0369	0.0371
	E_{lpath}	29.6107	31.1931
Rhand	E_{jerk}	0.0304	0.0367
	E_{lpath}	25.3052	29.8871
Lelbow	E_{jerk}	0.0135	0.0162
	E_{lpath}	11.6434	13.4594
Relbow	E_{jerk}	0.0102	0.0161
	E_{lpath}	9.0550	13.9018
Head	E^{ψ}_{jerk}	0.0515	0.0478
	E^{ϕ}_{jerk}	0.0301	0.0248

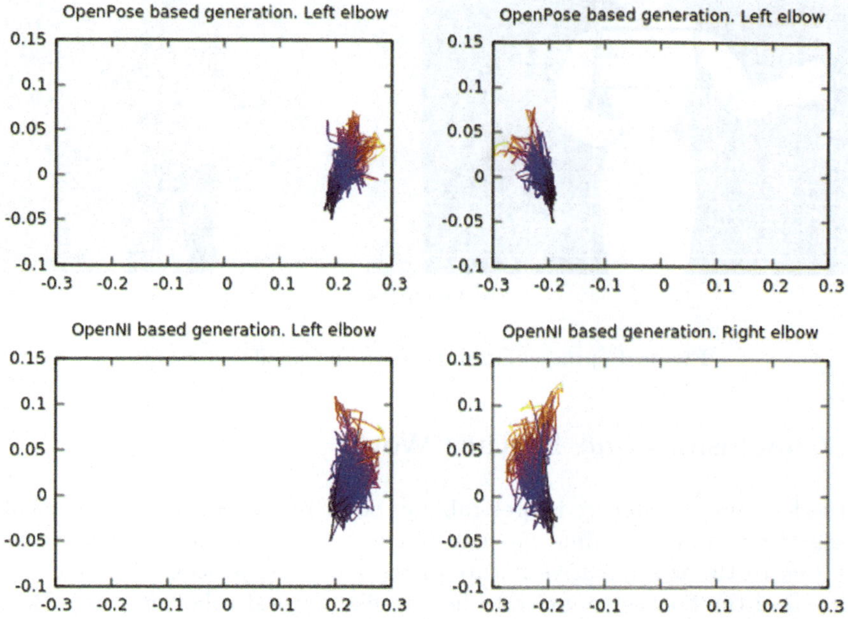

Fig. 5. Elbows' motion paths

Figures 6(a) and 6(b) show the dissimilarities between the two capture systems. Note the differences in head, hand opening and elbow positions compared to the model pose.

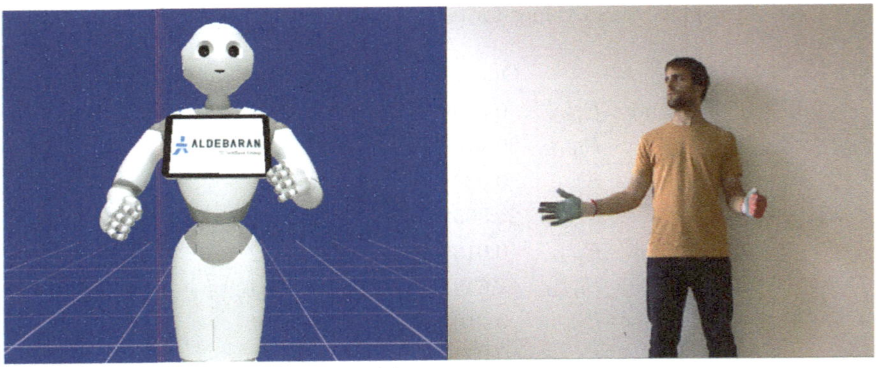

(a) OpenNI

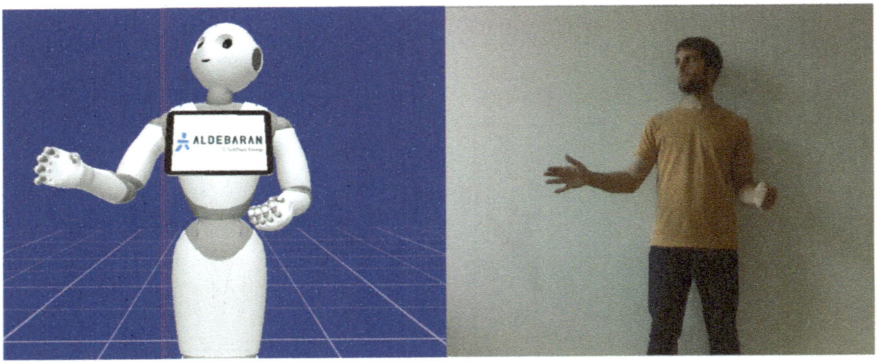

(b) OpenPose

Fig. 6. Reproduction of poses in the simulated robot

7 Conclusions and Further Work

Markerless motion capture is preferable in order not to constraint the speakers during the motion recording process. It allows to the human to behave naturally due to the non-intrusive nature of the recording process. This fact is not petty since the goal is to generate human like natural robot gestures. Beyond this, OpenPose captures motion more faithfully, i.e. the skeleton obtained better gathers the pose of the human model. OpenPose based approach induces better replication and, as a consequence, better motion generation as it has been empirically measured.

A more expressive database would broaden the variety of gestures being generated by the robot. Before exposing the robot to a large audience, a wider set of human models need to be recorded. However, the most interesting point in our opinion is that the developed system offers the basis for extending the beats to a generator conditioned on other kind of movements as emotion-based or context related gestures.

We could conclude with an optimistic answer to the question presented in the title. There is still a long way for a fully functional system able to learn the wide spectrum of human gestures just by observation, but this research has shown one of the steady steps in that direction.

References

1. Alibeigi, M., Rabiee, S., Ahmadabadi, M.N.: Inverse kinematics based human mimicking system using skeletal tracking technology. J. Intell. Rob. Syst. **85**(1), 27–45 (2017)
2. Augello, A., Cipolla, E., Infantino, I., Manfrè, A., Pilato, G., Vella, F.: Creative robot dance with variational encoder. CoRR **abs/1707.01489** (2017)
3. Beck, A., Yumak, Z., Magnenat-Thalmann, N.: Body movements generation for virtual characters and social robots. Soc. Signal Process. **20**, 273–286 (2017)
4. Breazeal, C.: Designing Sociable Robots. Intelligent Robotics and Autonomous Agents. MIT Press, Cambridge (2004)
5. Calinon, S., D'halluin, F., Sauser, E.L., Cakdwell, D.G., Billard, A.G.: Learning and reproduction of gestures by imitation. In: International Conference on Intelligent Robots and Systems, pp. 2769–2774 (2004)
6. Cao, Z., Hidalgo Martinez, G., Simon, T., Wei, S., Sheikh, Y.A.: Openpose: realtime multi-person 2D pose estimation using part affinity fields. IEEE Trans. Pattern Anal. Machi. Intell. 1 (2019). https://doi.org/10.1109/TPAMI.2019.2929257
7. Cao, Z., Simon, T., Wei, S.E., Sheikh, Y.: Realtime multi-person 2D pose estimation using part affinity fields. In: CVPR (2017)
8. Enrique Castillo, J.M.G., Hadi, A.S.: Learning bayesian networks. In: Expert Systems and Probabilistic Network Models. Monographs in Computer Science, Springer, New York (1997)
9. Everitt, B., Hand, D.: Finite Mixture Distributions. Chapman and Hall, Dordrecht (1981)
10. Fadli, H., Machbub, C., Hidayat, E.: Human gesture imitation on NAO humanoid robot using Kinect based on inverse kinematics methodn. In: International Conference on Advanced Mechatronics, Intelligent Manufacture, and Industrial Automation (ICAMIMIA). IEEE (2015)
11. Goodfellow, I.: NIPS Tutorial: Generative adversarial networks. ArXiv e-prints (2017)
12. Goodfellow, I., Pouget-Abadie, J., Mirza, M., Xu, B., Warde-Farley, D., Ozair, S., Courville, A., Bengio, Y.: Generative adversarial nets. In: Advances in Neural Information Processing Systems, pp. 2672–2680 (2014)
13. Gupta, A., Johnson, J., Fei-Fei, L., Savarese, S., Alahi, A.: Social GAN: socially acceptable trajectories with generative adversarial networks. CoRR **abs/1803.10892** (2018). http://arxiv.org/abs/1803.10892
14. Kofinas, N., Orfanoudakis, E., Lagoudakis, M.G.: Complete analytical forward and inverse kinematics for the NAO humanoid robot. J. Intell. Rob. Syst. **77**(2), 251–264 (2015). https://doi.org/10.1007/s10846-013-0015-4
15. Kwon, J., Park, F.C.: Using hidden markov models to generate natural humanoid movement. In: International Conference on Intelligent Robots and Systems (IROS). IEEE/RSJ (2006)
16. LeCun, Y., Bengio, Y., Hinton, G.: Deep learning. Nature **521**(7553), 436–444 (2015)

17. Manfrè, A., Infantino, I., Vella, F., Gaglio, S.: An automatic system for humanoid dance creation. Biologically Inspired Cogn. Architect. **15**, 1–9 (2016)
18. McNeill, D.: Hand and Mind: What Gestures Reveal about Thought. University of Chicago press, Chicago (1992)
19. Mehta, D., Sridhar, S., Sotnychenko, O., Rhodin, H., Shafiei, M., Seidel, H.P., Xu, W., Casas, D., Theobalt, C.: VNect: real-time 3D human pose estimation with a single RGB camera. ACM Trans. Graph. **36**(4), 1–14 (2017)
20. Poubel, L.P.: Whole-body online human motion imitation by a humanoid robot using task specification. Master's thesis, Ecole Centrale de Nantes–Warsaw University of Technology (2013)
21. Rabiner, L.R.: A tutorial on hidden Markov models and selected applications in speech recognition. Proc. IEEE **77**, 257–286 (1989)
22. Rodriguez, I., Astigarraga, A., Jauregi, E., Ruiz, T., Lazkano, E.: Humanizing NAO robot teleoperation using ROS. In: International Conference on Humanoid Robots (Humanoids), IEEE-RAS (2014)
23. Rodriguez, I., Astigarraga, A., Ruiz, T., Lazkano, E.: Singing minstrel robots, a means for improving social behaviors. In: IEEE International Conference on Robotics and Automation (ICRA), pp. 2902–2907 (2016)
24. Rodriguez, I., Martínez-Otzeta, J.M., Irigoien, I., Lazkano, E.: Spontaneous talking gestures using generative adversarial networks. Robot. Auton. Syst. **114**, 57–65 (2019)
25. Schubert, T., Eggensperger, K., Gkogkidis, A., Hutter, F., Ball, T., Burgard, W.: Automatic bone parameter estimation for skeleton tracking in optical motion capture. In: International Conference on Robotics and Automation (ICRA). IEEE (2016)
26. Shotton, J., Fitzgibbon, A., Cook, M., Sharp, T., Finocchio, M., Moore, R., Kipman, A., Blake, A.: Real-time human pose recognition in parts from single depth images. In: Proceedings of the 2011 IEEE Conference on Computer Vision and Pattern Recognition, CVPR 2011, pp. 1297–1304. IEEE Computer Society, Washington, DC, USA (2011). https://doi.org/10.1109/CVPR.2011.5995316
27. Tanwani, A.K.: Generative models for learning robot manipulation. Ph.D. thesis, École Polytechnique Fédéral de Laussane (EPFL) (2018)
28. Tits, M., Tilmanne, J., Dutoit, T.: Robust and automatic motion-capture data recovery using soft skeleton constraints and model averaging. PLOS ONE **13**(7), 1–21 (2018)
29. Wold, S., Esbensen, K., Geladi, P.: Principal component analysis. Chemometr. Intell. Lab. Syst. **2**(1–3), 37–52 (1987)
30. Zabala, U., Rodriguez, I., Martínez-Otzeta, J.M., Lazkano, E.: Learning to gesticulate by observation using a deep generative approach. In: 11th International Conference on Social Robotics (ICSR), pp. 666–675. Springer (2019). https://doi.org/10.1007/978-3-030-35888-4_62
31. Zhang, Z., Niu, Y., Kong, L.D., Lin, S., Wang, H.: A real-time upper-body robot imitation system. Int. J. Robot. Control **2**, 49–56 (2019). https://doi.org/10.5430/ijrc.v2n1p49
32. Zhang, Z., Niu, Y., Yan, Z., Lin, S.: Real-time whole-body imitation by humanoid robots and task-oriented teleoperation using an analytical mapping method and quantitative evaluation. Appl. Sci. **8**(10) (2018). https://www.mdpi.com/2076-3417/8/10/2005

Evaluation of a Multi-speaker System for Socially Assistive HRI in Real Scenarios

Antonio Martínez-Colón[1], Raquel Viciana-Abad[1](✉), Jose Manuel Perez-Lorenzo[1], Christine Evers[2], and Patrick A. Naylor[3]

[1] Universidad de Jaén, Jaén, Spain
{amcolon,rviciana,jmperez}@ujaen.es
[2] University of Southampton, Southampton, UK
c.evers@soton.ac.uk
[3] Imperial College London, London, UK
p.naylor@imperial.ac.uk

Abstract. In the field of social human-robot interaction, and in particular for social assistive robotics, the capacity of recognizing the speaker's discourse in very diverse conditions and where more than one interlocutor may be present, plays an essential role. The use of a mics. array that can be mounted in a robot supported by a voice enhancement module has been evaluated, with the goal of improving the performance of current automatic speech recognition (ASR) systems in multi-speaker conditions. An evaluation has been made of the improvement in terms of intelligibility scores that can be achieved in the operation of two off-the-shelf ASR solutions in situations that contemplate the typical scenarios where a robot with these characteristics can be found. The results have identified the conditions in which a low computational cost demand algorithm can be beneficial to improve intelligibility scores in real environments.

Keywords: Beamforming · ASR · Array · Masking · Intelligibility

1 Introduction

One of the main current concerns, both of the central and regional administrations, and of society in general, is to analyse the capacity to attend to and cover the growing needs of people in a situation of dependence, and in particular the elderly. In response to this challenge, in recent years, a great effort is being made to develop Socially Assistive Robots (SAR) as tools to support and leverage resources in different tasks in retirement homes [1]. If the development of an interaction mechanism adapted to the needs and tastes of the people is already complex, it is even more of a challenge when it is conceived for older people, in which the digital gap and the possible sensory limitations are one more handicap to be considered. In this sense, Automatic Speech Recognition

L. M. Bergasa et al. (Eds.): WAF 2020, AISC 1285, pp. 151–166, 2021.
https://doi.org/10.1007/978-3-030-62579-5_11

(ASR), together with voice synthesis, is a very powerful form of interaction to provide naturalness and social engagement to human robot interaction (HRI), as stated for other examples of computer based service systems [2]. However, different studies within the SAR field have concluded about the preferences for other interaction modalities that a priori are not so straight forward or natural, mainly due to technological limitations of these systems [3] nowadays. These are associated with the ASR engines difficulties to adapt well to dynamic environments with very diverse acoustic properties, together with the challenge of recognising one person's voice among others while talking simultaneously, or the a.k.a. 'cocktail party' situation and the effect of noise of very different nature.

On the other hand, there are great advances in the field of commercial ASR integrated in VoIP value-added services and "voice bots", thanks to deep learning strategies and the extensive corpus used for training, favour their increasing use and success rate. However, these systems still tend to be based on a one-to-one interaction where it is also understood more for the use of specific commands than for an open speech dialog. In this regard, as described in Sect. 3, research is focused on mainly two approaches: dealing with the changing acoustic properties of the environments by employing de-reverberation algorithms to enhance the voice signal used as an input for these systems and using arrays of microphones (mics) to exploit their space filtering capabilities. There are also recent approaches based on combining these techniques by employing deep learning with complete training procedures as recently reviewed in [4]. However, there is an important trade-off to consider between responsiveness and intelligibility in order to keep interactivity and naturalness perception high in HRI systems.

Regarding the use of beamforming techniques, this can be compared as an artificial way to provide attention to a specific source of acoustic information, in a similar way as humans attentional mechanism focuses on a specific conversation, but just exploiting the spatial filtering capabilities. One approach would be that reactive and deliberative agents of an attentional mechanism govern the signal enhancement module to improve the ASR input signal. This study analyses the extent to which beamforming algorithms may improve a commercial ASR engine from audio signals captured by a mic. array that can be mounted on SARs.

2 The Use of ASR for Socially Assistive HRI

Voice synthesis and ASRs have been widely used in SARs as they are one of the most natural interaction modalities for HRI. However, these systems have been mainly considered in one-to-one interaction. The development of commercial service robots, such as Pepper (Softbank Robotics) and XR1 (CloudMinds), designed to provide services in public places has led to greater requirements in terms of voice recognition in multispeakers environments and with more challenging and dynamic conditions (reverberation, noise, interference, distances).

Thus, recent research projects such as EARS (FP7- 609465, Embodied Audition for RobotS) have focused on improving the auditory perception system of social robots and in the development of open-source robot audition [5] systems further integrated in different robots (Honda ASIMO, SIG2, Robovie-R2 and HRP-2). A similar approach to our proposal, but with post filtering and masking integrated together with the MFCC features of an ASR embedded in SIG2 robotic head has been evaluated with a MIMO (multiple input multiple output) approach [6], and further research [7] is still being developed in this line.

The scope of the robotic research into ASR systems is very broad, due to additional problems such as internal noises and the effect of the mic. integration in the robot structure. Nowadays, APIs such as those of Google, Microsoft, IBM, Nuance, etc. are being integrated in the robot system due to the improvement of multimedia distribution in networks and rapid access to SaaS in the Cloud. Thus, recent studies [8] have evaluated in real or "out of the lab" conditions, the performance of these ASR systems in terms of WER (Word Error Rate). In particular, Jankowski et al. have recently established that in Chinese most of them exhibit a degradation not bearable with SNR lower than 15 dB.

Until recently, the use of systems that hinge on effective verbal communication in scenarios where the dialogue requires more than a command, has caused the discouragement of the users or even their adaptation to the robot system. This is also typical in HCI for VR, the Cyborg's dilemma [9] that refers to the paradoxical situation in which the development of increasingly "natural" and embodied interfaces leads to "unnatural" adaptations or changes in the user. In [10], commercial ASR engines (Google, Bing, Sphinx, Nuance) have been evaluated using fixed, all spontaneous, and clean spontaneous kids speech utterances recorded during their interaction with a NAO robot, in typical conditions of social HRI scenarios. Their performance metrics not only consider WER, but also matches with "relaxed accuracy", as they can be enough for an ASR system to recognise sentences. Their results highlighted that Google outperformed the others with three types of utterances.

The differences in the gender representation in broadcast corpora used to train ASR engines is being also an aspect considered [11], as in the case of children and women, it may have a negative impact in HRI while being used with general population, and together with voice differences should be considered.

3 Implementation of an Acoustic Signal Enhancement Module for the Source of Interest in Real Time

In relation to the separation of sources of interest (SOI), when they are of different nature, and not only associated with speakers, there are many studies that propose to work with statistical models and deep learning strategies (neural networks, Gaussian mixture models, support vector machines) that contemplate the different nature of the signals present. As reviewed in [4], deep learning strategies, based on supervised training, have improved performance of data-driven approach in speech processing. However, these techniques require defining a proper training target and binaural or monoaural features as an input. Under the framework of Auditory Scene Analysis (ASA), the main benefits of the statistical techniques is that is not necessary to know "a priori" aspects such as the geometry of the room, number of sources, number of mics., which is referred to as Blind Source Separation (BSS) systems. In this sense, PCA (Principal Component Analysis), ICA (Independent component Analysis) and NMF (Non-negative Matrix Factorization) methods [12] are widely used, even though they are prepared for signals that vary mainly in amplitude and not in phase.

3.1 Beamforming Techniques

Another approach under ASA highlights the role of auditory attention in stream segregation, based on knowledge about the target sound, such as previous knowledge of a speaker's presence (utterance including the name), source location or type of sound (alarm, crying, etc.). Considering location, beamforming techniques may be a gross separation technique previous to other attention-related parameters. These other parameters that allow a more refined separation are commonly based in applying algorithms in time-frequency (T-F) representations.

Different beamforming alternatives have been analyzed for separation purposes [12], and performance it is usually compared to that obtained from Delay and Sum Beamforming (DSB) by measuring the improvement due to attenuating interference from other directions, which is the more basic and less computational demand algorithm. DSB adds multiple mic. signals for target direction in phase, and its generic output for a beamformer focal point, $\mathbf{r}_p$, is given by:

$$y\left(\mathbf{r}_p\right)[n] = \frac{1}{M}\sum_{m=1}^{M} x_m\left[n - \tau_{pm} \cdot f_s\right] \tag{1}$$

where x_m is the mth mic. response for sample n, f_s is the sample frequency, $1/M$ (M the number of mics.) is the traditionally selected weights magnitude, and τ_{pm} are delays due to sound signals propagation. This delay is due to the distance between $\mathbf{r}_p$ and the mics. array at the sound speed c, thus being τ_{pm}:

$$\tau_{pm} = \frac{d_{pm}}{c} = \frac{\sqrt{\left(x_p - x_{mic}\right)^2 + \left(y_p - y_{mic}\right)^2 + \left(z_p - z_{mic}\right)^2}}{c} \tag{2}$$

However, DSB does not usually allow sufficient reduction of the interfering signal where the signal-to-interference ratio (SIR) is low, in the case of interfering signals energy greater than that of the signal of interest [12]. In particular, dynamic beamformers, such as MVDR (Minimum Variance Distortionless Response), LCMV (Linearly Constrained Minimum Variance) and GSC (Generalized Sidelobe Canceller) are an alternative in situations with moving speakers and unknown room conditions. In the case of MVDR, the objective is to calculate the direction vectors to minimize the output energy from not target directions, which requires the calculation of the co-variance matrix of the captured signals, processing it for each frequency bins and dealing with the complexity of non-inverted matrix computation. Therefore, it has a high computational cost and also requires knowing how much the SOI signal should be amplified to minimize the interfering energy, being difficult its implementation in real time environments. The possible use of LCMV (an evolution of MVDR) has been also analysed, because in addition to minimizing the interfering signal energy at the output, interfering signals can be cancelled if they come from known directions. This solution is interesting because the reactive module already implemented in [13] is able to localize an "a priori" unknown number of interference sources. However, previous studies [14] have indicated that despite the same demands in terms of computational cost, the recalibration required for changes in the power of the SOI signal is not so limiting, but it is necessary to have a very precise computation of the directions of the interfering signals, which is not the case for a very dynamic scenario, as the once considered with the array mounted in a robot. A GSC implementation has been therefore considered, which is basically like LCMV but cancelling all directions that don't come from the SOI direction, thus simplifying the calculation of the conditioned part. GSC is based on the application of a fixed beamformer with the purpose of estimating the SOI output signal, having therefore part of the interfering signals eliminated, and on the other hand, making an estimation of the noise and interference through the use of a blocking matrix that eliminates the target source. The implementation of the interference modeling part can be done by calculating the adaptive filter weights based on the use of a least squares algorithm, in particular NLMS (Normalized Least Mean Square) [15]. Thus, achieving that in a dynamic way it is possible to model the present interference and directional noise sources. To avoid the computational cost of the null space calculation of the condition matrix, a time domain implementation of the classical adaptative beamformer Griffiths-Jim Beamformer (GJBF) [16] has been programmed. This approach consists in using a DSB in the fixed part and a blocking matrix associated with the subtraction of adjacent signals. The GSC output beamformer is given by:

$$y\left[n\right] = y_s\left[n\right] - \sum_{k=1}^{M-1} \boldsymbol{w}_k^T\left[n\right]\boldsymbol{z}_k\left[n\right] \tag{3}$$

where $y_s\left[n\right]$ represents a fixed beamformer computed with weights $(w_1, .., w_M)$ to each M mic., keeping its behaviour constant with time, $\boldsymbol{z}_k\left[n\right]$ is the kth Blocking Matrix output of O total samples, and $\boldsymbol{w}_k\left[n\right]$ is the kth column of the NLMS filter tap weight matrix $\mathbf{W}$ of length O. The adaptive filters are updated with the NLMS as follows:

$$\boldsymbol{w}_k\left[n+1\right] = \beta\boldsymbol{w}_k\left[n\right] + \mu y\left[n\right]\frac{\boldsymbol{z}_k\left[n\right]}{\left\|\boldsymbol{z}_k\left[n\right]\right\|^2} \tag{4}$$

β is the forgetting factor ($0 < \beta < 1$), the $\|\|^2$ is the calculation of the Euclidean norm, and μ is the step size parameter ($\mu > 0$) determining how much the filter tap changes with each iteration. Large values of μ result in fast convergence toward a steady-state signal with large misadjustment, and small values result on the opposite situation. β modifies the influence of previous calculated tap weights on the future weights. Therefore, both parameters affect the stability of the NLMS filters and they have been set to $\mu = 0.1$ and $\beta = 0.9$ as in [17].

3.2 Masking Techniques

Masking techniques are used together with beamformers to treat speech separation as a supervised learning problem based on computing two-dimensional masks in T-F. Recent studies have evaluated a basic implementation of a DSB and a binary mask, featured by its low computational demands, for distributed array of mics. in both simulated and real scenarios. In particular, this study [17] showed the benefits of a masking implementation in terms of the intelligibility with a distributed mic. array. In this experiment, instead of a distributed approach, we have proposed the use of a small circular MEMS array and the evaluation has considered environments with multiple stationary sound sources, such as human speakers, an air-conditioner (AC) machine, TV noise, etc.

Given an environment with Q sound sources distributed through the room, the T-F masking algorithm consists of a short-time windowing (20–50 ms) and the spectrum computation of the signal after being steered to each source position by a beamformer. Thus, the discrete function Y represents the T-F of a beamformed signal as follows:

$$Y\left[k,i,\mathbf{r}_p\right] = \sum_{q=1}^{Q} G_{pq}\left[k\right] \cdot X\left[k,i,\mathbf{r}_q\right] \tag{5}$$

where i is the index of a particular window, k is the frequency index (frequency bin), $X\left[\cdot\right]$ is a time frequency representation of an audio source signal located at position $\mathbf{r}_q$, and $G_{pq}\left[k\right]$ is the discrete beamformer transfer function for the sound source located at position $\mathbf{r}_q$ with the beamformer pointing to $\mathbf{r}_p$.

Even though the beamformer has its highest gain at the focal point, a T-F window can be dominated by an interferer in particular moments when the SOI doesn't speak or when the interferer speaks louder than the SOI. A spectral power ratio is used to determine T-F windows where the SOI is the dominant source and those in which the interferer is the dominant source:

$$S_{pq}\left[k,i\right] = \frac{\left|Y\left[k,i,\mathbf{r}_p\right]\right|^2}{\left|Y\left[k,i,\mathbf{r}_q\right]\right|^2} \tag{6}$$

being $\mathbf{r}_p$ the position where the SOI is located and $\mathbf{r}_q$ the position of an interferer. A binary mask is chosen as:

$$T_{pq} = \begin{cases} 1, & if\ S_{pq}\left[k,i\right] \geq 1 \\ 0, & if\ S_{pq}\left[k,i\right] < 1 \end{cases} \tag{7}$$

If several interference sources are present in the environment, a mask is chosen as a multiplication (or binary "AND" operation) of each mask corresponding to individual sources:

$$T_p\left[k,i\right] = \prod_{q=1,q\neq p}^{Q} T_p\left[k,i\right] \tag{8}$$

Thus, the output of a signal spectrum for a specific T-F window is given as:

$$Y^{'}\left[k,i,\mathbf{r}_p\right] = T_p\left[k,i\right] \cdot Y\left[k,i,\mathbf{r}_p\right] \tag{9}$$

Finally, the time domain signal can be reconstructed processing its inverse FFT. Once T-F areas where predominate interferent sound sources are masked, the intelligibility of the SOI improves.

3.3 Simulated Evaluations of Intelligibility

STOI (short-time objective intelligibility) [18], which measures the correlation between the short-time temporal envelops of a reference utterance and a separated utterance, is the most commonly used metric in recent years [4] in studies with pre-recorded signals and simulated scenarios. The outcomes of these studies have been later contrasted with other intelligibility measurements associated with the use of ASR and scenarios where real noise sources are considered. Studies [19] have evaluated word intelligibility scores with respect to SNR from different kinds of interference (multiple voices, broadband noise and complex tone). Results have proved a Speech Reception Threshold (SRT) of around -2 dB for two speakers and a SRT of 2 dB for broadband noise. However, when the interference is due to voices, the word intelligibility score considering a threshold of 50% has a more negative dependence on the SNR.

4 Overview of the General System and Implementation

The algorithms evaluated have been implemented as part of a system conceived to endorse a robot with capabilities of exhibiting a socially-accepted behaviour. For that purpose, three main modules have been designed, a perception system supported by sensors (visual, audio, odometer, laser, etc.) that can be included within a robot housing; a module that implements the agents of an attentional mechanism in charge of focusing the robot's attention to different points of interests, depending on the task to develop and the world's state (environment model), and a sensorial processing system. The attentional mechanism is commanded by two planners: a reactive planner in charge of allowing fast reactions in the robot as a response to the detection of new events (i.e. an alarm, or the a arrival of a new person) and a deliberative one that focuses the robot's attention to specific points commanded by rules that consider the task to perform and the environment state. One of the tasks included in the sensorial processing system is the capability of detecting, tracking and identifying the existing people in the surroundings. As part of this module it is vital to recognise the content of a conversation with the robot in typical 'cocktail party' situations, where multiple speakers must be talking simultaneously. In this situation, the attentional mechanism based on the information obtained from the sensorial processing module and the environment model must indicate who is the SOI, based on their identification, relative positions, speech content and the task being performed.

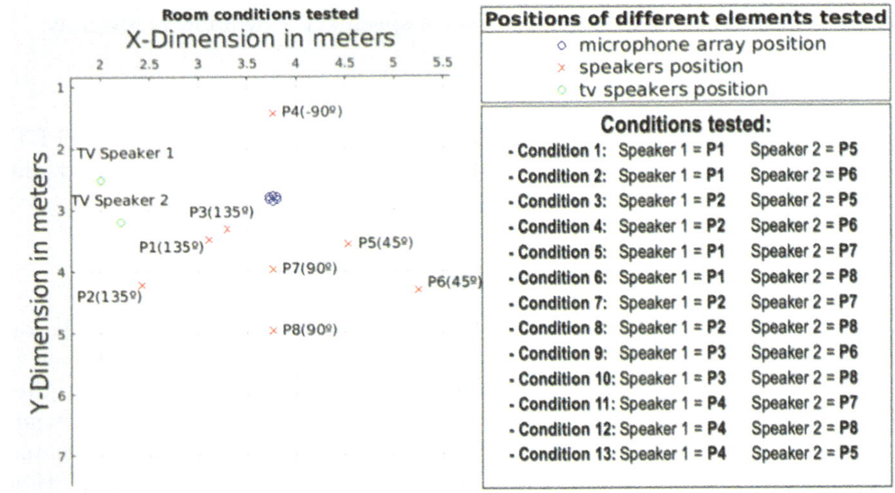

Fig. 1. Conditions tested in a research lab that model some of the most challenging situations in terms of relative positions between the target speaker and the interferer, that could take place on a real environment in an AAL.

In particular, the module of the acoustic signal enhancement is fed with the position information (cylindrical coordinates) of the target speaker and in the case of the masking system the position of the main disrupting speaker is also provided. The implementation is therefore made through three components: acquisition and recording of audio captured; the real time enhancement process (beamforming and masking) and the ASR agent implemented by calls to the REST API of an external ASR engine. Moreover, although still not connected to the enhancement agent, two algorithms for the source localization have been already implemented in C++ and evaluated: KDS-EM (Kurtosis-driven split-Expectation Maximization) binaural algorithm [13] and a SRP-PHAT algorithm with multichannel audio captures [20] obtained from a circular array, in particular a XMOS array based on 6 Micro Electrical-Mechanical System (MEMS) mics. These localization modules have been implemented to allow their execution in real time in C++ and through the use of Robocomp, a robotic middleware [3] that among other aspects allows the connection of different agents and components as well as the data exchange through ICE communication middleware[1].

The implementation of DSB, GSC and DSB+Masking algorithms has been carried out considering a capturing time of 15 s, with a processing window of 80 ms and an overlap of 50%.

The experiment considers the performance analysis in terms of recognition accuracy of two ASR engines, Google Speech API[2] and Watson Speech to Text

[1] https://zeroc.com/products/ice.
[2] https://cloud.google.com/speech-to-text?hl=es-419.

API from IBM[3], that can be accessed in the ASR component with externals calls through their APIs. In particular, for this study both engines are used through the corresponding web browser utility. In the case of Google Speech API, the model with higher fit in each case has been chosen (Default and Command).

Table 1. Average and standard deviation of STOI measurements obtained simulating the 13 different situations showed in Fig. 1 in an anechoic room using Matlab

Intelligibility	raw	DSB	GSC	Masking
STOI	0.69 (0.08)	0.73 (0.07)	0.71 (0.05)	0.84 (0.04)

5 Evaluation

5.1 Off-line Analysis of Intelligibility

The three algorithms (DSB, GSC and DSB+Masking) have been evaluated in Matlab by simulating the lab conditions without any noise source and the cocktail party effect with anechoic recordings of two speakers (one male and one female) by using the "Image method for efficiently simulating small-room acoustics"[4], for the specific case of a 192.4 m^3 rectangular room. The speakers have been placed in the 13 conditions shown in Fig. 1. The outcomes of the enhancement process have been evaluated in terms of STOI in average for all the conditions and results are shown in Table 1. These results indicate that in these challenging conditions (close interferer and use of small array of mic) the improvement in terms of STOI of the binary masking applied is high (DSB+Masking 21,7%) but as the expense of knowing the interferers' position. In contrast, the improvement due to the beamformers is low (DSB: 5,8%; GSC: 2,9%).

5.2 Experimental Procedure and Variables

The experiments consider an emulation of a common room of a retirement home, where a social robot may be interacting with residents. In these rooms, it is normal the presence of caregivers talking among themselves and typical noise sources associated with TV and AC systems. In particular, the emulated scenario considers two simultaneous speakers where the goal of the ASR is to recognise the open commands of the resident speaker, which is more difficult than to work with fix or template provided grammars.

Both speakers talk in Spanish, to properly measure ASR performance. One of them emulates a resident asking the robot for help regarding five basic daily life matters (taking food or medicines, luminosity and temperature conditions,

[3] https://speech-to-text-demo.ng.bluemix.net/.

[4] http://web.engr.uky.edu/~donohue/audio/Arrays/MAToolbox.htm.

TV volume, visits). Sentences of different lengths were chosen and including in certain cases the name of the robot (Felipe). The other speaker was emulating typical sentences of a worker or caregiver, that may also have common key words with the one associated to the sentences addressed to the robot. They are longer sentences as the goal is to analyse the system's behaviour in the worst scenario, that is to say, when most of the time both speakers are talking simultaneously.

The objective of evaluating two ASRs is to analyse whether the enhancement module behaves in the same way with ASRs with different performances under the optimal conditions of a one-to-one interaction. Google ASR has been chosen because it has already been labelled as one of the best options [10], and IBM because it has a lower performance. The experiment has been completed first with just one speaker (ideal one-to-one interaction) and after that with multiple speakers. Testing in both cases ASR engines in two conditions: with and without the presence of noise. Regarding the positions of the speakers, they were chosen with the goal of analysing the worst possible conditions in terms of speakers closeness (positions and angles) to the array for both, the target and the interfering speaker. Thus, the 13 conditions showed in Fig. 1 have been evaluated. For each condition, 4 tests were carried out: two in which the woman acts as the target speaker saying two phrases from the list of target phrases while the male says long disruptive sentences, and two tests in which they change their roles. This procedure gives rise to 52 tests carried out under conditions without noise and another 52 with background noise. The noise conditions include the AC system of the laboratory working and a TV programme (news with reporters talking).

The preliminary one-speaker test was conducted by placing the less average power voice speaker (a woman) in the four positions around the array considered in the experimental set-up. In each position, the woman has said five different sentences and the sentences analysed with the two ASR systems were analysed for the raw data and after applying the three enhancement techniques.

ASR systems usually under-perform with voices associated with young users or women in the simultaneous presence of male voices, most of the time due to the lower power voice and pitch differences, but also as stated in [11] due to the existence of less corpora for children and/or women used to generate models. For this reason, one of the speakers was a man (speaker 2) and the other a woman (speaker 1) so as to evaluate the behaviour of the ASR engines after applying the enhancement algorithms in this complex situation.

In order to evaluate the ASR intelligibility with the goal of analysing the extent to which results correspond with the STOI measurements, two measurements have been employed:

- **WER** (Word Error Rate): Ratio between the number of wrong recognised words and the total number of words in the target speaker's sentence.
- **RAR** (Relaxed Accuracy Rate): Ratio between the number of key words (adjectives, nouns, verbs and adverbs) recognised between the total amount of key words in the target speaker's sentence.

5.3 Setup for the Real Environment Conditions

The real scenario is a research lab with similar dimensions to the common rooms typically used in some residences. Its dimensions are: 5.83 m × 10 m × 3.34 m. The XMOS mic. array has been placed on a table in coordinates ($x = 3.76$, $y = 2.81$, $z = 1.21$) expressed in metres. Three walls are smooth and covered with plaster and one of them is formed by large windows. The room is featured with a reverberation of 1 s (RT_{60}).

Table 2. WER ratio [% RAR] measurements obtained with Google and IBM ASR engines for only one speaker (the woman as she is the speaker with lower voice power) in conditions with and without noise.

ASR	Non noisy scenario			Noisy scenario-SNR (dB) [−0.8, 3.7]		
	raw	DSB	GSC	raw	DSB	GSC
Google	**0.02 [97]**	**0 [100]**	**0 [100]**	**0 [99]**	**0 [100]**	**0 [100]**
IBM	**0.37 [59]**	**0.18 [82]**	**0.29 [69]**	0.64 [32]	**0.42 [56]**	0.65 [34]

Table 3. WER ratio [% RAR] measurements with two simultaneous speakers obtained with Google and IBM ASR for the 13 conditions tested with and without noise.

ASR	Non noisy scenario				Noisy scenario [(dB) [−0.8, 3.7]]			
	raw	DSB	GSC	Masking	raw	DSB	GSC	Masking
Google	**0.44 [55]**	**0.40 [60]**	**0.47 [53]**	**0.12 [88]**	0.50 [49]	**0.48 [51]**	0.60 [41]	**0.14 [86]**
IBM	0.8 [22]	0.62 [38]	0.78 [23]	0.52 [49]	0.95 [5]	0.85 [16]	0.94 [6]	0.72 [28]

The noise condition considers real noise sources (AC system and TV) placed in the positions represented in Fig. 1. As there is utterance by utterance variability in power, the SNR in dB has been computed with the average values of the one-speaker audio recordings in the different positions, obtaining SNR values within a range of [−0.8 dB, 3.7 dB]. In these conditions, [19] has specified that the word recognition accuracy for a person with broadband noise is around 50%. The SIR (Signal to Interference Ratio) has been measured considering the relative speakers' voice power in the 4 positions where they have been placed, by recording 5 different sentences in each position. An average SIR has been obtained by considering all the conditions within a range of [−11 dB, 3 dB] when the target source is a woman and within a range of [−3 dB, 11 dB] when the target source is a man. Miler's study [19] has also established that for these ranges, considering one voice interference, the word intelligibility would be between 50%–80% when the woman is the target speaker, and between 70%–90% when the target speaker is a man.

6 Results

The performance of the two ASR engines in terms of WER and RAR, in the ideal situation with one-to-one interaction (with only a woman), is shown in Table 2. As can be seen, results indicate that, Google's ASR always reaches the highest intelligibly scores with and without noise, while the IBM's performance fails mainly due to conditions where the speaker is at 2 m. from the array. In the case of IBM's ASR engine the WER reaches 0.37 (59% for RAR) in the condition without noise, and the degradation not detected in Google's ASR due to the ambient noise, in this case has been higher (WER score of 0.64). The effect of beamformers is positive but most evident for DSB, with a reduction in the WER scores of: 52% without noise and 35% with noise. Thus, as expected, in these noise conditions (TV, AC-system), that is to say broadband noises or background interferences, DSB outperforms GSC in terms of ASR scores improvement. The same tests have been carried out for the other speaker (a man) detecting the same performance.

Average ASR results for two simultaneous speakers in the 13 conditions shown in Fig. 1, with and without ambient noise, are shown in Table 3. As expected, results indicate a clear degradation of performance without any signal enhancement (nearly 40% increase of WER from raw data), for both engines, as

Table 4. Results with two simultaneous speakers in terms of WER ratio and [% RAR] using Google ASR for the most challenging situations (A: both speakers close; B: distant speakers; C: interferer closer than the target speaker), with and without noise.

Cat	Angle	SOI	Non noisy scenario			Noisy scenario		
			DSB	GSC	Masking	DSB	GSC	Masking
A	45°	female	1 [0]	1 [0]	**0.14 [83]**	1 [0]	1 [0]	0.67 [33]
		male	**0.11 [83]**	**0 [100]**	**0 [100]**	**0.11 [83]**	**0.11 [83]**	**0.11 [83]**
	90°	female	1 [0]	1 [0]	**0.25 [75]**	1 [0]	1 [0]	**0.11 [88]**
		male	**0.07 [90]**	**0 [100]**	**0 [100]**	**0 [90]**	0.59 [35]	**0 [100]**
	180°	female	**0.31 [83]**	**0.27 [69]**	**0 [100]**	0.61 [38]	0.75 [27]	**0 [100]**
		male	**0 [100]**	**0 [100]**	**0 [100]**	**0 [100]**	**0.07 [90]**	**0 [100]**
	225°	female	**0.5 [50]**	**0.43 [62]**	**0 [100]**	1 [0]	1 [0]	**0 [100]**
		male	**0.1 [87]**	**0.05 [100]**	**0 [100]**	**0 [100]**	**0 [100]**	**0 [100]**
B	45°	female	0.84 [19]	1 [0]	**0.11 [92]**	1 [0]	1 [0]	0.66 [42]
		male	**0 [100]**	**0.18 [87]**	**0 [100]**	**0.12 [87]**	**0.5 [50]**	**0.18 [75]**
	90°	female	0.75 [25]	1 [0]	**0 [100]**	1 [0]	1 [0]	**0.15 [85]**
		male	**0.14 [80]**	**0.57 [50]**	**0.15 [80]**	0.57 [40]	0.88 [20]	**0.03 [100]**
C	45°	female	1 [0]	1 [0]	0.69 [30]	1 [0]	1 [0]	1 [0]
		male	**0.47 [51]**	0.62 [37]	**0.4 [62]**	0.75 [25]	1 [0]	**0.25 [75]**
	90°	female	1 [0]	1 [0]	0.57 [37]	1 [0]	1 [0]	0.57 [37]
		male	1 [0]	1 [0]	**0.07 [91]**	1 [0]	1 [0]	**0.07 [91]**
	180°	male	**0.35 [66]**	**0.5 [50]**	**0.07 [91]**	**0.5 [50]**	**0.21 [74]**	**0 [100]**

they employ training models considering one person speech recognition. In conditions, where the target and interferer are both relatively close to each others and the array, just the T-F masking algorithm has improved the scores, which agrees with the simulations performed in terms of STOI, but indicating that the intelligibly measured with STOI underestimated the improvement of the masking in real conditions. Regarding the differences between engines in the effect of the target voice enhancement, for Google the masking has increased the performance in more than 70% of the cases with and without noise. In the case of the IBM ASR engine, this improvement could bring performance to a bearable value (0.5 WER and 50% RAR) only for conditions without environment noise, with an increase of 35% (WER reduction from 0.8 to 0.52).

From the 52 tests performed, the results obtained in the most challenging situations have been analysed in detail, classifying them according to: the difference in metres between SOI and interferer, the azimuth angle separation, whether the target speaker was a man or a woman, resulting 3 categories associated with the experimental conditions of Fig. 1, as follows:

A: Both speakers close: 1 m. (C5-45°, C1-90°, C11-180°, C13-225°)
B: Both speakers far away: 2 m. (C8-45°, C4-90°)
C: The interferer closer (1 m) than the target speaker (2 m) (C7 female-45°, C6 male-45°, C10 male-45°, C3 female-90°, C2 male-90°, C12 male-180°).

When analysing the results by category, it becomes clear that in the case of a good ASR engine like Google's (See results in Table 4), the obvious need for masking starts to be evident for a SRI below 0 dB. In our experiment, in the case of a man being the target speaker (−3 dB, +11 dB) this is observed in category C and in the case of a woman (−11 dB, +3 dB) in all the categories analyzed in detail. Even in category C the masking no longer allows the WER to be below 0.5 or the RAR to exceed 50%. In the case of an engine like IBM's, performance was worse due to the higher effect of a low SNR and SIR. The beamformers effect has not been able to improve sufficiently the signal in these challenging cases and only the masking has been able to improve performance when the SIR is higher than zero. In our experiment this happens only in category A and only when the target speaker is a man.

7 Conclusions and Future Research

The study presented has made it possible to evaluate the extent in which a simple signal conditioning algorithm with a low computational cost makes it possible to improve the degree of use in terms of intelligibility of a low-cost array that can be easily integrated into a robot. There are numerous studies that evaluate different proposals for a module that allows speech signal enhancement of a target speaker in a multi-speaker scenario with simulated environments and/or standardized test-benches. However, the experiment performed in this research has evaluated the proposed system in challenging conditions and closer to real scenarios for a robot equipped with an ASR. The outcomes have highlighted the feasibility of improving efficiency of this natural interaction modality by using a basic DSB+Masking approach with a small mic. array mounted on a robot, which is a more flexible solution that an array distributed in a room.

The time domain implementation of a dynamic GSC beamformer has exhibited very low results being outperformed by a fixed DSB beamformer mainly due to the problem of the SOI signal cancellation of the blocking matrix. This is in accordance with previous studies, which have proposed as a solution implementations based on adaptive filter coefficients in linear arrays [21] or other types of null-steering algorithms [22]. Thus, these approaches avoid the need of accurately detecting the interferer position, which is required in the masking approach, but should be tested in dynamic situations and with small arrays.

On the other hand, assuming that the localization module has a certain dynamic precision determination of at least one speaker and one interferer, the basic masking proposal evaluated has been addressed as beneficial in terms of intelligibility as previously obtained with a distributed array. The 'cocktail party' situation with more than two participants has been emulated with a realistic noise condition and intelligibility performance has remained with feasible scores. However, further research is needed once the integration among the localization, signal enhancement, and ASR modules is successfully completed, in order to address the behaviour not only in terms of intelligibility but keeping under certain threshold recognition delays that may interfer with interactivity perception.

Considering tha the final objective is to include a robot as part of an AAL that serves to as tools that facilitate the development of certain monotonous but important tasks, such as actively monitoring an older person's degree of autonomy or assisting them in everyday tasks, it is necessary to evaluate the capacity of the system overcoming age-impairment interaction capabilities and gender limitations, in order to draw conclusions about the feasibility of assuring ASR performance with limited dialogues in real conditions.

Acknowledgements. This work has been funded by the National Research Project TEST-RTI2018-099522-A-C44': "Test-beds for the Evaluation of Social Awareness in Assistance Robotics" and thanks to the collaboration with CSP group at Imperial College London, funded by the Spanish Ministry of Science, Innovation and University through the lectures mobility programme (Jose Castillejo's 2018 grant). Most of the information about the typical life in a retirement home and Felipe's robot name have been gathered from the experiences during the work developed in Vitalia Teatinos and supported by the Regional Project AT17-5509-UMA 'ROSI'.

References

1. Kriegel, J., Grabner, V., Tuttle-Weidinger, L., Ehrenmüller, I.: Socially assistive robots (SAR) in in-patient care for the elderly. Stud. Health Technol. Inform. **260**, 178–185 (2019). https://doi.org/10.3233/978-1-61499-971-3-178
2. Beckert, E., et al.: Event-based experiments in an assistive environment using wireless sensor networks and voice recognition. In: 2nd International Conference on PErvasive Technologies Related to Assistive Environments (PETRA09), pp. 1–8. ACM, Corfu (2009). https://doi.org/10.1145/1579114.1579131

3. Martínez, J., et al.: Towards a robust robotic assistant for comprehensive geriatric assessment procedures: updating the CLARC system. In: 27th IEEE International Symposium on Robot and Human Interactive Communication (RO-MAN), pp. 820–825. IEEE Press, Nanjing (2018). https://doi.org/10.1109/ROMAN.2018.8525818
4. Wang, D., Chen, J.: Supervised speech separation based on deep learning: an overview. IEEE/ACM Trans. Audio Speech Lang. Process. **26**, 1702–1726 (2018). https://doi.org/10.3233/978-1-61499-971-3-178
5. Okuno, H.G., Nakadai, K., Kim, H.: Robot audition: missing feature theory approach and active audition. In: Springer Tracts in Advanced Robotics (14th Conference Robotics Research), vol. 70, pp. 227–244 (2009). https://doi.org/10.1007/978-3-642-19457-3_14
6. Valin, J., et al.: Robust recognition of simultaneous speech by a mobile robot. IEEE Trans. Robot. **23**, 742–752 (2007). https://doi.org/10.1109/TRO.2007.900612
7. Chang, X., et al.: MIMO-speech: end-to-end multi-channel multi-speaker speech recognition. In: IEEE Automatic Speech Recognition and Understanding Workshop (ASRU), pp. 237–244 (2020). https://doi.org/10.1109/ASRU46091.2019.9003986
8. Jankowski, C., Mruthyunjaya, V., Lin, R.: Improved robust ASR for social robots in public spaces (2020)
9. Biocca, F.: The cyborg's dilemma: embodiment in virtual environments. In: Second International Conference on Cognitive Technology Humanizing the Information Age, pp. 12–26, Japan (1997). https://doi.org/10.1109/CT.1997.617676
10. Kennedy, J., et al.: Child speech recognition in human-robot interaction: evaluations and recommendations. In: 12th ACM/IEEE International Conference on Human-Robot Interaction (HRI), pp. 82?90. IEEE/ACM, Vienna (2017). https://doi.org/10.1145/2909824.3020229
11. Garnerin, M., Rossato, S., Laurent, B.: Gender representation in french broadcast corpora and its impact on ASR performance. In: 1st International Workshop on AI for Smart TV Content Production, Access and Delivery (AI4TV 2019), pp. 3–9. ACM, New York (2019). https://doi.org/10.1145/3347449.3357480
12. Nikunen, J., Diment, A., Virtanen, T.: Separation of moving sound sources using multichannel NMF and acoustic trackings. IEEE/ACM Trans. Audio Speech Lang. Process. **26**, 281–295 (2018). https://doi.org/10.1109/TASLP.2017.2774925
13. Reche, P.J., et al.: Binaural lateral localization of multiple sources in real environments using a kurtosis-driven split-EM algorithm. Eng. Appl. Artif. Intell. **69**, 137–146 (2018). https://doi.org/10.1016/j.engappai.2017.12.013
14. Souden, M., Benesty, J., Affes, S.: A study of the LCMV and MVDR noise reduction filters. IEEE Trans. Signal Process. **58**, 4925–4935 (2010). https://doi.org/10.1109/TSP.2010.2051803
15. Yu, Z.L., Er, M.J.: An extended generalized sidelobe canceller in time and frequency domain. In: 2004 IEEE International Symposium on Circuits and Systems, pp. 629–633. IEEE, Vancouver (2004). https://doi.org/10.1109/ISCAS.2004.1328825
16. Griffiths, L., Jim, C.: An alternative approach to linearly constrained adaptive beamforming. IEEE Trans. Antennas Propag. **30**, 27–34 (1982)
17. Morgan, J.P.: Time-frequency masking performance for improved intelligibility with microphone arrays. Master Thesis in the College of Engineering at the University of Kentucky (2017)

18. Taal, C.H., et al.: A short-time objective intelligibility measure for time-frequency weighted noisy speech. In: IEEE Transactions on Antennas and Propagation, pp. 4214–4217. IEEE, Dallas (2010). https://doi.org/10.1109/ICASSP.2010.5495701
19. Miller, G.A.: The masking of speech. Psychol. Bull. **44**, 105–129 (1947)
20. Martinez-Colon, A., et al.: Attentional mechanism based on a microphone array for embedded devices and a single camera. In: Workshop of Physical Agents, pp. 165–178. Springer. Madrid (2018)
21. Ni, F., Zhou, Y., Liu, H.: A robust GSC beamforming method for speech enhancement using linear microphone array. In: IEEE 21st International Workshop on Multimedia Signal Processing (MMSP), pp. 1–5. IEEE. Kuala Lumpur (2019)
22. Li, C., Benesty, J., Chen, J.: Beamforming based on null-steering with small spacing linear microphone arrays. J. Acoust. Soc. Am. **143**, 2651–2664 (2018)

Graph Neural Networks for Human-Aware Social Navigation

Luis J. Manso[1(✉)], Ronit R. Jorvekar[2], Diego R. Faria[1], Pablo Bustos[3], and Pilar Bachiller[3]

[1] Department of Computer Science, College of Engineering and Physical Sciences, Aston University, Birmingham, UK
l.manso@aston.ac.uk

[2] Department of Computer Engineering, Pune Institute of Computer Technology, Pune, India

[3] Robotics and Artificial Vision Laboratory, University of Extremadura, Badajoz, Spain

Abstract. Autonomous navigation is a key skill for assistive and service robots. To be successful, robots have to comply with social rules, such as avoiding the personal spaces of the people surrounding them, or not getting in the way of human-to-human and human-to-object interactions. This paper suggests using Graph Neural Networks to model how inconvenient the presence of a robot would be in a particular scenario according to learned human conventions so that it can be used by path planning algorithms. To do so, we propose two automated scenario-to-graph transformations and benchmark them with different Graph Neural Networks using the SocNav1 dataset [1]. We achieve close-to-human performance in the dataset and argue that, in addition to its promising results, the main advantage of the approach is its scalability in terms of the number of social factors that can be considered and easily embedded in code in comparison with model-based approaches. The code used to train and test the resulting graph neural network is available in a public repository.

Keywords: Social navigation · Graph Neural Networks · Human-robot interaction

1 Introduction

Human-aware robot navigation deals with the challenge of endowing mobile social robots with the capability of considering the emotions and safety of people nearby while moving around their surroundings. There is a wide range of works studying human-aware navigation from a considerably diverse set of perspectives. Pioneering works such as [2] started taking into account the personal spaces of the people surrounding the robots, often referred to as proxemics. In addition to proxemics, human motion patterns were analysed in [3] to estimate whether

L. M. Bergasa et al. (Eds.): WAF 2020, AISC 1285, pp. 167–179, 2021.
https://doi.org/10.1007/978-3-030-62579-5_12

humans are willing to interact. Semantic properties were also considered in [4]. Although not directly applied to navigation, the relationships between humans and objects were used in the context of ambient intelligence in [5]. Proxemics and object affordances were jointly considered in [6] for navigation purposes. Two extensive surveys on human-aware navigation can be found in [7] and [8].

Despite the previously mentioned approaches being built on well-studied psychological models, they have limitations. Considering new factors programmatically (*i.e.*, writing additional code) involves a potentially high number of coding hours, makes systems more complex, and increases the chances of including bugs. Additionally, with every new aspect to be considered for navigation, the decisions made become less *explainable*, which is precisely one of the main advantages of model-based approaches over data-driven ones. Besides the mentioned model scalability and explainability issues, model-based approaches have the intrinsic and rather obvious limitation that they only account for what the model explicitly considers. Given that these models are manually written by humans, they cannot account for aspects that the designers are not aware of.

Approaches leveraging machine learning have also been published. The parameters of a social force model (see [9]) are learned in [10] and [11] to navigate in human-populated environments. Inverse reinforcement learning is used in [12] and [13] to plan navigation routes based on a list of humans in a radius. Social norms are implemented using deep reinforcement learning in [14], again, considering a set of humans. An approach modelling crowd-robot interaction and navigation control is presented in [15]. It features a two-module architecture where single interactions are modelled and then aggregated. Although its authors reported good qualitative results, the approach does not contemplate integrating additional information (*e.g.*, relations between humans and objects, structure and size of the room). The work in [16] tackles the same problem using Gaussian Mixture Models. It has the advantage of requiring less training data, but the approach is also limited in terms of the input information used.

All the previous works and many others not mentioned have achieved outstanding results. Some model-based approaches such as [4] or [6] can leverage structured information to take into account space affordances. Still, the data considered to make such decisions are often handcrafted features based on an arbitrary subset of the data that a robot would be able to work with. There are many reasons motivating to seek a learning-based approach not requiring feature handcrafting or manual selection. Their design is time-consuming and often requires a deep understanding of the particular domain (see discussion in [17]). Additionally, there is generally no guarantee that a particular hand-engineered set of features is close to being the best possible one. On the other hand, most end-to-end deep learning approaches have important limitations too. They require a big amount of data and computational resources that are often scarce and expensive, and they are hard to explain and manually fine-tune. Somewhere in the middle of the spectrum, we have proposals advocating not to choose between hand-engineered features or end-to-end learning. In particular, [18] proposes Graph Neural Networks (GNNs) as a means to perform learning that allows combining raw data with hand-engineered features, and most importantly, learn from structured information. The relational inductive bias of GNNs is specially well-suited to learn about structured data and the relations between different types of entities, often requiring less training data than other approaches. In this line, we argue that using GNNs for human-aware navigation makes possible integrating new social cues in a straightforward fashion, by including more data in the graphs that they are fed.

In this paper, we use different GNN models to estimate social navigation compliance, *i.e.*, given a robot pose and a scenario where humans and objects can be interacting, estimating to what extent a robot would be disturbing the humans if it was located in such a pose. GNNs are proposed because the information that social robots can work with is not just a map and a list of people, but a more sophisticated data structure where the entities represented have different relations among them. For example, social robots can have information about who a human is talking to, where people are looking at, who is friends with who, or who is the owner of an object in the scenario. Regardless of how this information is acquired, it can be naturally represented using a graph, and GNNs are a particularly well-suited and scalable machine learning approach to work with these graphs.

2 Graph Neural Networks

Graph Neural Networks (GNNs) are a family of machine learning approaches based on neural networks that take graph-structured data as input. They allow classifying and making regressions on graphs, nodes, edges, as well as predicting link existence when working with partially observable phenomena. Except for few exceptions (*e.g.*, [19]) GNNs are composed by similar stacked blocks/layers operating on a graph whose structure remains static but the features associated to its nodes are updated in every layer of the network (see Fig. 1).

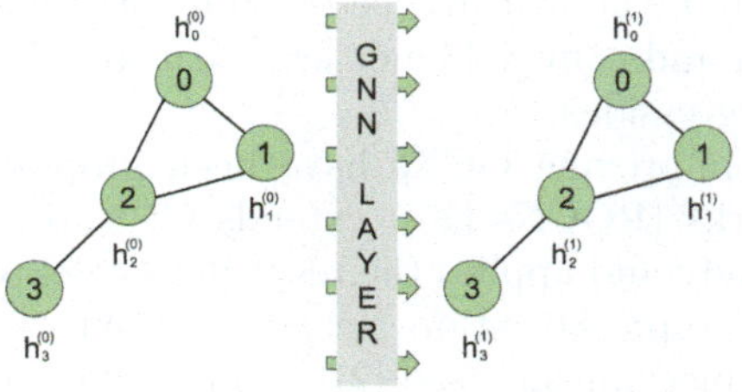

Fig. 1. A basic GNN block/layer. GNN layers output updated versions of the input graph. These updated graphs have the same nodes and links, but the feature vectors of the nodes will generally differ in size and content depending on the feature vectors of their neighbours and their own vectors in the input graph. A GNN is usually composed of several stacked GNN layers. Higher level features are learnt in the deeper layers, to that the output of any of the nodes in the last layer can used for classification or regression purposes.

As a consequence, the features associated to the nodes of the graph in each layer become more abstract and are influenced by a wider context as layers go deeper. The features in the nodes of the last layer are frequently used to perform the final classification or regression.

The first published efforts on applying neural networks to graphs date back to [20]. GNNs were further studied and formalised in [21] and [22]. However, it was with the appearance of Gated Graph Neural Networks (GG-NNs, [23]) and especially Graph Convolutional Networks (GCNs, [24]) that GNNs gained traction. The work presented in [18] reviewed and unified the notation used in the GNNs existing to the date.

Graph Convolutional Networks (GCN) [24] are one of the most common GNN blocks. Because of its simplicity, we build on the GCN block to provide the reader with an intuition of how GNNs work in general. Following the notation proposed in [18], GCN blocks operate over a graph $G = (V, E)$, where $V = \{v_i\}_{i=1:N^v}$ is a set of nodes, being v_i the feature vector of node i and N^v the number of vertices in the graph. E is a set of edges $E = \{(s_k, r_k)\}_{k=1:N^e}$, where s_k and r_k are the source and destination indices of edge k and N^e is the number of edges in the graph. Each GCN layer generates an updated representation v_i' for each node v_i using two functions:

$$\overline{e}_i = \rho^{e \rightarrow v}(E) = \sum_{\{k:r_k=i\}} e_k,$$

$$v_i' = \phi^v(\overline{e}_i, v_i) = NN_v([\overline{e}_i, v_i]).$$

For every node v_i, the first function ($\rho^{e \rightarrow v}(E)$) aggregates the feature vectors of other nodes with an edge towards it and generates a temporary aggregated feature $\overline{e}_i$ which is used by the second function. In a second pass, the function $\phi^v(\overline{e}_i, v_i)$ is used to generate updated v_i' feature vectors from the aggregated feature vectors using a neural network (usually a multi-layer perceptron, but the framework does not make any assumption on this). Such a learnable function is the same for all the nodes. By stacking several blocks where features are aggregated and updated, the feature vectors can carry information from nodes far away in the graph and convey higher level features that can be finally used for classification or regressions.

Several means of improving GCNs have been proposed. Relational Graph Convolutional Networks (RGCNs [25]) extends GCNs by considering different types of edges separately and applies the resulting model to vertex classification and link prediction. Graph Attention Networks (GATs [26]) extend GCNs by adding self-attention mechanisms (see [27]) and applies the resulting model to vertex classification. For a more detailed review of GNNs and the generalised framework, please refer to [18].

3 Formalisation of the Problem

The aim of this work is to analyse the scope of GNNs in the field of human-aware social navigation. Our study has been set up using the *SocNav1* dataset (see [1]) which provides social compliance labels for specific scenarios. It contains scenarios with a robot in a room, a number of objects and a number of people that can potentially be interacting with other objects or people. Each sample is labelled with a value from 0 to 100 depending on to what extent the subjects that labelled the scenarios considered that the robot is disturbing the people in the scenario. The dataset provides 16336 labelled samples to be used for training purposes, 556 scenarios as the development dataset and additional 556 for final testing purposes.

As previously noted, GNNs are a flexible framework that allows working somewhere in the middle of end-to-end and feature engineered learning. Developers can use as many data features as desired and are free to structure the input graph data as they please. The only limitations are those of the particular GNN layer blocks used. In particular, while GCN and GAT do not support labelled edges, RGCN and GG-NN do. To account for this limitation, two automated scenario-to-graph transformations were used in the experiments, depending on the GNN block to be tested: one without edge labels and one with them.

The first version of the scenario-to-graph transformation used to represent the scenarios does not use labelled edges. It uses 6 node types (the features associated to each of the types are detailed later in this section):

- **robot:** The dataset only includes one robot in each scenario, so there is just one robot symbol in each of the graphs. However, GNNs do not have such restriction.
- **wall:** A node for each of the segments defining the room. They are connected to the room node.
- **room:** Used to represent the room where the robot is located. It is connected to the robot.
- **object:** A node for each object in the scenario.
- **human:** A node for each human. Humans might be interacting with objects or other humans.
- **interaction:** An interaction node is created for every human-to-human or human-to-object interaction.

Figure 2 depicts two areas of a scenario where four humans are shown in a room with several objects. Two of the humans are interacting with each other, another human is interacting with an object, and the remaining human is not engaging in interaction with any human or object. The structure of the resulting non-labelled graph is shown in Fig. 3a.

The features used for **human** and **object** nodes are: distance, the relative angle from the robot's point of view, and its orientation, from the robot's point of view too. For **room** symbols the features are: the distance to the closest human and the number of humans. For the **wall** segments and the **interaction** symbols, the features are the distance and orientation from the robot's frame of reference. For wall segments, the position is the centre of the segment and the orientation is the tangent. These features geometrically define the room, which is relevant to characterise the density of people in the room and the distance from the robot to each of the walls. For interactions, the position is the midpoint

(a) An area of a scenario where two humans interacting in a room are depicted.

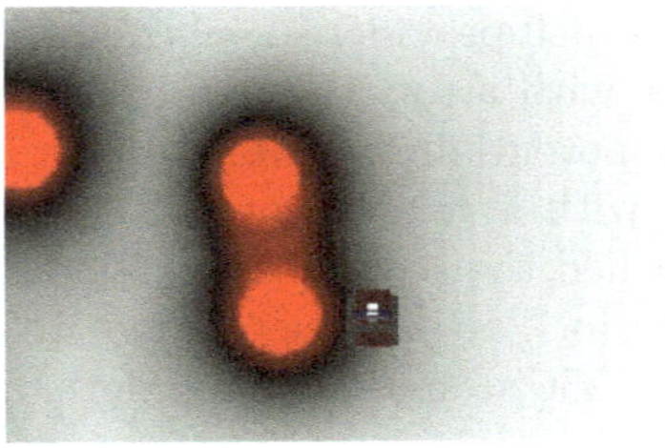

(b) Heat map of the social compliance estimation for the area shown in Fig. 2a in the different positions in the environment.

(c) An area of the same scenario with two humans. The human on the left is not engaged in any interaction. The human on the right is interacting with the object in front of her.

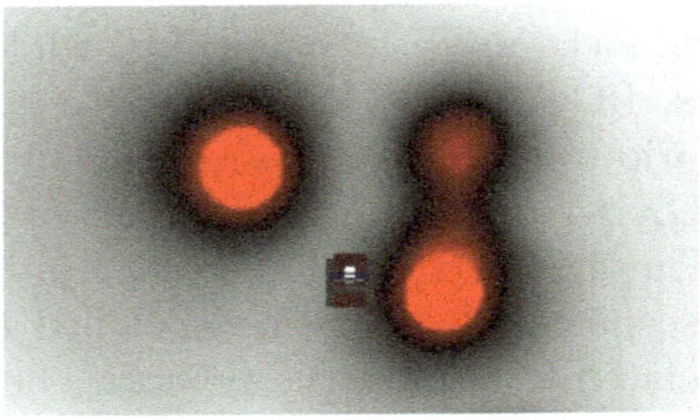

(d) Heat map of the social compliance estimation for the area shown in Fig. 2c in the different positions in the environment.

Fig. 2. Different areas of a scenario where social interactions are being held and their corresponding estimated heat map of *"social inconvenience"*.

between the interacting symbols, and the orientation is the tangent of the line connecting the endpoints. Features related to distances are expressed in meters, whereas those related to angles are actually expressed as two different numbers, $sin(\alpha)$ and $cos(\alpha)$. The final features of the nodes are built by concatenating the one-hot encoding that determines the type of the symbol and the features for the different node types. It is worth noting that by building feature vectors this way, their size increases with every new type. This limitation is currently being studied by the GNN scientific community.

For the GNN blocks that can work with labelled edges, a slightly different version of the scenario-to-graph transformation is used. The first difference is that in this version of the scenario-to-graph model there are no interaction nodes. The elements interacting are linked to each other directly. Robot, room, wall, human and object nodes are attributed with the same features as in the previous model. The second difference is related to the labelling of the edges. In this domain, the semantics of the edges can be inferred from the types of the nodes being connected. For example, *wall* and *room* nodes are always connected by the same kind of relation "composes". Similarly, *humans* and *object* nodes are always connected by the relation "interacts_with_human". The same holds the other

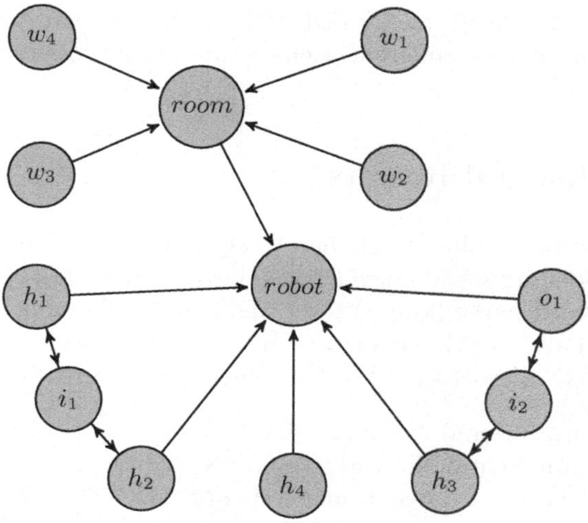

(a) Graph without labelled edges.

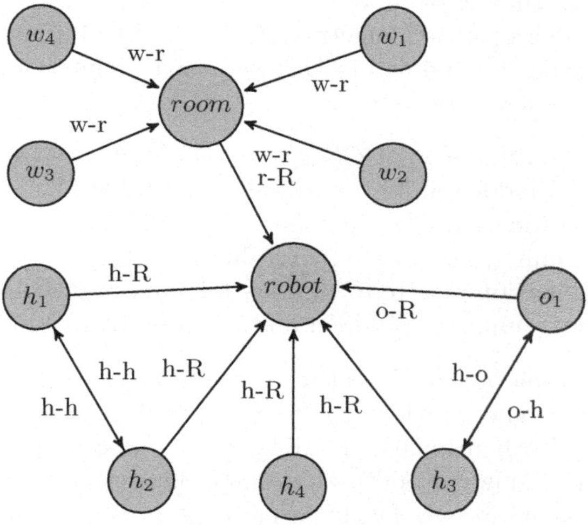

(b) Graph with labelled edges.

Fig. 3. Examples of how the scenario-to-graph transformation work based on the scenario depicted in Fig. 2.

way around: "composes" relations only occur between *wall* and *room* nodes, and "interacts_with_human" relations only occur with *humans* and *object* nodes. Therefore, for simplicity, the label used for the edges is the concatenation of the types involved. The structure of the resulting labelled graph for the scenario depicted in Fig. 2 is shown in Fig. 3b.

Because all nodes are connected to the robot, the GNNs were trained to perform the regression on the feature vector of the *robot* node in the last layer.

Using the previously mentioned dataset and the two proposed scenario-to-graph transformations, different architectures using different GNN blocks were compared.

4 Experimental Results

The experimental results are aligned with the contributions of the paper, which deal with modelling social conventions. Therefore, we assume that we can build on top of a third party body tracker and a path planning system and proceed with the evaluation of the algorithm against the dataset.

The four GNN blocks used in the experiments were:

- Graph Convolutional Networks (GCN) [24].
- Gated Graph Neural Networks (GG-NN) [23].
- Relational Graph Convolutional Networks (RGCNs) [25].
- Graph Attention Networks (GAT) [26].

If available, each of these GNN blocks was benchmarked using Deep Graph Library (DGL) [28] and PyTorch-Geometric (PyG) [29]. In addition to using several layers of the same GNN building blocks, alternative architectures combining RGCN and GAT layers were also tested:

1. A sequential combination of n RGCN layers followed by n GAT layers with the same number of hidden units (alternative 8 in Table 1).
2. An interleaved combination of n RGCN and n GAT layers with the same number of hidden units (alternative 9 in Table 1).
3. A sequential combination of n RGCN layers followed by n GAT layers with a linearly decreasing number of hidden units (alternative 10 in Table 1).

As a result, 10 framework-architecture combinations were benchmarked. Table 1 describes them and provides their corresponding performance on the development dataset. To benchmark the different architectures, 5000 training sessions were launched using the SocNav1 training dataset and evaluated using the SocNav1 development dataset. The hyperparameters were randomly sampled from the range values shown in Table 2.

The results obtained (see Table 1) show that, for the dataset and the framework/architecture combinations benchmarked, DGL/GAT delivered the best results, with a training loss of 0.01701 for the development dataset. The parameters used by the DGL/GAT combination were: batch size: 273, number of hidden units: 129, number of attention heads: 2, number of attention heads in the last layer: 3, learning rate: 5e−05, weight decay regularisation: 1e−05, number of layers: 4, no dropout, alpha parameter of the ReLU non-linearity 0.2114. After selecting the best set of hyperparameters, the network was compared with a third *test* dataset, **obtaining an MSE of 0.03173**. Figures 2b and 2d provide an intuition of the output of the network for the scenarios depicted in Figs. 2a and 2c considering all the different positions of the robot in the environment when looking along the y axis.

Table 1. A description of the different framework/architecture combinations and the experimental results obtained from their benchmark for the SocNav1 dataset.

Alt. #	Framework	Network architecture	Training loss (MSE)
1	DGL	GCN	0.02283
2	**DGL**	**GAT**	**0.01701**
3	DGL	GAT2	0.01740
4	PyG	GCN	0.29778
5	PyG	GAT	0.01804
6	PyG	RGCN	0.02827
7	PyG	GG-NN	0.02718
8	PyG	RGCN||GAT 1	0.02238
9	PyG	RGCN||GAT 2	0.02147
10	PyG	RGCN||GAT 3	0.0182

Table 2. Hyperparameter values. Only applicable to Graph Attention Network blocks.

Hyperparameter	Min	Max
Epochs	1000	
Patience	5	
Batch size	100	1500
Hidden units	50	320
Attention heads	2	9
Learning rate	1e−6	1e−4
Weight decay	0.0	1e−6
Layers	2	8
Dropout	0.0	1e−6
Alpha	0.1	0.3

It is worth noting that, due to the subjective nature of the labels in the dataset (human feelings are utterly subjective), there is some level of disagreement even among humans. To compare the performance of the network with human performance, we asked 4 subjects to label all the scenarios of the development dataset. The mean MSE obtained for the different **subjects was 0.022**. This means that the network performs close to human accuracy. Figure 4 shows an histogram comparing the error made by the GNN-based regression in comparison to humans.

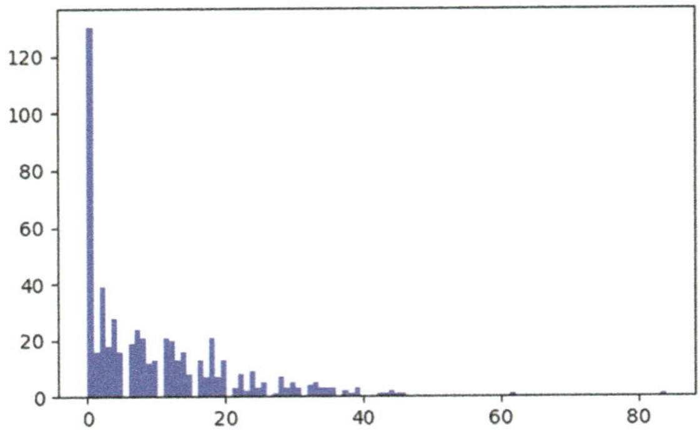

Fig. 4. Histogram of the absolute error in the test dataset for the network performing best in the development dataset.

Most algorithms presented in Sect. 1 deal with modelling human intimate, personal, social and interaction spaces instead of social inconvenience, which seems to be a more general term. Keeping that in mind, the algorithm proposed in [6] was tested against the test dataset and got a MSE of 0.12965. The relative bad performance can be explained by the fact that other algorithms do not take into account walls and that their actual goal is to model personal spaces instead of feelings in general.

Regarding the effect of the presence of humans, we can see from Fig. 2d that the learnt function is slightly skewed to the front of the humans, but not as much as modelled in other works such as [30] or [6]. One of the possible reasons why the "personal space" is close to being circular is the fact that, in the dataset, humans appear to be standing still. It is still yet to be studied, probably using more detailed and realistic datasets, how would the personal space look like if humans were moving.

The results obtained using GNN blocks supporting edge labels were inferior to those obtained using GAT, which does not support edge labels. Two reasons might be the cause of this phenomena: **a)** as mentioned in Sect. 3 the labels edges can be inferred from the types of the nodes, so that information is to some extent redundant; **b)** the inductive bias of GATs is strong and appropriate for the problem at hand. This does not mean that the same results would be obtained in other problems where the label of the edges cannot be inferred.

5 Conclusions

Although there have been attempts to predict people's trajectories using graph neural networks, to our knowledge, this paper presented the first graph neural network modelling human-aware navigation conventions. The scenario-to-graph transformation model and the graph neural network developed as a result of the work presented in this paper achieved a performance comparable to that of humans. The final model can be used as part of a social navigation system to predict the degree of acceptance of the robot position according to social conventions. Thus, given the graph representation of a scenario, the GNN provides a quantification of how good a specific robot location is for humans. Even though the results achieved are remarkable, the key fact is that this approach allows to include more relational information. This will allow to include more sources of information in our decisions without a big impact in the development.

There is room for improvement, particularly related to: **a)** personalisation (different people generally feel different about robots), and **b)** movement (the inconvenience of the presence of a robot is probably influenced by the movement of the people and the robot). Still, we include interactions and walls, features which are seldom considered in other works. As far as we know, interactions have only been considered in [6] and [31].

The code to test the resulting GNN has been published in a public repository as open-source software[1], as well as the code implementing the scenario-to-graph transformation and the code to train the models suggested.

References

1. Manso, L.J., Nuñez, P., Calderita, L.V., Faria, D.R., Bachiller, P.: SocNav1: a dataset to benchmark and learn social navigation conventions. Data **5**(1), 7 (2020). https://www.mdpi.com/2306-5729/5/1/7
2. Pacchierotti, E., Christensen, H.I., Jensfelt, P.: Human-robot embodied interaction in hallway settings: a pilot user study. In: IEEE International Workshop on Robot and Human Interactive Communication, vol. 2005, pp. 164–171. IEEE (2005). https://doi.org/10.1109/ROMAN.2005.1513774
3. Hansen, S.T., Svenstrup, M., Andersen, H.J., Bak, T.: Adaptive human aware navigation based on motion pattern analysis. In: Proceedings - IEEE International Workshop on Robot and Human Interactive Communication, pp. 927–932 (2009). https://doi.org/10.1109/ROMAN.2009.5326212
4. Cosley, D., Baxter, J., Lee, S., Alson, B., Nomura, S., Adams, P., Sarabu, C., Gay, G.: A tag in the hand: supporting semantic, social, and spatial navigation in museums. In: Proceedings of the 27th International Conference on Human Factors in Computing Systems (CHI 2009), pp. 1953–1962 (2009). https://doi.org/10.1145/1518701.1518999
5. Bhatt, M., Dylla, F.: A qualitative model of dynamic scene analysis and interpretation in ambient intelligence systems. Int. J. Robot. Autom. **24**(3), 1–18 (2010). https://doi.org/10.2316/journal.206.2009.3.206-3274
6. Vega, A., Manso, L.J., Macharet, D.G., Bustos, P., Núñez, P.: Socially aware robot navigation system in human-populated and interactive environments based on an adaptive spatial density function and space affordances. Pattern Recogn. Lett. **118**, 72–84 (2019). https://doi.org/10.1016/j.patrec.2018.07.015
7. Rios-Martinez, J., Spalanzani, A., Laugier, C.: From proxemics theory to socially-aware navigation: a survey. Int. J. Soc. Robot. **7**(2), 137–153 (2015). https://doi.org/10.1007/s12369-014-0251-1
8. Charalampous, K., Kostavelis, I., Gasteratos, A.: Recent trends in social aware robot navigation: a survey. Robot. Auton. Syst. **93**, 85–104 (2017). https://doi.org/10.1016/j.robot.2017.03.002

[1] https://github.com/robocomp/sngnn.

9. Helbing, D., Molnár, P.: Social force model for pedestrian dynamics. Phys. Rev. E **51**(5), 4282–4286 (1995). https://doi.org/10.1103/PhysRevE.51.4282
10. Ferrer, G., Garrell, A., Sanfeliu, A.: Social-aware robot navigation in urban environments. In: 2013 European Conference on Mobile Robots, ECMR 2013 - Conference Proceedings, pp. 331–336 (2013). https://doi.org/10.1109/ECMR.2013.6698863
11. Patompak, P., Jeong, S., Nilkhamhang, I., Chong, N.Y.: Learning proxemics for personalized human-robot social interaction. Int. J. Soc. Robot. (2019). https://doi.org/10.1007/s12369-019-00560-9
12. Ramon-Vigo, R., Perez-Higueras, N., Caballero, F., Merino, L.: Transferring human navigation behaviors into a robot local planner. In: IEEE RO-MAN 2014 - 23rd IEEE International Symposium on Robot and Human Interactive Communication: Human-Robot Co-Existence: Adaptive Interfaces and Systems for Daily Life, Therapy, Assistance and Socially Engaging Interactions, pp. 774–779 (2014). https://doi.org/10.1109/ROMAN.2014.6926347
13. Vasquez, D., Okal, B., Arras, K.O.: Inverse reinforcement learning algorithms and features for robot navigation in crowds: an experimental comparison. In: IEEE International Conference on Intelligent Robots and Systems (IROS), pp. 1341–1346 (2014). https://doi.org/10.1109/IROS.2014.6942731
14. Chen, Y.F., Everett, M., Liu, M., How, J.P.: Socially aware motion planning with deep reinforcement learning. In: IEEE International Conference on Intelligent Robots and Systems, September 2017, pp. 1343–1350 (2017). https://doi.org/10.1109/IROS.2017.8202312
15. Chen, C., Liu, Y., Kreiss, S., Alahi, A.: Crowd-robot interaction: crowd-aware robot navigation with attention-based deep reinforcement learning. In: International Conference on Robotics and Automation (ICRA), pp. 6015–6022. IEEE (2019). http://arxiv.org/abs/1809.08835
16. Martins, G.S., Rocha, R.P., Pais, F.J., Menezes, P.: Clusternav: learning-based robust navigation operating in cluttered environments. In: 2019 International Conference on Robotics and Automation (ICRA), pp. 9624–9630. IEEE (2019)
17. Lecun, Y., Bengio, Y., Hinton, G.: Deep learning. Nature **521**(7553), 436–444 (2015). https://doi.org/10.1038/nature14539
18. Battaglia, P.W., Hamrick, J.B., Bapst, V., Sanchez-Gonzalez, A., Zambaldi, V., Malinowski, M., Tacchetti, A., Raposo, D., Santoro, A., Faulkner, R., Gulcehre, C., Song, F., Ballard, A., Gilmer, J., Dahl, G., Vaswani, A., Allen, K., Nash, C., Langston, V., Dyer, C., Heess, N., Wierstra, D., Kohli, P., Botvinick, M., Vinyals, O., Li, Y., Pascanu, R.: Relational inductive biases, deep learning, and graph networks, pp. 1–40 (2018). https://doi.org/10.1017/S0031182005008516, http://arxiv.org/abs/1806.01261
19. Ying, Z., You, J., Morris, C., Ren, X., Hamilton, W., Leskovec, J.: Hierarchical graph representation learning with differentiable pooling. In: Advances in Neural Information Processing Systems, pp. 4800–4810 (2018)
20. Sperduti, A., Starita, A.: Supervised neural networks for the classification of structures. IEEE Trans. Neural Netw. **8**(3), 1–22 (1997)
21. Gori, M., Monfardini, G., Scarselli, F.: A new model for learning in graph domains. In: Proceedings of the International Joint Conference on Neural Networks, vol. 2, pp. 729–734 (2005). https://doi.org/10.1109/IJCNN.2005.1555942
22. Scarselli, F., Gori, M., Tsoi, A.C., Hagenbuchner, M., Monfardini, G.: The graph neural network model. IEEE Trans. Neural Netw. **20**(1), 61–80 (2009). https://doi.org/10.1109/TNN.2008.2005605
23. Li, Y., Tarlow, D., Brockschmidt, M., Zemel, R.: Gated graph sequence neural networks, no. 1, pp. 1–20 (2015). arXiv: 1511.05493

24. Kipf, T.N., Welling, M.: Semi-supervised classification with graph convolutional networks, pp. 1–14 (2016). arXiv:1609.02907
25. Schlichtkrull, M., Kipf, T.N., Bloem, P., van den Berg, R., Titov, I., Welling, M.: Modeling relational data with graph convolutional networks. Lecture Notes in Computer Science (Including Subseries Lecture Notes in Artificial Intelligence and Lecture Notes in Bioinformatics). LNCS, vol. 10843, no. 1, pp. 593–607 (2018). https://doi.org/10.1007/978-3-319-93417-4_38
26. Veličković, P., Cucurull, G., Casanova, A., Romero, A., Liò, P., Bengio, Y.: Graph attention networks. In: Proceedings of the International Conference on Learning Representations 2018, 2015, pp. 1–11 (2018). http://arxiv.org/abs/1710.10903
27. Vaswani, A., Shazeer, N., Parmar, N., Uszkoreit, J., Jones, L., Gomez, A.N., Kaiser, Ł., Polosukhin, I.: Attention is all you need. In: Advances in Neural Information Processing Systems, pp. 5998–6008 (2017)
28. Wang, M., Yu, L., Zheng, D., Gan, Q., Gai, Y., Ye, Z., Li, M., Zhou, J., Huang, Q., Ma, C., Huang, Z., Guo, Q., Zhang, H., Lin, H., Zhao, J., Li, J., Smola, A., Zhang, Z.: Deep graph library: towards efficient and scalable deep learning on graphs. In: ICLR 2019 Workshop on Representation Learning on Graphs and Manifolds (RLGM), pp. 1–7 (2019)
29. Fey, M., Lenssen, J.E.: Fast graph representation learning with PyTorch geometric. In: ICLR Workshop on Representation Learning on Graphs and Manifolds (2019)
30. Kirby, R., Simmons, R., Forlizzi, J.: Companion: a constraint-optimizing method for person-acceptable navigation. In: RO-MAN 2009-The 18th IEEE International Symposium on Robot and Human Interactive Communication, pp. 607–612. IEEE (2009)
31. Cruz-maya, A.: Enabling socially competent navigation through incorporating HRI, pp. 9–12. arXiv: 1904.09116v1 (2019)

A Toolkit to Generate Social Navigation Datasets

Rishabh Baghel[1], Aditya Kapoor[2], Pilar Bachiller[3], Ronit R. Jorvekar[4], Daniel Rodriguez-Criado[5(✉)], and Luis J. Manso[5]

[1] Department of Computer Science and Engineering, Indian Institute of Information Technology Guwahati, Guwahati, India
[2] Department of Computer Science and Information Systems, Birla Institute of Technology and Science Goa, Pilani, India
[3] Robotics and Artificial Vision Laboratory, University of Extremadura, Badajoz, Spain
[4] Department of Computer Engineering, Pune Institute of Computer Technology, Pune, India
[5] Department of Computer Science, College of Engineering and Physical Sciences, Aston University, Birmingham, UK
{190229717,l.manso}@aston.ac.uk

Abstract. Social navigation datasets are necessary to assess social navigation algorithms and train machine learning algorithms. Most of the currently available datasets target pedestrians' movements as a pattern to be replicated by robots. It can be argued that one of the main reasons for this to happen is that compiling datasets where real robots are manually controlled, as they would be expected to behave when moving, is a very resource-intensive task. Another aspect that is often missing in datasets is symbolic information that could be relevant, such as human activities, relationships or interactions. Unfortunately, the available datasets targeting robots and supporting symbolic information are restricted to static scenes. This paper argues that simulation can be used to gather social navigation data in an effective and cost-efficient way and presents a toolkit for this purpose. A use case studying the application of graph neural networks to create learned control policies using supervised learning is presented as an example of how it can be used.

Keywords: Social navigation · Robot simulation · Social robotics · Navigation dataset

1 Introduction

To boost robots' prevalence in the service industry, health and home settings, their efficiency and social acceptance must be improved. This lack of social

R. Baghel and A. Kapoor—Contributed equally to this work.

L. M. Bergasa et al. (Eds.): WAF 2020, AISC 1285, pp. 180–193, 2021.
https://doi.org/10.1007/978-3-030-62579-5_13

acceptance requires a seamless interaction with humans that cannot be achieved without considering humans' goals, emotions, and predicting their future behaviour. These skills are essential to successfully work with humans and avoid user rejection [1].

Social acceptance is particularly important for mobile robots. For instance, while moving, robots should estimate human trajectories to avoid getting in their way. Robots should gather information about the subjects' current activity, emotions, and intentions to provide an efficient and natural interaction while offering them help. Let us consider a scenario where a guest arrives at the robot's owner's apartment, and they start having a heated argument just after that. Should the robot approach the guest to offer a cup of tea as it would usually do? Approaching humans in those conditions is probably something that roboticists would like to avoid. The problem in this scenario is that the robot might not be able to interpret the social context correctly and assess which actions are acceptable. The general challenge while programming robots with some degree of ability to understand humans is to cope with the complexity and breadth of factors involved.

The algorithms in charge of controlling robots and those estimating humans' emotions and aims are usually manually programmed, but they can also be based on Machine Learning (ML) models or a combination of both approaches. Handcrafted algorithms require a considerable amount of time and are hard to debug, expensive to develop, and even expert developers might disregard variables that could be of use. ML models are very often able to consider factors that go beyond experts' intuition. However, until recently, we lacked ML algorithms to work with structured data such as graphs, so most ML-based algorithms tend to discard structural information or make poor use of it. Recent non-Euclidean ML techniques such as Graph Neural Networks (GNNs) [2] can work with arbitrarily complex structured data in the form of graphs, as opposed to the Euclidean structures that conventional ML approaches are limited to.

In social navigation, datasets are crucial to a) assess algorithms' appropriateness and; b) to train ML models for social navigation. Although this holds regardless of whether structured or conventional ML models are used, the toolkit presented in this paper focuses on -but is not limited to- structured (*i.e.*, graph-like) data. The motivation is that structured data can naturally be represented as a graph [2], but the opposite does not always hold.

Given the subjective and highly-interactive nature of social navigation interactions and Human-Robot Interaction (HRI), generating large datasets for social navigation in real-life scenarios would be prohibitively expensive and time-consuming. Previous works have used a tool to generate a social navigation dataset, SocNav1. The dataset, presented in [3], was designed to be used to train and benchmark models to assess the social compliance of robots in a given scenario. Although SocNav1 is limited in many aspects (covered in the next section), it is one of the few datasets considering not only metric properties but also relationships, which can be of high importance and most datasets neglect.

The contributions of the paper at hand are a new open-source modular toolkit based on CoppeliaSim and PyRep [4,5] to generate social navigation datasets, and a use-case where the toolkit is applied to create a dataset. The tool aims to enable third-party researchers to develop their datasets while reducing the time spent developing the required tools. The use-case provided demonstrates that the data generated using the toolkit can be used to train robots in social navigation tasks.

2 Navigation Datasets

There are many publicly available datasets and tools that can be used for social navigation. These datasets can be classified as video recordings, geometric data, symbolic data, audio, or multimodal data -or a combination of them. Video datasets, which are the most common type, can be divided into real or simulated. EIPD [6], a video dataset, was used to learn the behaviour of pedestrians in [7]. The video datasets ETH [8] and UCY [9] were used in [10] to train a social attention model. ETH contains two sets of videos that were obtained from a bird's eye view and were manually annotated. It has a total of 650 human trajectory tracks of over 25 min. A dataset for public space surveillance tasks was also presented in [11]. It consists of 28 videos from 6 in different scenarios. Another dataset of videos from CCTV cameras involving pedestrians is presented in [12]. In addition to pedestrians, [13] contains videos of other agents in a social scenario such as bicyclists, cars, buses, and even skateboarders. Hence, [6,8,9,11–13] belong to the category of video datasets. Recording videos is relatively easy and inexpensive, but the videos need to be annotated, and even if there is some form of person detector to generate the annotations, these need to be supervised by humans. Similarly, videos have no semantic features *per se* (*e.g.*, who is talking to whom), so this kind of information is missing or has to be added by hand.

The Carnegie Mellon University (CMU) dataset [14] is a 3D dataset which was used in [15] to encode the context-dependent spatio-temporal interactions into social-maps. This dataset contains 2605 pre-recorded action sequences, and 55 of these sequences correspond to social interactions between two subjects. It involves two subjects, whereas in real-life, the social scenarios contain more entities and are more complex in nature. A multimodal dataset with 64 min of multimodal sensor data is presented in [16]. It includes omnidirectional and regular RGB video streams, 3D point cloud data from a LIDAR sensor, audio signals, and encoder values from a robot's wheels.

Simulators are also a useful tool while collecting data as they are cost-effective, and because the state of the world is fully observable, there is no need for manual annotations. On the other hand, one of the main drawbacks of simulations is that they will always yield approximations to real data, and that might introduce biases. Pedestrian simulations were used in [17] to fit a social force model where the model parameters were learned using thousands of simulations. The UCY dataset [9], which was mentioned in video datasets, could also fall in this category because it introduces a data-driven simulation approach where agents' behaviour is based on thousands of previous examples in the database. The SocNav1 dataset [3] combines geometric and semantic relations. Instead of containing subjects moving, SocNav1 provides a series of scenes with humans and a robot. The labelled information corresponds to the *social score* of the robot in that given position. One of the advantages of SocNav1 is that it combines geometric with symbolic information, so social relations can be considered to estimate the score. The dataset's main limitation is that the scenes are static, so it does not consider human movement.

From the reviewed datasets, only SocNav1 provides information related to how humans want robots to behave. Instead, most datasets provide human movement information, expecting that robots should mimic their behaviour. Mimicking human navigation is not a bad alternative but in some cases robots are expected to behave differently. For instance, it could be interesting to make robots especially careful regarding social distances (*e.g.*, *"when moving around older people or people handling fragile objects"*), or make them extra cautious when it comes to causing interruptions (*e.g.*, *"be extra careful not to go in the way of people walking or having a conversation"*). However, SocNav1 is limited to static scenes, which neglects the importance of the robots' trajectories and the people around them. The gap identified during the review is the lack of datasets targeting how people would like robots to move, which might differ from how pedestrians move. The next section describes our proposal, whose aim is not to provide a dataset but a tool to create them according to the users' needs.

3 Proposal

No one-size-fits-all approach will suffice for generating datasets, as different datasets will work with different kinds of data. Therefore, instead of creating an off-the-shelf solution, our SOcial NAvigation Toolkit for data Acquisition (SONATA) provides software components that can be used as scaffolding to implement different dataset generation tools. The general structure of the toolkit is shown in Fig. 1. It is composed of three software components (rounded squares) and a Python module. The module is built on top of PyRep [5], which is in turn used to communicate with CoppeliaSim [4]. The module's goal is twofold: a) to provide a high-level API to create manual or randomly-generated social navigation scenes, and b) to control the simulation and record data.

The toolkit architecture has been designed as a network of components to reduce software coupling and ensure that the computational load can be distributed among different cores or even different computers. The *simulator* component instantiates the *CoppeliaSim* simulation, publishes the information about the simulated entities (people, walls, objects, and goals), and acts as an interface with the simulator, handling robot control and scene regeneration commands from a *controller* component. The *event-publisher* component receives events from a joystick or mouse and publishes the corresponding data. These two software elements are common to every specific dataset generation tool built using

SONATA. The only problem-specific component is the *controller*, so the toolkit has been designed to facilitate forking the project and modifying the controller component.

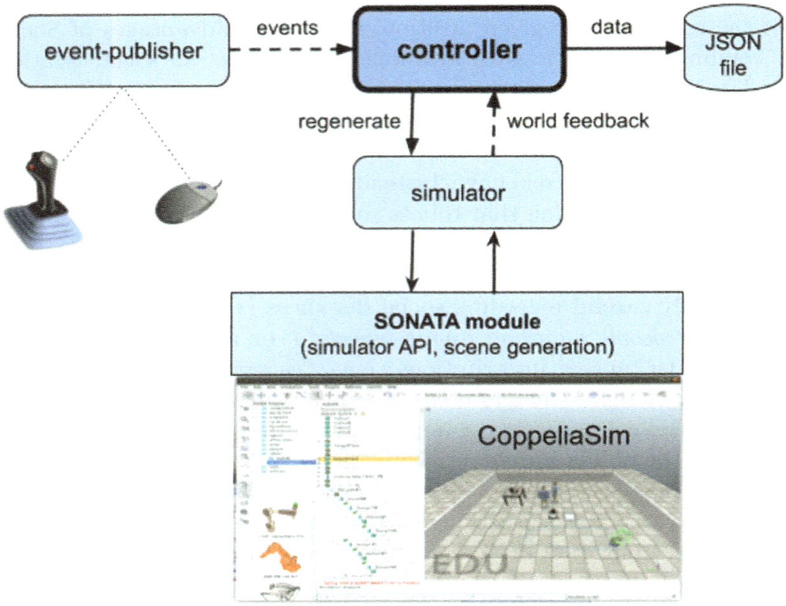

Fig. 1. General structure of the proposed toolkit for dataset generation.

The *controller* is subscribed to the event publisher and the simulator feedback. It is therefore in charge of gathering data from the simulation and the user input, as well as controlling what happens in the simulation by sending commands to the *simulator* component (*e.g.*, moving the robot, the humans, restarting the simulation once a sample has been acquired). Its default GUI contains a view of the simulation and options related to the generation of scenarios (see Fig. 2). The controller obtains data related to the simulation from the subscription to different topics provided by the *simulator* component. Specifically, in a given scene, the simulator publishes geometrical attributes of each entity included in it (*e.g.*, humans, objects, walls) as well as semantic information about relationships between entities (*e.g.*, two humans talking, two humans walking together). Besides this information, the simulator component publishes an image of the scene captured by a virtual camera, which provides a third-person view of the robot. This image, which is shown in the main GUI of the controller, provides a general view of the scene to help the user label it or provide feedback according to the specific problem. The controller also interacts with the simulation using two additional interfaces implemented in the *simulator* component, which is accessible through proxy objects. One of these interfaces provides an operation to regenerate the scenario. For each type of entity (objects and relations), the user can set a minimum and maximum number of instances and randomly generate a new scenario according to the specified ranges. The second interface is used to control the base of the simulated robot. This control can be used as part of the labelling process or to change the user's point of view. Figure 3 summarises the topics that the controller component has access to.

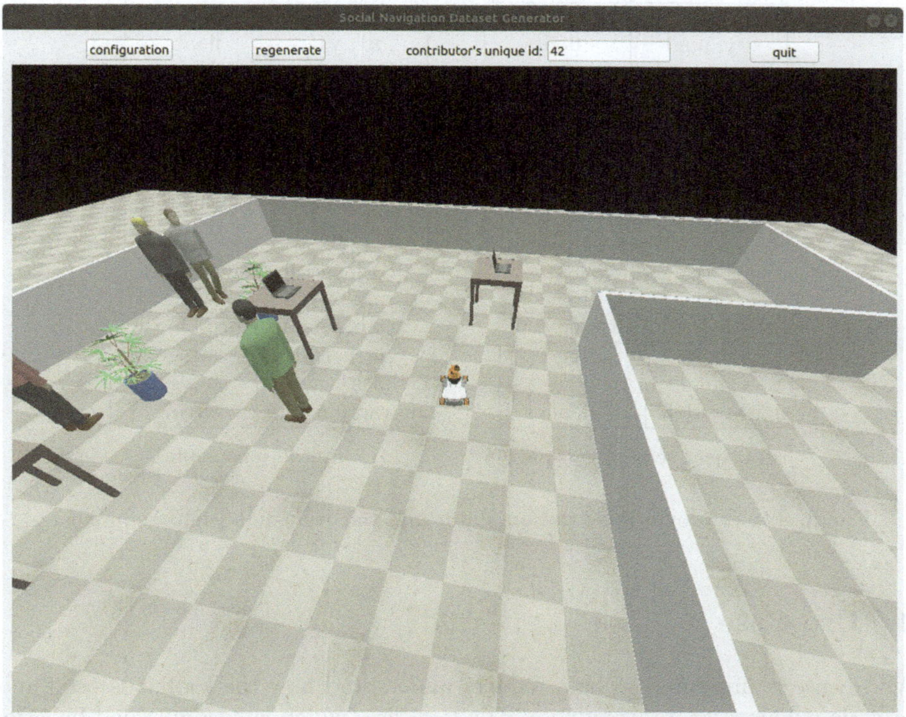

Fig. 2. GUI example for the controller component.

Along with the simulation data, the controller may access the user input to assign a label to each scene. The user input reaches the controller through its subscription to a topic published by the *event-publisher* component. The received device events can be transformed into labels for a particular dataset. Thus, to name some examples, an event can be associated to a binary (true/false) user response to a given question about the scene, a degree of acceptance of the user to a particular situation, or an action performed by the robot.

The main control loop of the default *controller* component periodically associates the current state of the scene with the label specified by the user and updates a list containing labelled data of the scenario. The list is stored in a

JSON file on user demand. By default, the name of the file contains a user identifier and the timestamp to identify the file uniquely. The choice of using JSON over other formats is motivated by its readability and ease to export to other formats. Nevertheless, using other formats is possible and only depends on the developer's preferences for the dataset generation tool.

variable name	variable type
id	integer
x, y, angle	float
ix, iy, iangle	float

(a) human sequence topic

variable name	variable type
wall id	integer
x1, y1	float
x2, y2	float

(b) wall sequence topic

variable name	variable type
identifier	integer
x, y	float

(c) goal topic

variable name	variable type
axis id	integer
value	float

(d) joystick event topic

variable name	variable type
id	integer
x, y, angle	float
sideX, sideY	float

(e) object sequence topic

variable name	variable type
entity1 id	integer
entity2 id	integer
interaction type	string

(f) interactions topic

Fig. 3. Topics provided by the simulator and the event-publisher components

4 Use Case

A use case is presented to help readers understand how the toolkit is used and provide evidence of its usefulness. The use case, which is accompanied by brief experimental results, covers the design and implementation of a dataset generation tool and how the data was gathered.

4.1 Use Case Description

The application of the dataset obtained from the use case is to train a supervised social navigation machine learning model to move a robot. The particular model described later in this section is a Relational Graph Neural Network (R-GCN) [18]. The implementation details and the model's experimental results are not deeply described as the aim of the use case is only to provide an example of use. However, the model serves the purpose of studying the limitations of supervised learning for robot control.

Given that the model developed is used to control a robot using supervised learning, the model's output is the control commands that are meant to move the robot. Thus, the desired outputs in the dataset are such control commands. The input data are sequences of the different states of the environment where the robot is asked to move and the goal and the robot itself. Given that the goal would be to achieve social navigation, the users providing the input-output relation (*i.e.*, the sequence of robot commands for a given scenario) were asked to make the robot move towards the goal according to the following guidelines: a) avoid invading people's personal space; b) avoid getting in the way of people when they walk; c) avoid getting in the interaction area of people when they are interacting with an object (*e.g.*, looking at a computer screen) or another person; and d) adjust the speed not to have a big impact on people's comfort. Examples of these situations to avoid are shown in Fig. 4.

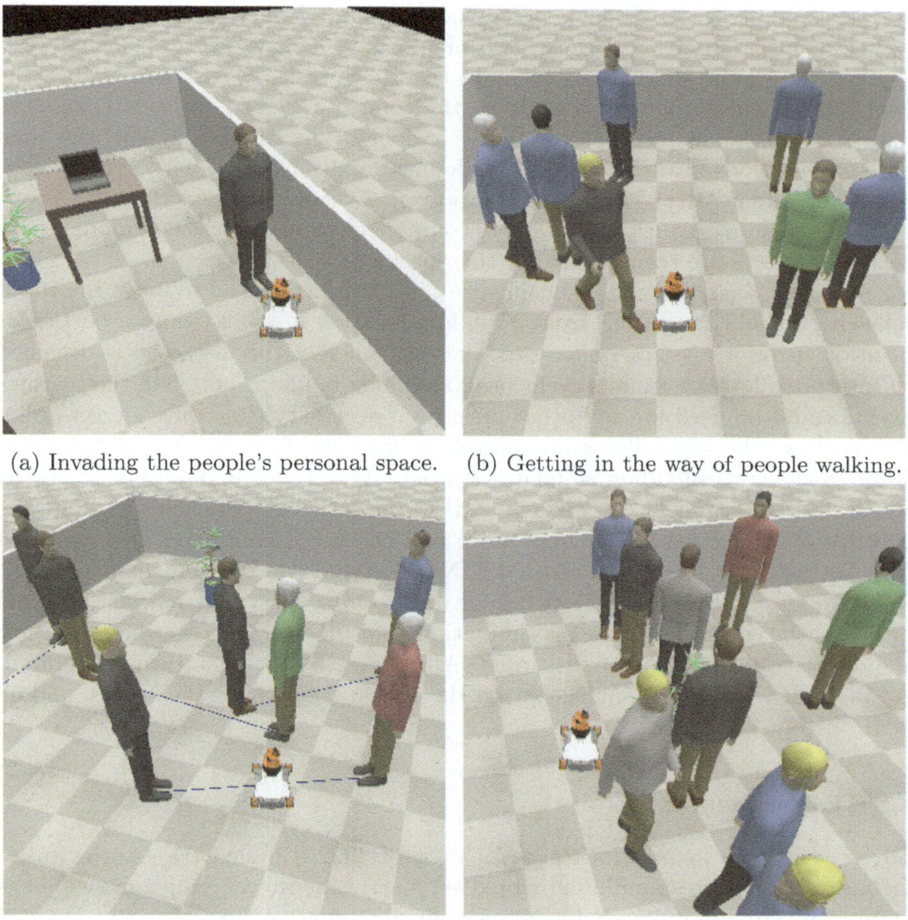

(a) Invading the people's personal space.

(b) Getting in the way of people walking.

(c) Interrupting people's interactions.

(d) Situation where the robot would have to move slowly.

Fig. 4. Examples of robot behaviour that should be avoided.

The dataset scenarios are rectangular and L-shaped rooms, with a randomised number of humans, tables, laptops, and plants (the numbers and types of objects do not have a significant impact on this task). Some humans are static, while the rest randomly walk to places in the room. Static humans might or might not be looking at one of the laptops in the scenario or talking to each other. Moving humans might or might not be walking, accompanied by other fellow humans.

The subjects contributing to the datasets have a third-person view of the robot and the environment, such as the one shown in Fig. 2.

4.2 Use Case Design and Implementation

As discussed in Sect. 3, developers only need to modify the *controller* component. It has to be adapted to perform two main tasks: a) save the data of interest, and b) control the simulation. To make the above tasks as easy as possible, the controller is subscribed by default to the topics described in Sect. 3, and the data is accessible as instance variables. To control the simulation, the users can access operations available through different proxies charged with synchronous communication with the simulator component. Every call to a method of a proxy invokes a remote operation in a remote component. Thus, all the communication-related subjects are transparent for the developer.

For every scenario in this specific use case, users are asked to move the robot following social conventions from a random location in the scene to a particular position using a joystick or mouse. The controller of this use case is designed to interpret device events as velocity commands of the robot, but the joystick events could be used for any other purpose if necessary (*e.g.*, provide rewards). In this case, the controller sends movement commands to the simulated robot through the corresponding simulator's proxy. Also, the command is also stored as the desired output for the current state of the scenario. For each time step, the description of the world's current state and the robot command are appended to a list which is serialised into a JSON file at the end of each sequence (*i.e.*, a JSON file is generated for each goal met). Once the robot reaches the goal position, the controller stops the simulation, and the user decides whether or not the data should be kept or discarded. After that, a new scenario is generated. Additionally, the user can regenerate the scenario by using the corresponding GUI button.

4.3 Experimental Results

To test the data acquisition process, a set of 3092 sampled trajectories (6184 including their corresponding mirrored versions) generated by the authors of the paper were stored, and an R-GCN was trained to control the robot. The R-GCN has 5 layers with 40, 30, 20, 15, and 3 units, respectively. The 3 units' output in the last layer is used to control the robot as advance, lateral, and rotation speed.

The graphs used as input are composed of room representations for several consecutive time frames. This is done to let the network gather information about how the room's entities evolve through time. Therefore, the creation of the final graph is done in three steps: Firstly, graphs representing the room are created for two or three frames in a row. Secondly, these graphs are bonded, creating temporal edges from one node to the same one in the next time frame. Finally, a feature vector is associated with each node adding valuable information for training the network.

The graphs have 5 different types of nodes, one for each element in the scenes: people, objects, walls, the goal, and the room node acting as a global node. Thus, there are bidirectional edges from the room node to every other node in the graph. Additionally, edges are also used for describing interactions between entities (*e.g.*, a human talking to another human, a human interacting with an object). All feature vectors have the same dimensions since it is a limitation of the network architecture. They are composed of different fields concatenated one after the other making a total of 42 elements. The following equation shows the structure of the feature vector for node number i in layer 0 (the input layer):

$$h_i^{(0)} = (OH_{t_i}|OH_{f_i}|ts_i|p_i|o_i|r_i|w_i|g_i)$$

The first two fields (OH_{t_i} and OH_{f_i}) are one-hot encodings. The first one is for the type of nodes, so it has 5 elements (one for each type of node). The second one encodes the time frame, which has a length of 3, given that the graphs are composed of a maximum of 3 frames. ts_i encodes the time passed since the beginning of the sequence, and has a dimension of 2. The remaining fields store normalised geometric information for each of the entities of the graph. For instance, if the node is a person, the p_i field will store its coordinates, velocity, orientation. The rest of the geometric feature fields will be filled with zeros. All the geometric features are computed from the robot's reference frame.

The hyperparameters of the R-GCN were:

- Number of layers: 5
- Hidden units: 50 (first hidden layer), 40, 30 and 15 (last hidden layer)
- Output units: 3 (velocity components of the robot)
- Activation functions: *leaky ReLU* for the hidden layers and *tanh* for the output layer
- Learning rate: 0.0005
- Weight decay: 10^{-9}
- Batch size: 40.

We obtained an MSE of 0.0230 for the training set, and 0.0328 and 0.0331 for the development and test sets. Results of the execution of the robot using the network output in different scenarios can be found in the following link: http://shorturl.at/etLOW. Figures 5, 6 and 7 show some snapshots of three of these executions. Each sequence is ordered from top to bottom and left to right. The model trained can reach the goal of 76.6% of the times (46 out of 60). Given the limited number of samples and the complexity of some scenarios, this percentage can be considered reasonable. Nevertheless, the behaviour of the robot is not always the most appropriate. Sometimes the robot gets too close to humans and objects (Fig. 6). Besides, in some scenarios, the robot interrupts people's interactions even though it would be possible to find an alternative path to reach the goal position (Fig. 7). These results are in line with the literature, which does not recommend using supervised learning for robot control because the samples' distribution tends to be biased towards safe situations. This bias makes robots unreliable in unexpected situations. Besides showing an example of data gathering using the proposed toolkit, reproducing these results was one of the purposes of the use case. The analysis of the influence of larger datasets in supervised learning and exploring alternative approaches to this problem, such as inverse reinforcement learning, is left for future research and lie out of the scope of the paper.

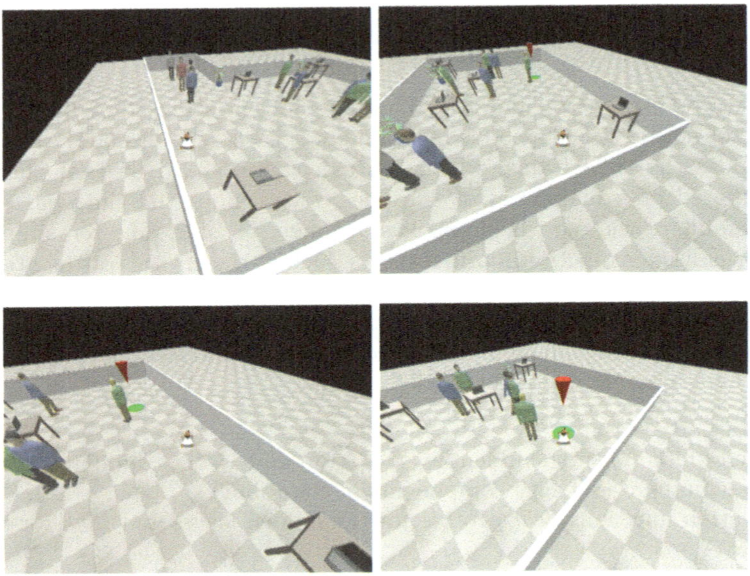

Fig. 5. Execution of the robot using the trained model (first scenario). The robot correctly orients its body to avoid a standing human on the way to the goal position.

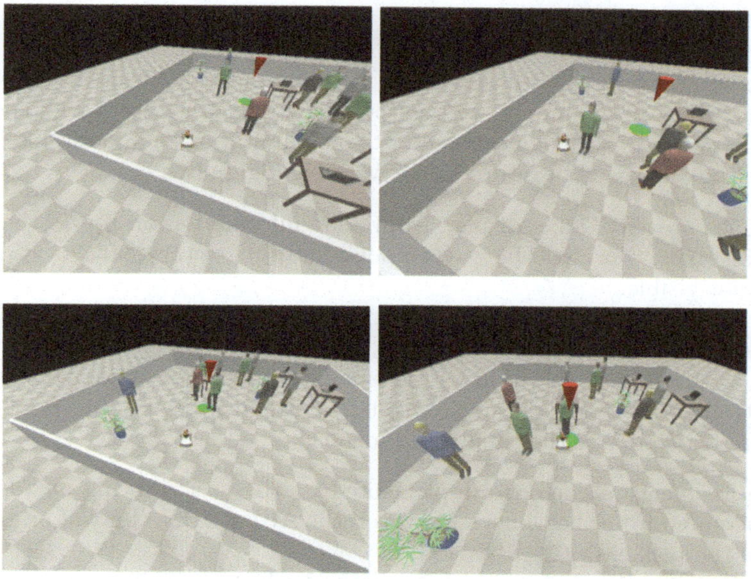

Fig. 6. Execution of the robot using the trained model (second scenario). The robot avoids a human walking in the opposite direction but gets too close to a human when it reaches the goal.

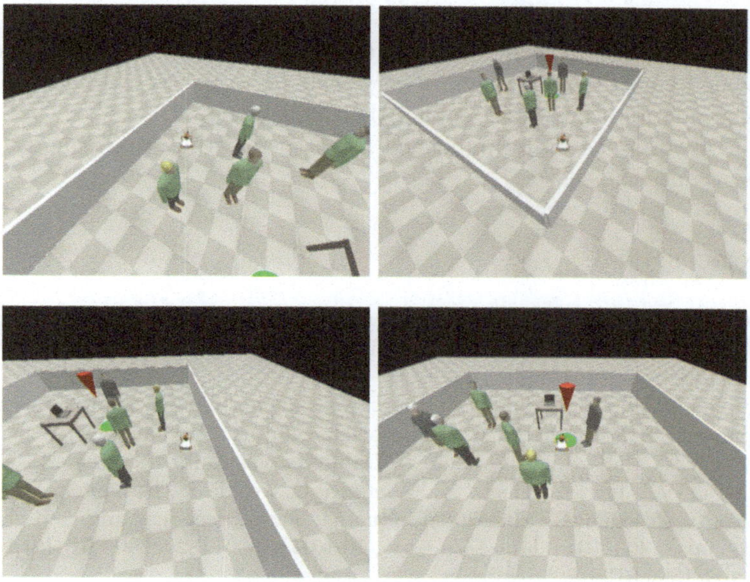

Fig. 7. Execution of the robot using the trained model (third scenario). The robot changes its trajectory to avoid a group of humans on the way to the target. It gets into an interaction area to finally reach the goal position.

5 Conclusions

The experience working with the toolkit to gather data for the use case is deemed satisfactory, as 3092 samples were gathered by only 6 subjects who spent less than 31 h in total. Although satisfactory, some limitations and desirable features that will be addressed in future works were highlighted:

- Absence of a HUD-like display to provide information on top of the image. For instance, this could be used to point subjects to the goal's location when it is not visible.
- The meshes of the simulated humans sometimes intersect and go through walls that are meant to be solid.
- The orientation of the head of the simulated humans cannot be controlled. Controlling the head orientation could be useful to hint subjects about the future trajectory of the simulated humans.

The first feature could be implemented by the user drawing on top of the image acquired from the camera. However, the API of SONATA will be extended to make the task easier.

The humans' behaviour, which is related both to the collision detection and the lack of ability to move their heads, depends on how they are simulated in CoppeliaSim. Enforcing collision detection could be solved by adding a control layer. To solve the second issue, it would be necessary to implement a new controllable simulated human in CoppeliaSim as an extension of the current *"path planning Bill"* model.

As a final remark, it must be noted that any simulation will differ from real life to some extent. Even if they are very realistic, subjects will know that they are in a simulation that might impact their behaviour.

The software has been published in https://github.com/ljmanso/sonata under a permissive license.

References

1. De Graaf, M.M., Allouch, S.B., Klamer, T.: Sharing a life with harvey: exploring the acceptance of and relationship-building with a social robot. Comput. Hum. Behav. **43**, 1–14 (2015)
2. Battaglia, P.W., Hamrick, J.B., Bapst, V., Sanchez-Gonzalez, A., Zambaldi, V., Malinowski, M., Tacchetti, A., Raposo, D., Santoro, A., Faulkner, R., Gulcehre, C., Song, F., Ballard, A., Gilmer, J., Dahl, G., Vaswani, A., Allen, K., Nash, C., Langston, V., Dyer, C., Heess, N., Wierstra, D., Kohli, P., Botvinick, M., Vinyals, O., Li, Y., Pascanu, R.: Relational inductive biases, deep learning, and graph networks, pp. 1–40 (2018). https://doi.org/10.1017/S0031182005008516, http://arxiv.org/abs/1806.01261
3. Manso, L.J., Nuñez, P., Calderita, L.V., Faria, D.R., Bachiller, P.: SocNav1: a dataset to benchmark and learn social navigation conventions. Data **5**(1), 7 (2020). https://www.mdpi.com/2306-5729/5/1/7

4. Rohmer, E., Singh, S.P., Freese, M.: Coppeliasim (formerly V-rep): a versatile and scalable robot simulation framework. In: Proceedings of the International Conference on Intelligent Robots and Systems, pp. 1321–1326 (2013)
5. James, S., Freese, M., Davison, A.J.: PyRep: bringing V-REP to deep robot learning. arXiv preprint arXiv:1906.11176 (2019)
6. Majecka, B.: Statistical models of pedestrian behaviour in the forum. Master's thesis, School of Informatics, University of Edinburgh (2009)
7. Luber, M., Spinello, L., Silva, J., Arras, K.O.: Socially-aware robot navigation: a learning approach (2012)
8. Pellegrini, S., Ess, A., Van Gool, L.: You'll never walk alone: modeling social behavior for multi-target tracking. In: RONIT, pp. 261–268 (2009). https://doi.org/10.1109/ICCV.2009.5459260
9. Lerner, A., Chrysanthou, Y., Lischinski, D.: Crowds by example. Comput. Graph. Forum **26**, 655–664 (2007). https://doi.org/10.1111/j.1467-8659.2007.01089.x
10. Vemula, A., Mülling, K., Oh, J.: Social attention: modeling attention in human crowds. CoRR abs/1710.04689 (2017). http://arxiv.org/abs/1710.04689
11. Fisher, R.: The pets04 surveillance ground-truth data sets. In: Proceedings of 6th IEEE International Workshop on Performance Evaluation of Tracking and Surveillance, pp. 1–5 (2004)
12. Benfold, B., Reid, I.: Guiding visual surveillance by tracking human attention. In: Proceedings of the 20th British Machine Vision Conference (2009)
13. Robicquet, A., Sadeghian, A., Alahi, A., Savarese, S.: Learning social etiquette: human trajectory understanding in crowded scenes. In: Leibe, B., Matas, J., Sebe, N., Welling, M. (eds.) Computer Vision - ECCV 2016, pp. 549–565. Springer, Cham (2016)
14. CMU: Carnegie Mellon University (CMU) dataset (2008). http://mocap.cs.cmu.edu/
15. Papadakis, P., Spalanzani, A., Laugier, C.: Social mapping of human-populated environments by implicit function learning. In: IEEE/RSJ International Conference on Intelligent Robots and Systems, pp. 1701–1706 (2013). https://doi.org/10.1109/IROS.2013.6696578
16. Martín-Martín, R., Rezatofighi, H., Shenoi, A., Patel, M., Gwak, J., Dass, N., Federman, A., Goebel, P., Savarese, S.: JRDB: a dataset and benchmark for visual perception for navigation in human environments (2019)
17. Ferrer, G., Garrell, A., Sanfeliu, A.: Social-aware robot navigation in urban environments. In: 2013 European Conference on Mobile Robots, ECMR 2013 - Conference Proceedings, pp. 331–336 (2013). https://doi.org/10.1109/ECMR.2013.6698863
18. Schlichtkrull, M., Kipf, T.N., Bloem, P., van den Berg, R., Titov, I., Welling, M.: Modeling relational data with graph convolutional networks. Lecture Notes in Computer Science (Including Subseries Lecture Notes in Artificial Intelligence and Lecture Notes in Bioinformatics). LNCS, vol. 10843, no. 1, pp. 593–607 (2018)

JBCA: Designing an Adaptative Continuous Authentication Architecture

Javier Junquera-Sánchez(✉), Carlos Cilleruelo-Rodríguez, Luis de-Marcos, and José Javier Martínez-Herráiz

Dpto. Ciencias de la Computación, Universidad de Alcalá, Alcala de Henares, Spain
javier.junquera@uah.es

Abstract. Several continuous authentication methods provide very accurate ways of detecting the impersonation of a user. However, even though they should be access control chain's links, they are commonly documented as algorithms unrelated to any system. In this paper, we propose a continuous authentication architecture to integrate several kinds of continuous authentication engines in a complete framework. The architecture covers different use cases where continuous authentication methods, placed in the user's device or in the cloud, collaborate to determine if the user's identity has changed. We use ensemble learning principles to combine the values generated by each CA model, and compute the trust level adaptively, addressing common errors caused by changes in the environments of continuous authentication systems.

Keywords: Continuous authentication · Ensemble learning · IT architectures · Authorization

1 Introduction

Nowadays mobile phones have become a critical asset for information security. There is a lot of personal information inside each mobile device and it can be the key to access to other systems (e.g. they are commonly used as a second factor for authentication). Furthermore, mobile devices can be carried everywhere, in contrast with other company devices that can improve their security through physical security policies, like video surveillance. And they are very lightweight, so a robber who steals one and transport it easily.

One way to offer better security to mobile devices is Continuous Authentication (CA). CA is the set of techniques and procedures destined to verify the users' identity after their authentication, and without their active participation. Most of these techniques are based on capturing behavioral metrics, or biometrics which not require an active implication. It is useful for situations where a user forgets unlocked his device somewhere (e.g. during a work break) and an adversary uses it without his consent.

L. M. Bergasa et al. (Eds.): WAF 2020, AISC 1285, pp. 194–209, 2021.
https://doi.org/10.1007/978-3-030-62579-5_14

CA can use more than one approach to create better detection results, especially taking into account that this CA process could happen in several different situations where some models could work better than others (e.g. photography could be a very precise approach, but in some poor visibility conditions, could be better to use a keystroke dynamics approach). As a result of this, there exist some initiatives to orchestrate or process multiple inputs to provide CA. However, most of them do not cover the data retrieving phase or lack the same problem: all the metrics have to be stored in the same place.

Besides collecting and storing all this behavioral user information arises privacy concerns. On one hand, because the information needed to run the system is sensible (e.g. biometrics). And on the other hand, because the model itself could be dangerous, as it could allow detecting the user's identity without his consent.

During our process researching CA systems and models, we design an architecture for retrieving and combining different kinds of CA metrics. To combine different decision models we implemented a voting system and the logic for delivering the results to an alert system (e.g. like a SIEM). Our architecture covers these three phases:

- Guidelines proposal for collecting, storing, and generating CA trust values from behavioral metrics
- Retrieving and combining partial trusts
- Delivering final trust alerts.

The present work aims to provide a complete architecture for a CA system. The goals are to cover the complete data flow, from metrics to trust values, allowing the combination of results from several CA decision models, but also being useful the simplest possible model (i.e. a model with just one CA provider). This paper contributes to knowledge by presenting the first holistic EDR architecture for CA.

2 Background

2.1 Continuous Authentication

Continuous authentication is the phase of an access control system which ensures the identity of the user remains the same that was in authentication process. Nowadays, there are several CA approaches with high detection accuracy, but none of them addressed the full architecture of a CA system. Almost every approach can be adapted to different systems (e.g. mouse pointer dynamics could have particularities if the user uses a touchpad, instead of a mouse [19]) or processed with different algorithms.

Traditionally, keystroke dynamics is one of the most primitive approaches [24], as it could be implemented mostly everywhere (almost all computers have had a keyboard). Many different features from keyboard writing sessions have been studied including writing speed, time between keystrokes or time pressing each key [22].

In mobile devices, the most significant evolutive step is touchscreen dynamics [13]. The touchscreen is the hegemonic peripheric device of smartphones like the keyboard is of the computers. The way users scroll a text provides lots of data [25]. Also, what is already known about keystroke dynamics' can be combined with the touchscreen to detect how the user writes in a virtual keyboard [14].

Finally, the HMOG research (based on Hand Movement, Orientation, and Grasp) condenses both models and enriches them with other phone-specific sensors (e.g. accelerometer) that can lead to knowing the specific circumstances in which the user is using it: lying down, running, sitting, etc. A lot of sensors can provide data to infer them, like light sensors [5], position sensors [16] or microphone [6].

2.2 Data Analysis

Each CA system needs to evaluate the data received. There are multiple methods for processing the information retrieved from each source [20].

Several of them are almost trivial when the features have been correctly extracted. The approaches which better illustrate the key point of a powerful CA system are those based on image recognition [27]. The process for extracting the color of the user's eyes can be difficult and entails several complex algorithms; but when done, determining if it is the user's eye color is as trivial as comparing two values [12].

However, other features easier to collect, are more difficult to evaluate. The most used algorithm for processing time-based metrics is SVM [3], as these kinds of metrics often fit well in regression models. For approaches related to tracking the user's behavior (i.e. based on the change of states of the system), the most effective algorithms are those suited for classification problems [17] (like k-nearest-neighbors method).

Also, the camera can be used as a rather effective input for a CA system. Approaches such as detecting the color of the user's clothes can very accurate during a working session. For instance, if the t-shirt color changes, another person could be using the mobile phone. CA should also consider the frequency of check or how often the system should try to identify the user. Moreover, another important parameter that must be defined, is how much time should session be hold (i.e. during a workday, a week, etc.).

According to [28], the main dimensions that define the validity of a CA model are universality, distinctiveness, permanence and collectability. These dimensions define, for example, whether a model is valid for a long time, just for a working session; or if it fits for any person (e.g. a hair colour based model requires all the population to have hair). Hence, the most important action to create a useful CA system is to find which features provide a most versatile and reliable way for evaluating the user's identity.

2.3 Multi-feature Approaches

Not all this data can be processed inside mobile devices, or in devices with limited capacity (for example, machine learning could be an demanding operation), and in most cases, it has to be processed outside of the users' device.

As HMOG combines multiple features for determining if the user's identity remains to be the genuine, other researches have tried to get better performance processing several features at the same time [4].

Some of this research starts from realizing that, even though the fusion of multiple features could deliver better results, not all of them have the same weight. The framework proposed in [18] implements a decision fusion system to combine several phone metrics with touchscreen gesture dynamics, in order to evaluate each user's gesture in a specific context.

There are also approaches that, instead of combining multiple features' results, process the same feature with different AI models. Combining them to obtain a more accurate evaluation is known as ensemble learning [9,23]. Our CA proposal uses ensemble learning as a method of combining several decision models.

2.4 Access Control

The authentication of the different components of the architecture should accomplish security, but also usability and scalability. It is necessary to understand the nature of the different architecture's actors.

OAuth2 [15] is an authorization protocol. This protocol allows the user to delegate its identity into a client, and let him interact with the resources server as it was the user itself.

There are multiple actors with different interests over the system. These actors correspond to the three components of an OAuth2 system: resource server (our server), clients (other service providers), and users (the final users).

The most common use case of OAuth2 is a process where a social network asks us if we grand access to some resources (like knowing our identity) to a third-party service, in order to login to this third party with the social network credentials. The background of this process is the authorization_code mode of OAuth2.

OAuth2 standard has several authorization modes [1], some of them deprecated. In order to understand our CA system architecture is it necessary to understand the following:

- Authorization code: The most common OAuth2 system. The user is redirected to the third party login page (in this case, to our's login page, from a third party application) and grants access to his data.
- Device code: Similar to authorization code, but implemented for devices without the capability of loading a web page.
- Client credentials: Used to authenticate clients instead of users. It is useful when clients should make use of the resource server for anything else than act as the user.

The OAuth2 standard has been largely studied, and the fruit of these studies has suffered several updates. As a result, the most secure configuration of OAuth2 has derived in a kind of new standard with some constraints concerning the original version [21].

While authorization code mode would allow both third party and self-programmed systems (respectively) to interact with a server, client credentials mode would also allow identifying clients (servers or systems where the user has no participation) in order to let them interact with our API when necessary.

OpenID Connect is a protocol constructed over OAuth2 to standardize also the process for retrieving user's data: username, mail, etc.

Also, the selected way for the identity is JWT as an access token, and a string refresh token (to allow the revocation without checking always during the JWT token validity time).

3 Architecture Overview

The architecture consists of three types of elements:

- Authorization server: Responsible of the authentication in the architecture.
- Resouces server (JBCA API): Manages the trust obtained from different features of the users' identity (generated by the CA clients) in order to create a global trust value.
- Clients: Act as the interface between the users and the architecture. The are several types of client, documented in Sect. 4.

The use cases proposed in Sect. 8 aims to provide a comprehensive example of the role that these components play in the architecture (also shown in Fig. 2).

4 Clients

As the clients would be not just the source of information about the users, but also the consumers of the architecture's products, it is necessary to start with them for explaining the architecture. There would be three types of client.

4.1 CA Clients

This category defines the clients which can generate trust values. Independently how they work, they would have their own CA processes to calculate the trustworthiness of each user identity, and send the results to the resources server. These types of clients should have a backend for protecting their own authentication secrets (i.e. the keys for authenticating themselves as clients against the authorization server), but they can implement their internal logic however they need it.

Nonetheless, it is necessary to mention that could be special kind of CA client which not always would have access to a backend: the offline clients.

Offline Clients. It is not a type of client, but a subtype of CA client. Generate models in the cloud, or storing this data out of the user's assets, could be a problem with certain types of data. This problem is especially serious when is specially protected data (like medical, biometrics, religion...) or there are cross-border connections (e.g. between an EU user to a USA server). So, sometimes, it is desirable that the user's data is not stored in a backend (e.g. very sensible data). For this purpose, we have designed what we have seen fit to call *offline client*: a client with CA capabilities, but without a backend.

Even though this could seem unnecessary (as they produce exactly the same type of data than common CA Clients), from the point of view of authorization, we have to take care of its particularities.

4.2 Trust Clients

The purpose of the architecture is to generate the most precise possible value for indicating trustworthiness in a user's identity. The consumers of this value are trust clients. They would ask anytime which is the result of the evaluation process, for example, to close a session of a medical application if the user might have been impersonated.

4.3 Partner Clients

This is the name given for those clients with special grants over the resources server. This type of client won't act as a specific user (i.e. won't be limited to interact with the information of the user which has granted them) but can retrieve the values generated for any user.

The aim of creating this role is not to bypass the users' grant, but allow the integration of the architecture in systems that already have their own purpose and need to provide CA capabilities to their tools.

5 Authorization Server

The authorization server will be in charge of authenticating the clients and generate the tokens to authorize them to use the server resources. It will also allow the whole system to maintain the user's identity unique.

5.1 Authorization Token

The authorization token would be codified using JWT, as it would provide enough information to the resource server for validating it without adding connection overload. This JWT token would be generated using the claims defined both in the JWT's RFC [7,26] and with the information that the OAuth2 introspection endpoint should provide [8].

The logic for revoking access to a client would work during the refresh process. If a user wants to revoke the rights granted to a client, the next time the client would try to refresh his access token (using OAuth2's refresh token), would be rejected, or his grants would be modified. Even though this approach could open a time slot where the user's revocation won't be effective, this time can be limited making the token's time of life shorter.

Finally, the server should be configured to be able to provide also information about the users' identity if required.

5.2 User Identity

To provide information about the user, an OpenID Connect layer will be implemented. The grants related to this functionality would be only given when they are necessary for the CA purpose (i.e. never like a social, or an identity validator feature). For example, the user's address could be necessary for a correlation with GPS data (if the authenticator makes this check), but to know the user's email address won't be necessary in this case.

The unique identifier all over the network would be placed in the subject claim (i.e. `sub`) of the JWT token, and should not apport any information about the user (would be, for example, an `UUID` value).

5.3 Client Roles

Those clients mentioned in Sect. 4 will have their particularities for authenticating.

- "Online" CA clients would retrieve OAuth2 tokens using the "auth_token" or "device_code" mode.
- "Offline" CA clients which have no possibility of authenticating themselves as clients, they should use the "device_code" mode.
- Partner clients should authenticate themselves using the `client_credentials` mode, as they will not directly interact with users.
- Trust clients will not need a particular authentication mode, and it will depend on the circumstances where they work (i.e. the capabilities for storing secretly a key or the type device where they work).

Each type of client should be allowed by the architecture administrators to use their authentication mode. For example, a regular CA client would be rejected if tries to get an OAuth2 token using the `client_credentials` mode.

5.4 Scope

As part of the OAuth2 protocol specification, there should be a list of defined grants, which will be set the `scope` field. The scope possibilities would be:

- ADD_MODEL_TRUST: Add trust of a particular model
- READ_GLOBAL_TRUST: Retrieve the trust generated by the combination of all partial trusts.
- OPENID and derivated: To obtain information about the user.

Table 1 shows the relation between the types of clients and their scope. These are the main groups, but can also be combined (for example, an EDR could want to write its own generated trust values and read the global trust result).

Table 1. Clients' configurations summary

Client type	OAuth2 modes	Scope
Online CA	auth_token, device_code	ADD_MODEL_TRUST
Offline CA	device_code	ADD_MODEL_TRUST
Partner	client_credentials	Specific for the partner
Trust clients	Any	READ_GLOBAL_TRUST

6 AI Models

Even though the architecture should be agnostic about how the clients calculate the trust values, at least it should have enough information to know how to combine them. The models should also comply with some quality requisites to be accepted in the system.

6.1 Quality

In contrast with the client, which should know how much data could be needed for training the models, or to evaluate the user's identity, the resources on server should only know about two quality parameters:

- Model's precision: It will be crucial for calculating the weight that each measure will have in the final trust value.
- Model's expiration: The time the value is valid to be part of the final trust value.

6.2 Requirements

The requirements that every registered model should meet are related to the nature of CA: rejecting intruders without interrupting the genuine user's work. So the models recall should be near 1.0 (i.e. practically total).

Also, as the strength of a CA system is to get a reliable value combining multiple input methods, so any data that contributes is valuable. Thus, the precision of the model should be the highest possible, but it would be enough with a precision greater than 0.5 (i.e. better than a random decision).

6.3 Privacy Concerns

The fact that all the process take place in the user's scope does not exempt from taking care of the information's security. On the one hand, securing this data is a crucial part of protecting the user's device; on the other hand, the responsibility remains with the data processor, regardless of where it performs its tasks.

As it could be necessary a significant amount of data about the user's behavior, it is necessary to take into account the possible problems it could generate. Every new model must follow a minimal risk policy, always choosing the alternatives that could make less impact on the user's privacy.

In order to fulfill with personal data protection principles [2], the next questions should be made in the design phase. Each time a CA decision model is created:

1. Can old data (i.e. the data used to generate the model) be removed? Or would it be necessary for creating future models?
2. Is it possible to recreate the original data from the model?

Those models where old data can be removed, to avoid its exposure, will be better from the privacy point of view. But if once created, the model makes it possible to generate again the data (or similar data), there should be implemented differential privacy countermeasures [11].

The data stored to generate the models also implies a risk (its leakage is the risk itself in both cases). There are also two questions to define before implementing a solution:

1. Can the data be aggregated?
2. Can it even be encrypted?

If the model consist on determining, for example, the progression followed by its individual metrics, sometimes will not be necessary the original dataset, but one whose elements follow the same pattern. In some other cases, it would be desirable at least to save it encrypted, and decrypt only when necessary.

There are also approaches based on secure multi-party computation or homomorphic encryption [10] that can be implemented in different models, to allow compute with them without decrypting.

6.4 Ethical Concerns

CA is, by definition, a process for validating the user's identity without his participation (i.e. without having to do any action). Even though this process aims to protect the user, this process could lead a malicious actor to identify the user not just without his participation, but without his knowledge or his consent.

Although the models capable of identifying the user would be more effective against intruders, those just focused on detecting when the user has changed (i.e. who is using the device is not the genuine user) would be more compliant with the user's privacy.

Moreover, those models which are valid during a short period of time (e.g. during a work session) would also be better, as others could use metrics from which the user cannot be detached.

7 Evaluation

To calculate the final trust we propose a two-step approach. The first task would be a simple combination based on a weighted average. The second one would consist of adding feedback to re-evaluate the models' precision with based on their agreement with the global trust.

Once at least one client has submitted a partial trust for a user, the evaluation process can be triggered.

7.1 Ensemble Learning

As knowing the precision of a model is a requirement for the architecture, it will determine the weight of each value in the final computation. We work with three variables: T_i (partial trust for model i), p_i (precision of model i) and r_i (reliability of model i, initialized to 1). The algorithm presented in Algorithm 1 shows how all trusts get merged.

Algorithm 1 Trust calculation

$T \leftarrow 0$ {Initializing trust}
$P \leftarrow \sum_0^i p_i$ {Sum of individual precisions}
for $n \leftarrow 0$ **to** i **do** {Weitghted trust sum}
  begin
  $T \leftarrow T + (T_i \times p_i \times r_i)$ {With this first $r_i = 1$}
  end
$T \leftarrow T/P$ {Normalize final trust}

7.2 Adaptative Model

Experiments done under very specific circumstances determined the initial values of precision for each model. The most accurate approach could fail under some conditions, and not all approaches will run on trusted environments (i.e. the local model's results are easier to be manipulated from an adversary than ones in the cloud). For this reason, the reliability of a model cannot be the same all time. For this reason, our model is designed to be adaptable using the reliability parameter.

On each evaluation process iteration, the reliability value of each model (r_i in 1 algorithm) will be updated following the decision diagram shown in Fig. 1. The

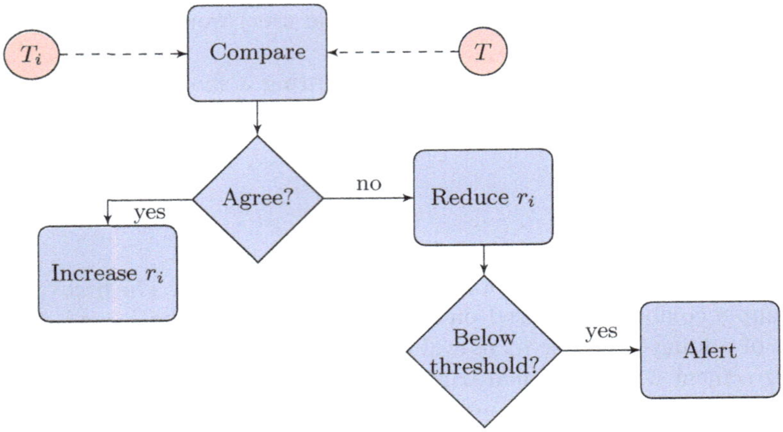

Fig. 1. Adaptative model

reason to determine if reliability is increased or decreased will not be quantitative (i.e. how near is its partial trust of the global trust), but qualitative (i.e. if agrees with the others about if the users are or not the genuine).

If after some iterations the rely value of a model goes below a threshold, this could have several meanings:

1. Its trust value is outdated
2. The circumstances make the model unprecise
3. Its trust value uploaded is being faked.

In this case the model will be disabled until a new trust value is added or an administrator checks which has been the reason of the fall.

8 Use Cases

The simplest way to illustrate the complete architecture is by observing its data flow through different use cases. In this section we document these three use cases that could be working independently:

- An "offline" model
- An EDR system
- A broker connected to the resource server sending metrics to a SIEM.

At the end of the section, we will show how they can be combined in a complete CA system (i.e. like the one shown in Fig. 2).

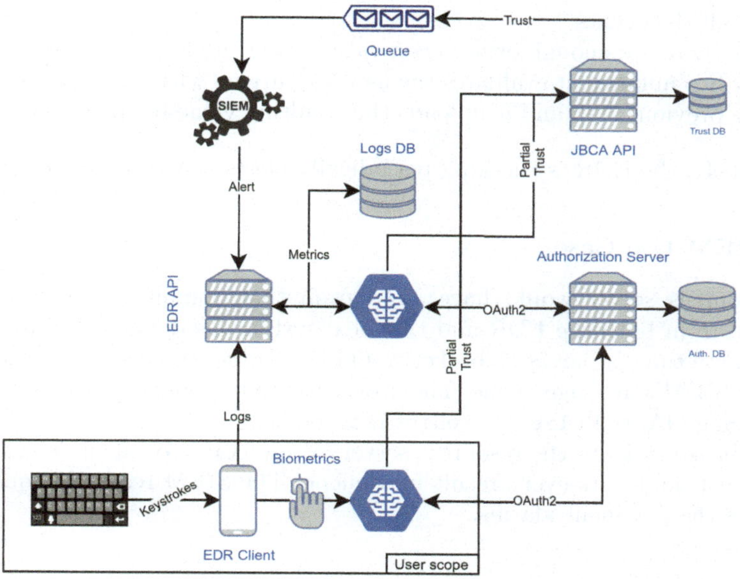

Fig. 2. Architecture overview

8.1 "Offline" Use Case

The "offline" models (i.e. those without a user-isolated backend) should authenticate using the OAuth2 mode `device code`. So, once the user is authenticated against the authorization server, the model would start working.

For example, let's imagine a model based on detecting the color of the user's clothes, and alert when it changes. The behavior would be:

1. When the user logs in, it takes a photograph, analyzes his clothes and saves locally which color they are.
2. Every minute a photograph is taken, colors are analyzed, and compared with the values saved in the first step, generating a similarity value between 0 and 1.
3. It upload the similarity result to the resource server.

8.2 EDR Use Case

Let's suppose that the EDR model consists on analyzing the user's typing dynamics in a cloud server (the backend). When the user logs in (using `authorization code` or `device code` method), the client sends to the EDR's backend all the typing events (e.g. key press, key release, deletion rate, etc.). The backend will:

1. Store all the events
2. When there are enough events, generates a user model with them
3. Every five minutes, it evaluates the new user events with the model generated in the previous step and it uploads the resulting value to the resource server.

Parallely, the EDR is checking periodically alerts generated by a SIEM.

8.3 SIEM Use Case

The resources server would have a program that combines the trust values uploaded from both the EDR and the local system (as documented in Sect. 7). At the same time, a partner's broker would be reading these global trust values, to feed a SIEM's message queue. The broker has to be connected to the resources server using OAuth2 `client credentials` mode.

The broker asks to the resources server, at intervals, to calculate each user's global trust, and puts every result in a queue. The SIEM reads this queue and generates the pertinent alarms.

8.4 Joint Use Case

Initially a user would unlock his device and start writing. At this moment, the local model will capture his features, and the EDR's agent will start sending information to his server. Ten minutes later, there would be a model of the user in both systems, and the resources server will be providing a trust value supporting his identity.

Now, let's suppose the user puts on a jacket and continues writing. The clothes-based model will start to send values that question the users' identity, but the keyboard model will counteract the result. Later, the user forgets the device unlocked in public place and leaves. If an unauthorized user (from now on, the attacker) takes the device and tries to use it, there would be three scenarios:

1. If the attacker wears clothes similar to the user's, and he uses the keyboard, the system will detect the impersonation
2. If the attacker wears clothes different to the user's, and he uses the keyboard, the same will happen
3. If the attacker wears clothes different to the user's, and he does not use the keyboard, he may eventually be able to use the device for a period of time. But when the keyboard's model trust value expires, the only available trust values will be related with the clothes, and this will prevail and reject the attacker.

9 Conclusions and Future Work

The CA architecture presented in this paper is a highly integrable system that covers all the stages necessary for evaluating a user's identity. The proposed authentication system, and the central trust system, make possible the use of different types of clients running in several kinds of environments. It also allows creating useful models without taking sensible data out of the user's device, granting the global result's integrity with an adaptative model against possible local tampering.

For future work we suggest analyzing ensemble learning methods to obtain the most accurate result. Determining the appropriate parameters for the adaptative strategy also requires further work. Finally, it is interesting to implement the architecture to continuously authenticate non-human users (e.g. validate the identity of sensors in a thermostat network).

Acknowledgements. This project has received funding from the European Union's Horizon 2020 Research and Innovation Program under grant agreement No. 826284 (ProTego).

References

1. OAuth Grant Types. https://oauth.net/2/grant-types/
2. REGULATION (EU) 2016/ 679 OF THE EUROPEAN PARLIAMENT AND OF THE COUNCIL - of 27 April 2016 - on the protection of natural persons with regard to the processing of personal data and on the free movement of such data, and repealing directive 95/ 46/ EC (general data protection regulation)
3. Alshehri, A., Coenen, F., Bollegala, D.: Iterative keystroke continuous authentication: a time series based approach. KI-Kunstliche Intelligenz **32**(4), 231–243 (2018). https://doi.org/10.1007/s13218-018-0526-z
4. Azzini, A., Marrara, S., Sassi, R., Scotti, F.: A fuzzy approach to multimodal biometric continuous authentication. Fuzzy Optim. Decis. Mak. **7**(3), 243 (2008). https://doi.org/10.1007/s10700-008-9034-1
5. Basar, O.E., Alptekin, G., Volaka, H.C., Isbilen, M., Incel, O.D.: Resource usage analysis of a mobile banking application using sensor-and-touchscreen-based continuous authentication. Procedia Comput. Sci. **155**, 185–192 (2019). https://doi.org/10.1016/j.procs.2019.08.028
6. Bonastre, J.F., Bimbot, F., Boe, L.J., Campbell, J.P., Reynolds, D.A., Magrin-Chagnolleau, I.: Person authentication by voice: a need for caution, p. 4 (2003)
7. Bradley, J., Sakimura, N., Jones, M.: JSON web token (JWT) (2015). https://tools.ietf.org/html/rfc7519
8. Campbell, B., Jones, M., Mortimore, C.: JSON web token (JWT) profile for OAuth 2.0 client authentication and authorization grants (2015). https://tools.ietf.org/html/rfc7523
9. Casale, P.: Approximate ensemble methods for physical activity recognition applications. ELCVIA: Electron. Lett. Comput. Vis. Image Anal. **13**, 22–23 (2014)
10. Chatterjee, A., Aung, K.M.M.: Fully homomorphic encryption in real world applications. In: Computer Architecture and Design Methodologies. Springer, Singapore (2019). https://doi.org/10.1007/978-981-13-6393-1

11. Dwork, C.: Differential privacy: a survey of results. In: Agrawal, M., Du, D., Duan, Z., Li, A. (eds.) Theory and Applications of Models of Computation, Lecture Notes in Computer Science, pp. 1–19. Springer, Heidelberg (2008). https://doi.org/10.1007/978-3-540-79228-4_1
12. Eberz, S., Rasmussen, K.B., Lenders, V., Martinovic, I.: Looks like eve: exposing insider threats using eye movement biometrics. ACM Trans. Priv. Secur. (TOPS) **19**(1), 1:1–1:31 (2016). https://doi.org/10.1145/2904018
13. Frank, M., Biedert, R., Ma, E., Martinovic, I., Song, D.: Touchalytics: on the applicability of touchscreen input as a behavioral biometric for continuous authentication. IEEE Trans. Inf. Forensics Secur. **8**(1), 136–148 (2012). https://doi.org/10.1109/TIFS.2012.2225048
14. Gascon, H., Uellenbeck, S., Wolf, C., Rieck, K.: Continuous authentication on mobile devices by analysis of typing motion behavior, p. 13 (2014)
15. Hardt, D.: The OAuth 2.0 authorization framework (2012). https://tools.ietf.org/html/rfc6749
16. Katevas, K., Haddadi, H., Tokarchuk, L.: Poster: SensingKit: a multi-platform mobile sensing framework for large-scale experiments. In: Proceedings of the 20th Annual International Conference on Mobile Computing and Networking, MobiCom 2014, pp. 375–378. Association for Computing Machinery, Maui (2014). https://doi.org/10.1145/2639108.2642910
17. Lin, F., Song, C., Zhuang, Y., Xu, W., Li, C., Ren, K.: Cardiac scan:a non-contact and continuous heart-based user authentication system. In: Proceedings of the 23rd Annual International Conference on Mobile Computing and Networking, MobiCom 2017, pp. 315–328. Association for Computing Machinery, Snowbird (2017). https://doi.org/10.1145/3117811.3117839
18. Mahfouz, A., Mahmoud, T.M., Sharaf Eldin, A.: A behavioral biometric authentication framework on smartphones. In: Proceedings of the 2017 ACM on Asia Conference on Computer and Communications Security, ASIA CCS 2017, pp. 923–925. Association for Computing Machinery, Abu Dhabi (2017). https://doi.org/10.1145/3052973.3055160
19. Mondal, S., Bours, P.: Continuous authentication using mouse dynamics. In: 2013 International Conference of the BIOSIG Special Interest Group (BIOSIG), pp. 1–12 (2013)
20. Morales, A., Fierrez, J., Tolosana, R., Ortega-Garcia, J., Galbally, J., Gomez-Barrero, M., Anjos, A., Marcel, S.: Keystroke biometrics ongoing competition. IEEE Access **4**, 7736–7746 (2016). https://doi.org/10.1109/ACCESS.2016.2626718
21. Parecki, A.: OAuth 2.1. IETF 106 (2019). https://www.ietf.org/proceedings/106/slides/slides-106-oauth-sessa-oauth-21-00.pdf
22. Pisani, P.H., Lorena, A.C.: A systematic review on keystroke dynamics. J. Braz. Comput. Soc. **19**(4), 573–587 (2013). https://doi.org/10.1007/s13173-013-0117-7
23. Polikar, R.: Ensemble learning. In: Zhang, C., Ma, Y. (eds.) Ensemble Machine Learning: Methods and Applications, pp. 1–34. Springer, US (2012). https://doi.org/10.1007/978-1-4419-9326-7_1
24. Shepherd, S.J.: Continuous authentication by analysis of keyboard typing characteristics, pp. 111–114 (1995). https://doi.org/10.1049/cp:19950480
25. Siirtola, P., Komulainen, J., Kellokumpu, V.: Effect of context in swipe gesture-based continuous authentication on smartphones (2019)
26. Solapurkar, P.: Building secure healthcare services using OAuth 2.0 and JSON web token in IOT cloud scenario. In: 2016 2nd International Conference on Contemporary Computing and Informatics (IC3I), pp. 99–104 (2016). https://doi.org/10.1109/IC3I.2016.7917942

27. Tsai, P., Khan, M.K., Pan, J., Liao, B.: Interactive artificial bee colony supported passive continuous authentication system. IEEE Syst. J. **8**(2), 395–405 (2012). https://doi.org/10.1109/JSYST.2012.2208153
28. Xu, H., Zhou, Y., Lyu, M.R.: Towards continuous and passive authentication via touch biometrics: an experimental study on smartphones. In: Proceedings of the Tenth USENIX Conference on Usable Privacy and Security, SOUPS 2014, pp. 187–198. USENIX Association, Menlo Park (2014)

Perception Systems

Optimization of a Robotics Gaze Control System

Jaime Duque Domingo[1(✉)], Jaime Gómez-García-Bermejo[1,2], and Eduardo Zalama[1,2]

[1] ITAP-DISA, University of Valladolid, Pl. Santa Cruz 8, 47002 Valladolid, Spain
jaime.duque@uva.es
[2] CARTIF Foundation, Boecillo, 47151 Valladolid, Spain

Abstract. Gaze control is an important issue in the interaction between a robot and humans. In particular, the decision of who to pay attention to in a multi-party conversation is a way of improving a robot's naturalness in human-robot interaction. A system based on a competitive neural network can decide who to look at with a smooth transition in the focus of attention when significant changes in stimuli are produced. One important aspect in this process is the configuration of the different parameters of such a neural network. The weights of the different stimuli have to be computed in order to achieve behavior similar to humans. This article explains how these weights can be obtained by solving an optimization problem. The experiments carried out and some results are also presented.

Keywords: Gaze control · Gaze engagement · Humanoid robot · ROS · HRI · Competitive network · Computer vision

1 Introduction

Gaze control represents an important discipline in the development of intelligent social robots so as to achieve higher evaluations of a robot's comprehension and naturalness in human-robot interaction [15]. When robots behave like a person, humans feel more comfortable. An approach to this problem consists in using a competitive neural network that receives different stimuli and returns a stable determination of the focus of attention that must be followed with the robot's eyes. One of the advantages of this approach is that the response of the competitive network is smooth and stable, avoiding erratic behavior. This approach also keeps, in the robot's memory, the information about people who previously interacted with the robot, regardless of whether they have left the robot's field of view. Different factors are taken into account: the human's gaze, who is speaking, pose, proxemics, visual focus of attention, hoarding conversation, habituation, etc. These factors produce stimuli which create a dynamic behavior, giving the

L. M. Bergasa et al. (Eds.): WAF 2020, AISC 1285, pp. 213–226, 2021.
https://doi.org/10.1007/978-3-030-62579-5_15

human interlocutors the feeling of speaking to another human. However, a problem that emerges is the need to determine the importance of the stimuli so as to be able to decide who should be the focus of attention. We have modeled this importance through a set of gain weights that are adjusted by optimization from an experimentation data set.

These gain weights represent how each stimulus contributes to a particular output of the neural network. In our experimentation, training is carried out by three people who interact with each other and with the robot. At this step, during an initial training, an external user (teacher) analyzes the sequence and establishes who the robot should pay attention to. This information is written down jointly with recorded stimuli. After the initial training, the gain weights are computed and the robot should pay attention to people as learned in the experiments. To validate the proposed methodology, a robotic head with a projected face that simulates the movement of the eyes has been developed. The result of the competition makes the robot set the focus on the person, using the combined movement of neck and eyes.

The present paper is structured as follows: Sect. 2 explores the state-of-the-art of the technologies considered in this paper. Section 3 briefly explains how the gaze control works with the competitive network. Section 4 presents the approach for computing the weights of the stimuli. In Sect. 5, the different experiments and results are reported. Finally, Sect. 6 notes the conclusions of the presented work and suggests future developments.

2 Overview of the Related Work

Human–Robot Interaction (HRI) is a discipline that allows robots that can communicate and respond to ongoing human communications and behavior to be improved [13]. Gaze control is an important issue concerning HRI. A recent survey of the state of the art in social eye gaze for HRI was presented by [1], distinguishing three different approaches to the problem: Human-focused, centered on understanding the characteristics of human behavior during interactions with robots; Design-focused, which studies how the design of a robot impacts on interactions with humans; and Technology-focused, centered on researching how to build tools to guide the robot's gaze in human interaction. The main challenges in a conversation are the management of attention and turn-taking between partners, controlling the gaze and adopting the right conversational roles [1].

According to [15], when a robot is a listener in a multi-party conversation and tracks the conversation with its gaze, it promotes higher evaluations of its comprehension and naturalness than a robot performing random gazing between speakers. Also, in [3], the authors showed how gaze control more effectively motivates users to repeatedly engage in therapeutic tasks. In addition, as seen in [8], the virtual agents which use turn-taking gaze during conversations are evaluated as more natural and pleasant than others that use none, or a random gaze control in their communication. At the same time, the robots with humanoid features positively influence people's behavior towards the machine and their expectations about its capabilities [24].

There have been different works on gaze control, as in [2], where the authors used a circular array of microphones in a social robot, named Maggie, to determine in which direction the robot should look. An infrared laser was used to obtain the distance with respect to the person so that the robot could move forward/backward. A Robot Assisted Therapy (RAT) method was presented in [20], where a robot perceived different stimuli (visual, auditory and tactile) to track a certain colored object, a face or a directional voice. These two previous works did not address the problem of who to pay attention to in a natural way. Different authors have indicated the benefits of fusing sensory information, such as [20,26] or [28]. In [26], a person is localized with a robotic head by simultaneously processing visual and audio data. However, these works do not focus on a conversation between multiple participants. In [28], the authors created a system to guide a robot's gaze at multiple humans who were interacting, by adding different stimuli: social features, proxemics values, orientation, and a memory component. This approach considered a limited number of stimuli and the maximum value could change abruptly and lead to erratic changes in the focus of attention. Our proposed method is based on a competitive neural network and fixes this problem, creating smooth transitions between participants.

The stimuli used by a robot can be diverse, but those based on computer vision, audio and memory represent some of the most commonly used. Visual information represents an important aspect of HRI [9]. In particular, the Robot's Field of View (FOV) has to be considered to detect visually people situated in front of the robot's camera. But when people are not situated in front of the robot, some information has to be kept in memory. A hypotheses generation was proposed by [21], inferring the people's position by using peripheral vision. Even though a person was not present in the foveal vision, the robot kept plausible hypotheses about the location. In [25], the authors proposed a dynamic visual memory to store information about objects from a moving camera on board a robot and created an attention system based on where to look to reobserve objects in memory and the need to explore new areas. The Visual Focus of Attention (VFOA) represents who or what people are looking at. As presented in [17], there is a relation between head poses and object locations. Audio represents an important stimulus, since microphone arrays can indicate the direction of the incoming sound. As explained previously, some works using microphone arrays, such as [2,26], have been used with social robots.

When a robot is interacting with people, their detection is required. Some methods are able to detect human bodies, such as *Haar filters* [27], HoG [5], or *Deep Convolution Neural Networks* (DCNN) [18]. However, in our approach, a face recognition algorithm has been preferred, since the participants are assumed to look directly at the robot or have a slightly turned position to look at other people interacting with it. As indicated for human body detections, Haar classifiers, HoG detectors or Deep Learning based solutions are widely used. Haar classifiers detect faces at different scales, but do not deal with non-frontal faces or occlusions.

It also returns a large number of false predictions. The HoG feature descriptor is fast, but does not work with small faces. The DLIB library [14] implements a CNN face detector using a Maximum-Margin Object Detector [7], which works for different face orientations and occlusions. Recognition can be implemented with Deep Residual Learning algorithms, which are very accurate. DLIB implements a ResNet network with 29 convolution layers and uses a pre-trained model which takes the 68 face landmarks obtained from an image [12].

Moreover, lip activity is important to detect whether a person is speaking or not. When several people are situated in front of the robot and some audio is detected, it is likely that the person who is moving the lips is currently speaking. Some works have explored different techniques to detect lip activity, such as [4], where the degree of disorder of pixel directions around the lips using the optical flow technique is measured; or [22], where a statistical algorithm using two detectors based on noise to characterize visual speech and silence in video sequences is created.

Competitive neural networks [10] can process different inputs and decide the winner in a dynamic and natural way. This kind of network can also have habituation capabilities. However, the problem is how to determine the importance of some stimuli over others. This can be done by a set of weights. These weights have to be configured or learned to replicate human behavior when the same stimuli are produced.

3 Gaze Control with a Competitive Network

Gaze control is usually implemented considering different stimuli [20,26,28]. Our approach for the stimuli integration is the use of an on-center off-surround competitive model [10]. Different stimuli are considered, which are the inputs of the competition at an instant of time t, I_{tkx}, where k is the number of persons who interact and x the number of stimuli. These stimuli have a binary value, indicating whether they are present or not, and are balanced by a weight, w_x. In some stimuli, such as speech detection, the value is activated if the different indicators exceed a threshold.

- I_{tk1} shows if a person k is situated in the robot's FOV at the instant of time t. People situated in front of the robot's FOV are usually candidates to interact with the robot. w_1 is the corresponding weight associated to that stimulus.
- I_{tk2} shows that a person k is considered to be speaking. This stimulus is obtained by performing lip movement detection, based on mouth landmarks. In addition, incoming audio has to be detected in the direction of the person k.
- I_{tk3} shows that a person k is gazing directly at the robot. The pose of a person is used to verify if that person is visually interacting with the robot, as shown in Fig. 1. Engagement in an interaction is increased by the *mutual gaze*, a kind of *shared looking* [23].

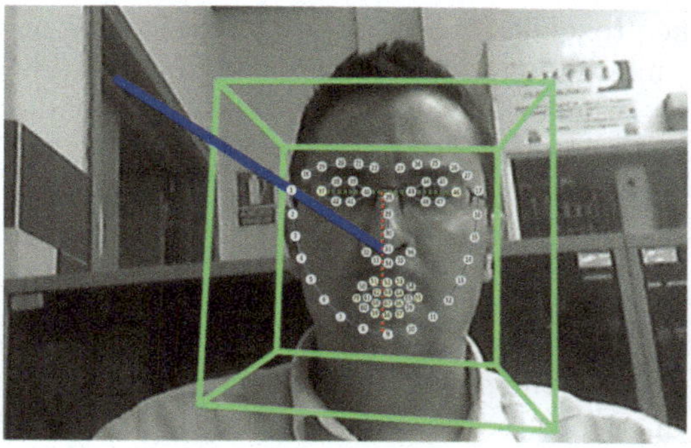

Fig. 1. Pose of a person given the DLIB face landmarks

- I_{tk4} shows that a person k is continuously moving. In a conversation with several people, an individual tends to look at another restless person. This stimulus requires the individual to be situated in the FOV of the robot. If the sum of the differences in a person's position between several frames is over a given threshold, the person is assumed to be restless.
- I_{tk5} shows that a person k is not situated in the robot's FOV, but for whom incoming audio could have been detected. When the robot does not see a person who has previously interacted, it keeps the previous position of the person in its memory and, if incoming audio is detected in his/her direction, the stimulus for that person k is activated. In a conversation between humans, when someone is speaking at their left/right side, a person tends to turn their head in the direction of the person who is speaking.
- I_{tk6} shows that a person k is in the VFOA of other people, but is not situated in the robot's FOV (see Fig. 2). When two or more people are looking at another person in a conversation, a stimulus is given to people in the direction of the gaze. Since the focus of attention is given to a concrete person, the corresponding stimulus is increased.

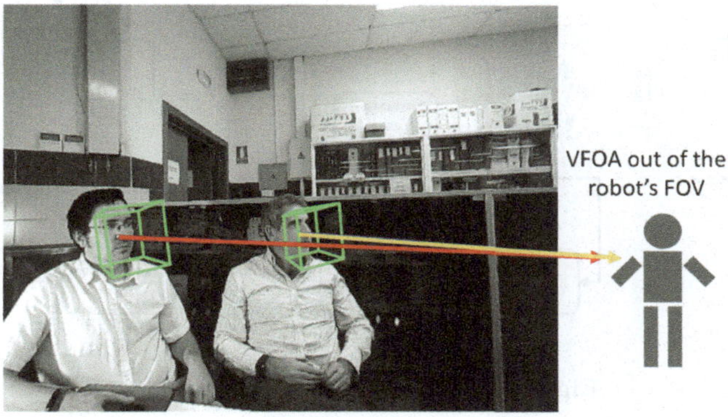

Fig. 2. Two people with their VFOA in another person

- I_{tk7} indicates the proxemics of a person. People situated at a certain distance are likely to be interacting with the robot [2,28]. In addition to the weight w_7, this stimulus is balanced by a tuning factor, depending on the distance between the robot and the person.

4 Analysis of the System

The different stimuli that influence a gaze control system must be weighted in such a way that it allows the system to behave similarly to humans. The gaze control system is implemented through a competitive neural network, where each stimulus competes with the others to provide a smooth dynamic result. However, the input of the competitive network, composed by the relation of stimuli and their weights, must be correctly assigned for the robot to have a realistic behavior. Figure 3 shows the system scheme, in which an optimization problem allows the weights that will be used in the gaze control system to be obtained. This optimization requires the data to be split into winners and losers.

The competitive network is shown in Fig. 4, where each input I_{tk} is obtained by adding the stimuli of the person k, I_{k1}, ..., I_{k7}, balanced by their respective weights, at time t. Thus, the input of the network produces an output O_{tk} that represents if a person k is the winner. The output with the highest value corresponds to the winner. The configuration of the weights, w_x, is not initially known, so a mechanism has been implemented to obtain it. The competitive network keeps information about preceding states and dynamically adapts the outputs depending on its configuration.

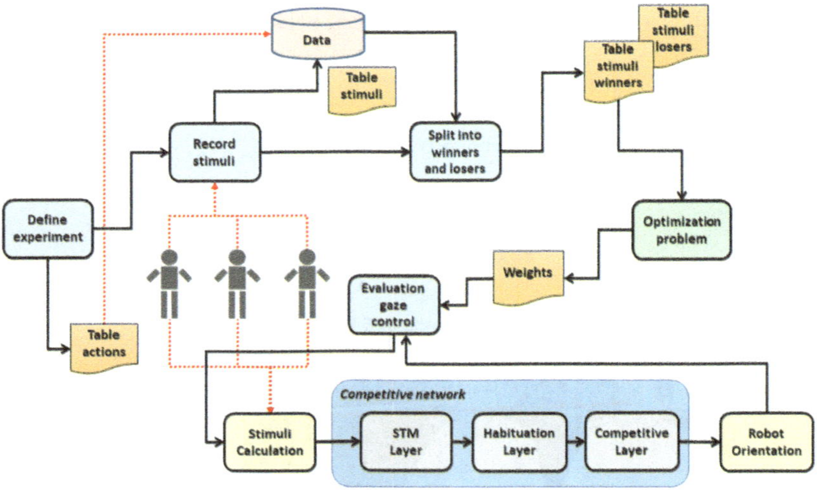

Fig. 3. Scheme of the training system

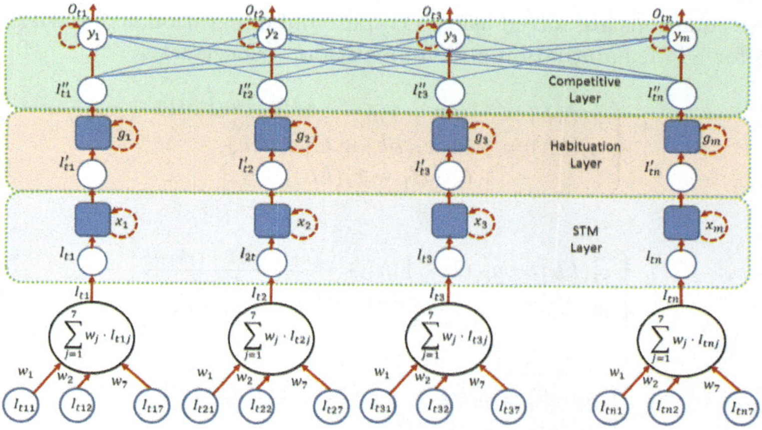

Fig. 4. Competitive neural network

At a given time t, there is an input vector of the network, I_t, composed by the value associated to each person (I_{t1}, ..., I_{tn}). I_{tk} is computed as shown in Eq. 1.

$$I_{tk} = \sum_{x=1}^{7} w_x \cdot I_{tkx} \tag{1}$$

The output of the network depends on the input vectors obtained previously, as seen in Eq. 2, where C, H and S are, respectively, the layers associated to the competitive, habituation and short time memory (STM) operations.

$$O_t = C(H(S(I_1, ..., I_t))) \tag{2}$$

The STM, habituation and competitive layers are based on the model of Grossberg [10] and use the differential Eqs. 3, 4 and 5, respectively.

$$\frac{dx_i}{dt} = -A_1 x_i + C_1(B_1 - x_i)[I_i w_i] \tag{3}$$

$$\frac{dg_i}{dt} = E(1 - g_i) - F I_i^{'} g_i \tag{4}$$

$$\frac{dy_i}{dt} = -A_2 y_i + C_2(B_2 - y_i)[I_i^{''} + D{y_i}^2] - y_i \sum_{i \neq j} D{y_i}^2 \tag{5}$$

These equations are solved using a trapezoidal integration, as shown in Eqs. 6, 7 and 8 for STM, habituation and competition, respectively.

$$\begin{cases} x_i(kh) = x_i((k-1)h) + \frac{g_i(kh)+g_i(k-1)h}{2} \\ g_i(kh) = -A_1 x_i(kh) + C_1 x_i(kh) \\ \qquad +C_1(B_1 - x_i(kh))[I_i w_i] \end{cases} \tag{6}$$

$$\begin{cases} g_i(kh) = g_i((k-1)h) + \frac{p(kh)+p((k-1)h))h}{2} \\ p(kh) = E[1 - g_i(kh)] - FI_i^{'}(kh)g_i(kh) \end{cases} \tag{7}$$

$$\begin{cases} y_i(kh) = y_i((k-1)h) + \frac{q_i(kh)+q_i(k-1)h}{2} \\ q_i(kh) = -A_2 y_i(kh) + C_2 y_i(kh) \\ \qquad +C_2(B_2 - y_i(kh))[I_i^{''} w_i] \\ \qquad - \sum_{i \neq j} D y_j((k-1)h)^2 \end{cases} \tag{8}$$

The STM layer, where A_1 corresponds to the decay rate, B_1 to the saturation and C_1 to the growth rate, receives the input stimuli, I_i, and produces an output where the duration of the stimuli is increased. The STM output is the input of the habituation layer, $I_i^{'}$, where the permanent stimuli lose interest through time. In the habituation layer, g_i is the gain for the stimuli. When a stimulus is active, the habituation gain decreases from a maximum value, 1, to the value given by $E/(E + FI_i^{'})$, where E and F correspond to the charge and discharge rates. The output of the habituation layer is the input of the competitive layer, $I_i^{''}$, where A_2 is the decay rate, B_2 the saturation value and C_2 marks the growth rate. D balances the parabolic function ${y_i}^2$, reinforcing the winner against the rest, which represents a lateral inhibition (off-surround). The output of the competitive layer shows the winner of the competition, as explained previously.

The exit from the network takes into account the dynamic nature of previous states by filtering spurious stimuli and producing a smooth change of the winning person focus of attention. However, the input of the network, I_i, corresponding to the stimuli, has to be optimized to obtain a group of weights which are optimal in the gaze control process. To this optimization, the training has to be carried out in a sequential way, following a list of steps previously recorded. The training is carried out by three people. At the same time, an external user (teacher) observes the interaction and annotates the time instants when a person should be the focus of attention.

When all data have been obtained, stimuli from expected winners and losers are separated based on the said manual annotation. At an instant t, there is an input value for the winner, $I_{t,winner}$, and an input value for every k loser, $I_{t,k}$. The process is modeled as an optimization problem, where the aim is to obtain the optimal weights that maximize the sum of the distances between the winners and the losers at each time instant t, as shown in Eq. 9. This procedure ensures that the weights are optimal to make the selected persons winners and separate them from the losers.

$$\begin{aligned} & \max \sum_{t=1}^{m} \left(\sum_{k \in losers} I_{t,winner} - I_{tk} \right) \\ & = \max \sum_{t=1}^{m} \left[\sum_{k \in losers} \left(\begin{array}{c} \sum_{x=1}^{7} w_x \cdot I_{t,winner,x} \\ - \sum_{x=1}^{7} w_x \cdot I_{t,k,x} \end{array} \right) \right] \end{aligned} \tag{9}$$

Some constraints have to be considered. First of all, the sum of all weights has to be equal to 1, as shown in Eq. 10. In addition, the weights range between 0 and 1, being bigger than 0 (Eq. 11).

$$\sum_{x=1}^{7} w_x = 1 \tag{10}$$

$$\forall x \in [1...7] : w_x \in [0,1] \wedge w_x > 0 \tag{11}$$

Secondly, there is a constraint for each case evaluated. The input of the winner, $I_{t,winner}$, is bigger than or equal to the losers at an instant t, as shown in Eq. 12. With this constraint, the problem is forced to behave as annotated during the training phase.

$$\forall t \in [1..m] \wedge k \in losers\ at\ t : I_{t,winner} \geq I_{t,k} \tag{12}$$

Another consideration is related to the proxemic stimulus. This is balanced by a factor depending on the distance between the person and the robot during the gaze control operation. During the training phase, a fixed value of 1 is assigned. The training is carried out by people who are situated at the *far phase* of the personal space (0.76 m to 1.22 m) or at the *close phase* of the social space (1.22 m to 2.10 m) [11], a region where this distance factor is 1. Beyond 2.10 m, this factor decreases and is not significant for the calculation of weights because, during our training, the participants have been situated within these distances. During normal operation time, an adjustment factor balances the weight of people situated beyond 2.10 m.

5 Experiments and Results Discussion

The experiment consisted in obtaining the optimal weights during training with three people interacting with the robot. The weights were not initially known and the persons followed a list of actions previously established, as shown in Table 1. The complete list of actions had 726 states. The table shows the losers

Table 1. Sequence of behavior of three people

State /	Person 1							Person 2							Person 3							Winner
Stimuli	1	2	3	4	5	6	7	1	2	3	4	5	6	7	1	2	3	4	5	6	7	
1	🟢						🟢	🔴		🔴				🔴								2
2	🟢		🟢				🟢	🔴		🔴				🔴	🟢						🟢	2
3	🔴		🔴	🔴			🔴	🟢		🟢				🟢	🟢		🟢				🟢	1
4	🟢		🟢				🟢	🔴	🔴	🔴	🔴			🔴	🟢		🟢				🟢	2
5	🟢		🟢				🟢	🟢		🟢				🟢	🔴			🔴			🔴	3
6	🔴		🔴	🔴			🔴	🟢		🟢				🟢	🟢						🟢	1
7	🟢						🟢	🔴	🔴	🔴				🔴	🟢		🟢				🟢	2
8	🔴		🔴				🔴	🟢						🟢	🟢	🟢					🟢	1
9	🟢						🟢	🟢		🟢				🟢	🔴						🔴	3
10	🟢		🟢				🟢	🔴		🔴				🔴								2
11	🟢						🟢	🟢		🟢				🟢					🔴	🔴		3
12	🔴	🔴	🔴				🔴	🟢						🟢						🟢		1

in green and the winners in red. Only 12 out of the 726 states are shown, but all of them were evaluated in the maximization problem.

The optimization problem was solved using the SLSQP algorithm [16], which obtained the results in 18 iterations and 0.23 s (in an intel i9-9900K, with 32 Gb of RAM). The obtained results were $w_1 = 0.06$, $w_2 = 0.25$, $w_3 = 0.06$, $w_4 = 0.16$, $w_5 = 0.16$, $w_6 = 0.25$ and $w_7 = 0.06$.

These weights were evaluated in the previously self-developed robotic head (see Fig. 5), where the competitive network had been deployed. The network created smooth transitions between the focus of attention, resulting in a natural behavior but, at the same time, considering properly which stimuli were more important according to the obtained weights. Figures 6 and 7 show the sum of input stimuli balanced by their weights, I_{tk}, and the output stimuli, O_{tk}, respectively.

As can be seen at the output of the neural network, the input stimuli are smoothed out to avoid sudden changes in the robot's focus. The results of the winning person are consistent with the choice made during the learning phase, being satisfactory in 99.03% of the test cases. The adjustment of these weights has allowed the creation of a robot that responds in a similar way to how a person behaves.

Fig. 5. Self-developed robot head

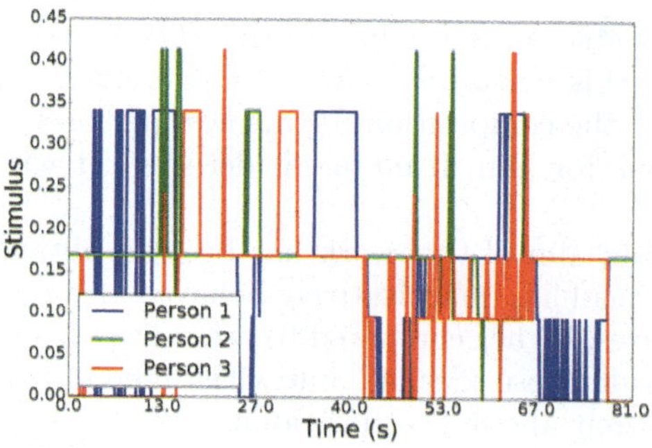

Fig. 6. Input stimuli of three people balanced by the weights

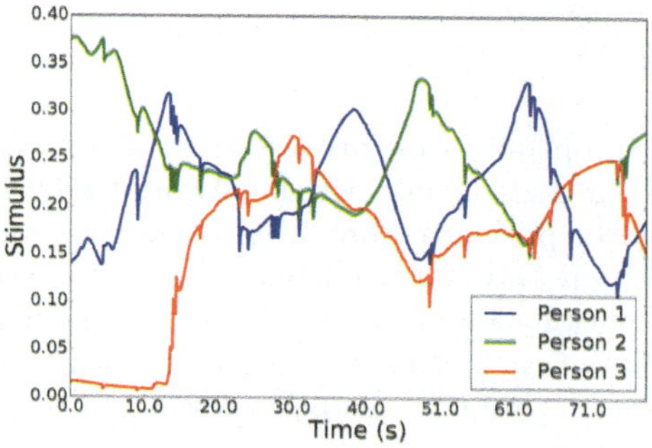

Fig. 7. Output stimuli of three people after the competitive network

Although the system has been trained with 3 people, the stimuli generated during the training are independent of the person. The methods of obtaining stimuli that have been used are generic and suitable for any person. The 726 cases used allow to refine the weights to a very wide range of situations. Training with more people may slightly vary these weights but not in a significant way. At normal operation time, the system works with 3 or more people, having even performed some test with more than 8 participants. Beyond the increase in computing time, the system was able to correctly gaze at the participants who were winning the competition. For performance reasons, our system has been limited to 10 participants.

When a participant disappears from the robot's FOV, the head is able to estimate the position using a Kalman filter [19]. If the robot interacts with 3 participants and one disappears, either because the person moves or because the head turns, when audio is detected in his direction or the other participants are looking at him (VFOA), stimuli will be provided to that participant and the head will gaze at this person in case of being winner, using the estimated Kalman position. When no stimuli are received from a participant for more than 20 s, the participant is removed. When a new person arrives on the scene, regardless of whether the robot is able to detect him in the FOV or not, if audio is detected in his direction or if this person is in the VFOA of other participants, the new participant will enter the competition. If this new person becomes a winner, the head will turn to look for him. If no one is detected, it will go back to the old winner.

Although the main aim of this work has been to obtain the weights of our neural network, the head has quantitatively behaved as we expected in operation time in 99.03% of cases, higher compared to other works, such as [28], where the authors obtained results close to 89.4% but with different experiment conditions. A complete explanation about the functioning of the neural network and the robotic head as well as the experiments performed and results obtained are presented in [6].

6 Conclusions

This work presents an optimization process for a robotic gaze control system. This gaze control system uses a competitive network which receives a large number of visual, auditory or presence stimuli. It allows a smooth transition, changing the focus of attention between participants, avoiding erratic movements. In addition, it has habituation capabilities to avoid someone from hoarding the conversation. The weights of each stimulus are not known a priori and a strategy based on an optimization problem has been developed to obtain them through experiential learning.

The computed weights were integrated into the gaze control system of a self-developed robotic head with combined body and eye movement, and the robot showed a natural interaction behavior similar to those annotated during the learning phase. The proposed method significantly improved the sensation of naturalness and realism of the robotic head. The movement of the robot joints and the expressions of the virtual agent projected on its 3D facial display were controlled by the system integrated in a ROS-based architecture. The design of the robot and the gaze control system creates a more realistic HRI system, which is more acceptable to interlocutors than other not-so-human robots.

The future objectives of the project will consist in integrating speech capabilities into the ROS architecture, offering a low-cost, intelligent robot with human-like behavior.

References

1. Admoni, H., Scassellati, B.: Social eye gaze in human-robot interaction: a review. J. Hum.-Robot Interact. **6**(1), 25–63 (2017)
2. Alonso-Martín, F., Gorostiza, J.F., Malfaz, M., Salichs, M.A.: User localization during human-robot interaction. Sensors **12**(7), 9913–9935 (2012)
3. Andrist, S., Mutlu, B., Tapus, A.: Look like me: matching robot personality via gaze to increase motivation. In: Proceedings of the 33rd Annual ACM Conference on Human Factors in Computing Systems, pp. 3603–3612. ACM (2015)

4. Bendris, M., Charlet, D., Chollet, G.: Lip activity detection for talking faces classification in tv-content. In: International Conference on Machine Vision, pp. 187–190 (2010)
5. Dalal, N., Triggs, B.: Histograms of oriented gradients for human detection (2005)
6. Duque-Domingo, J., Gómez-García-Bermejo, J., Zalama, E.: Gaze control of a robotic head for realistic interaction with humans. Front. Neurorobot. **14**, 34 (2020)
7. King, D.E.: Max-margin object detection. arXiv preprint arXiv:1502.00046 (2015)
8. Garau, M., Slater, M., Bee, S., Sasse, M.A.: The impact of eye gaze on communication using humanoid avatars. In: Proceedings of the SIGCHI Conference on Human Factors in Computing Systems, pp. 309–316. ACM (2001)
9. Gergle, D., Kraut, R.E., Fussell, S.R.: Using visual information for grounding and awareness in collaborative tasks. Hum.-Comput. Interact. **28**(1), 1–39 (2013)
10. Grossberg, S.: Contour enhancement, short term memory, and constancies in reverberating neural networks. In: Studies of Mind and Brain, pp. 332–378. Springer (1982)
11. Hall, E.T., Birdwhistell, R.L., Bock, B., Bohannan, P., Diebold Jr., A.R., Durbin, M., Edmonson, M.S., Fischer, J., Hymes, D., Kimball, S.T., et al.: Proxemics [and comments and replies]. Curr. Anthropol. **9**(2/3), 83–108 (1968)
12. Kazemi, V., Sullivan, J.: One millisecond face alignment with an ensemble of regression trees. In: Proceedings of the IEEE Conference on Computer Vision and Pattern Recognition, pp. 1867–1874 (2014)
13. Kiesler, S., Hinds, P.: Introduction to this special issue on human-robot interaction. Hum.-Comput. Interact. **19**(1–2), 1–8 (2004)
14. King, D.E.: DLIB-ML: a machine learning toolkit. J. Mach. Learn. Res. **10**(Jul), 1755–1758 (2009)
15. Kousidis, S., Schlangen, D.: The power of a glance: evaluating embodiment and turn-tracking strategies of an active robotic overhearer. In: 2015 AAAI Spring Symposium Series (2015)
16. Kraft, D., Schnepper, K.: Slsqp–A Nonlinear Programming Method with Quadratic Programming Subproblems. DLR, Oberpfaffenhofen (1989)
17. Massé, B.: Gaze direction in the context of social human-robot interaction. Ph.D. thesis (2018)
18. Ren, S., He, K., Girshick, R., Sun, J.: Faster R-CNN: towards real-time object detection with region proposal networks. In: Advances in Neural Information Processing Systems, pp. 91–99 (2015)
19. Rosales, R., Sclaroff, S.: Improved tracking of multiple humans with trajectory prediction and occlusion modeling. Tech. rep., Boston University Computer Science Department (1998)
20. Saldien, J., Vanderborght, B., Goris, K., Van Damme, M., Lefeber, D.: A motion system for social and animated robots. Int. J. Adv. Robot. Syst. **11**(5), 72 (2014)
21. Shiomi, M., Kanda, T., Miralles, N., Miyashita, T., Fasel, I., Movellan, J., Ishiguro, H.: Face-to-face interactive humanoid robot. In: 2004 IEEE/RSJ International Conference on Intelligent Robots and Systems (IROS) (IEEE Cat. No. 04CH37566), vol. 2, pp. 1340–1346. IEEE (2004)
22. Siatras, S., Nikolaidis, N., Krinidis, M., Pitas, I.: Visual lip activity detection and speaker detection using mouth region intensities. IEEE Trans. Circuits Syst. Video Technol. **19**(1), 133–137 (2008)
23. Sidner, C.L., Kidd, C.D., Lee, C., Lesh, N.: Where to look: a study of human-robot engagement. In: Proceedings of the 9th International Conference on Intelligent User Interfaces, pp. 78–84. ACM (2004)

24. Thrun, S.: Toward a framework for human-robot interaction. Hum.-Comput. Interact. **19**(1), 9–24 (2004)
25. Vega, J., Perdices, E., Cañas, J.: Robot evolutionary localization based on attentive visual short-term memory. Sensors **13**(1), 1268–1299 (2013)
26. Viciana-Abad, R., Marfil, R., Perez-Lorenzo, J., Bandera, J., Romero-Garces, A., Reche-Lopez, P.: Audio-visual perception system for a humanoid robotic head. Sensors **14**(6), 9522–9545 (2014)
27. Viola, P., Jones, M., et al.: Rapid object detection using a boosted cascade of simple features. CVPR **1**(1), 511–518 (2001)
28. Zaraki, A., Mazzei, D., Giuliani, M., De Rossi, D.: Designing and evaluating a social gaze-control system for a humanoid robot. IEEE Trans. Hum.-Mach. Syst. **44**(2), 157–168 (2014)

Open-Loop Sidescan Sonar Mosaic and ANN Velocity Estimation

José Manuel Bernabé Murcia(✉) and Humberto Martínez-Barberá

Department of Information and Communication Engineering,
University of Murcia, 30100 Murcia, Spain
{josemanuel.bernabe,humberto}@um.es

Abstract. This paper presents a system to generate open-loop mosaics from the images obtained by a sidescan sonar using dead reckoning as a positioning system in a low cost ROV, which does not include a Doppler velocity log (DVL). The ROV velocity is then estimated using an artificial neural network (ANN) for feeding the dead reckoning positioning. The training process of the neural network is also described. The sidescan sonar readings are used, based on the currently estimated position, to update the sonar mosaic. The process of the transformations and corrections of the sidescan sonar are also described. Finally, some results obtained in a thermal power plant are presented.

Keywords: Sidescan sonar mosaicing · Neuronal network velocity estimation · Dead reckoning

1 Introduction

The underwater exploration has increased in recent years, this increase is mainly caused by the appearance of ROVs and more recently AUVs. The ROVs are remotely operated underwater vehicles and AUVs are autonomous underwater vehicles, but they are still under development because they are not yet capable of performing complex tasks. Both ROVs and AUVs allow tasks to be carried out in greater depth, for longer and at a lower cost than those performed by divers. When we are performing underwater tasks that require the visualization of the seabed, the use of acoustic sonar is necessary because the spectrum of light in most situations is unable to reach the seabed, so traditional cameras are not useful.

An acoustic signal or sound pulse, often called a ping, is transmitted in the water by a transducer. If the seafloor or other object is in the track of the sound pulse, sound bounces off object and echoes back to transducer. The time between the emission of the sound pulse and the reception of the echo is what is used to calculate the distance to the object. Some sonar systems also measure the intensity of the echo, this information can be used to infer the features of some of the objects it reflects.

L. M. Bergasa et al. (Eds.): WAF 2020, AISC 1285, pp. 227–240, 2021.
https://doi.org/10.1007/978-3-030-62579-5_16

The acoustic sonar can be divided into three main categories [1]: SingleBeam Echosounder, MultiBeam Echosounder, SideScan Sonar. This paper focuses in the sidescan Sonar which are the preferred tool of choice for creating high-resolution maps of the seabed, they work at very high frequencies and are capable of covering large areas like multibeam echosounders. Unlike multibeam, sidescan sonar measures the strength of echoes instead of depth data. Therefore the scanning face produces black and white images of the seafloor.

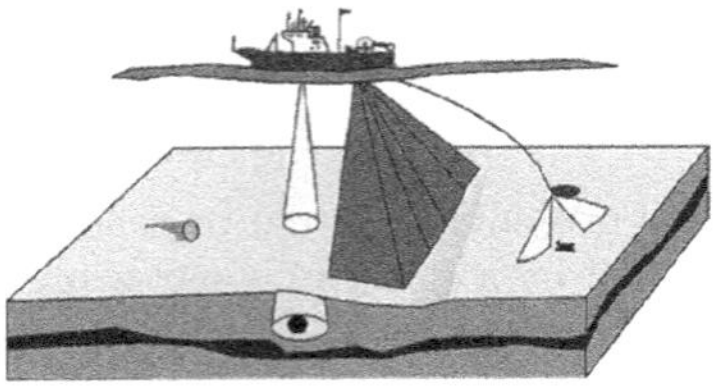

Fig. 1. SingleBeam, MultiBeam and SideScan (from left to right) [4].

The Sidescan sonar is used to map the seabed, so the creation of mosaics is necessary due to the way these devices work, since they are only capable of capturing a line from the seabed as we can see in the Fig. 1. The creation of these mosaics allow us to study the seabed but their usefulness goes beyond the study, since together with an echo sounder it gives us the possibility of creating highly detailed maps of the seabed and using them to locate the ROV within the generated map (SLAM), this being an important step for the construction of autonomous vehicles (AUV) [5,6].

The process of producing the mosaic of the seabed is known as mosaicing, individual sidescan traces from different traverses are pieced together in adjacent positions to form a complete image. Originally this process was done by printing the images on long sheets of paper, and the generation of mosaics was literally cut and pasted. Currently, mosaicing is made up of lines of adjacent pixels [3]. The process is simple, but carrying out the different corrections before and after the mosaicing as the ones proposed in the book [4] increases its complexity. These corrections are normally necessary due to distortions caused by oscillations in the scanning track. The most important thing to obtain a quality mosaic is the track, there are no corrections to solve large distortions caused by a bad track. There are two groups of corrections, radiometric and geometric. On the other hand, we have the problem of overlapping that is normally found in all the mosaic.

Radiometric corrections try to make corrections to the pixel values of an image obtained by the sidescan sonar, to obtain homogeneous intensity values by balancing histograms or correcting distortions present in the pixels. Examples covered by these corrections are: Requantization, Across-track Corrections, Along-track Corrections [4,7].

Geometric corrections try to correct those angular and depth variations of the sonar that degrade the quality of the sidescan sonar images causing effects such as image rescaling or loss of information. Examples covered by these corrections are: Slant-range Correction, Anamorphosis [4].

Overlapping is a very common problem and occurs normally, overlapping is the intersection between the different scanning tracks, a good solution to this problem translates into a high-quality mosaic [4]. In the last decades researchers have been working with methods from the field of artificial vision showing this very good results, the methods that stand out the most are non-symbolic and symbolic methods [2,12].

Non-symbolic Methods are methods based on a low level of information, such as pixel intensity. A non-symbolic method involves establishing a global relationship between the pixels of two images using only their gray levels. Within these methods we can find: Optical Flow, Phase Correlation, Mutual Information, Correlation Ratio [8,9].

Symbolic Methods are primarily based on feature matching and extraction. Features include: shapes, edges, corner detection. Within these methods we can find: SIFT, SURF, ORB [10,11].

Non-symbolic Methods vs Symbolic Methods. Both methods are good in different situations, for example, when there is a comparison of obvious features, the symbolic methods obtain very good results, but when this does not happen, the best are the non-symbolic ones [11].

The main objective of this paper is to present a system capable of generating mosaics of the seabed composed of the readings provided by sidescan sonar incorporated in a low cost ROV, applying to these images the necessary corrections and transformations for the position and pose data provided by the ROV. Furthermore, both the spatial and temporal consistency are sought, that is, a real relationship between the pixels and the distance travelled. In order to achieve these consistencies a positioning system is necessary, which is based on dead reckoning is also discussed in this work.

2 Positioning

We have previously discussed the importance of geocoding the images obtained by the sidescan sonar, for this work the sidescan sonar model BR-ROV from company Deep Vision is incorporated into a ROV model Sibiu Pro from company Nido Robotics. The first step to be able to geocoded the images that we obtain is to have an underwater positioning system, the knowledge of where the ROV is located in such a hostile environment for the propagation of signals such as water is a challenge even today, what it seeks is to obtain the position with the least possible error. The widely used technique is dead reckoning which is used on this work.

Dead reckoning estimates the navigation position from a initial position, inertial systems and velocity. In this work, the initial position is given by the GPS, the gyro of the inertial system is used to obtain the orientation of the ROV.

Both sensors are VectorNav VN-200. We use velocity and time to obtain the distance traveled. The more time passes without obtaining a reference position as the initial one, the greater the positioning error, since the error is cumulative, this error depends on how good the inertial systems are and the way to obtain the velocity [15]. The systems that best provide the velocity accurately are DVLs. DVL (Doppler Velocity Logger) is an acoustic sonar that sends at least three pulses in different directions towards the seabed, thus providing an estimate of the velocity relative to the seabed within an XYZ coordinate system. The main disadvantage is they are very expensive [16], there are other alternatives such as ship log that are widely used by ships, but they have a lower precision than a DVL in especially in vehicles with much greater mobility than that of a ship such as that of an ROV. The ROV used in this work has not a DVL device so other ways are used to get the velocity. In other fields, such as conventional vehicles, attempts are being made to predict or estimate velocity with the use of neural networks.

Neuronal networks are not new techniques from machine learning, but their use has increased in recent years mainly due to the computational increase of computers and the large amount of data that we are able to generate or find. We can find works carried out by researchers who have chosen to develop a neural network model to predict the velocity using time series, the objective in the case of electric vehicles is the optimization of batteries and in conventional vehicles both fuel savings and emissions of particles. In paper [13], an Multilayer Perceptron (MLP) model is carried out using the velocities at each instant of time as input, and the output it returns is the prediction of the velocity at the next instant of time. On the other hand, in thesis [14] it also performs an MLP, but also uses current consumption as input. Both works are oriented to the traffic conditions in the cities, causing constant changes in speed, therefore the objective of these works is to optimize consumption. No papers have been found related to the prediction of velocity in underwater vehicles (ROV or AUV).

The ROV is controlled using a joystick, which produces a control action with values in the range [0..1]. Moreover, depending on the user-defined controller gain in the range [0%..100%], the joystick action produces more or less thrust in the vehicle. In this case, the actual velocity depends not only on the joystick value but also the controller gain. It is important to note that the relationship between the joystick action plus gain and the current velocity is highly nonlinear and unknown. In order to train a neural network, we have collected a large amount of navigation data which includes the joystick values, the current gain and the current velocity. This velocity is estimated with the aid of a Nortek DVL-1000, which is only used during the neural network training and is not available in the production vehicle due to its extremely high cost (higher than the ready-to-sail vehicle itself).

The proposed neural network is an MLP, which given the large amount of data collected and based on the good results obtained by the works [14] and [13] it seems to be a good candidate. The training is carried out with the keras library in R, the length of the dataset is 24796, composed of three variables

<*joystick*, *gain*, *DVLvelocity*>, of which <*joystick*, *gain*> are the input variables and <*DVLvelocity*> the output variables. The neural network is composed of two nodes in the input layer, four hidden layers with 80, 60, 10 and 5 nodes respectively and an output layer with a node that is compared with the output variable which is the velocity of the DVL. The result is quite good, considering the limitations of the approach: the neural network result is not able to determine the sea current, which in some situations may negatively affect the estimations. Figure 2 shows the comparison of the neural network velocity estimation and the DVL velocity readings for a 50% gain.

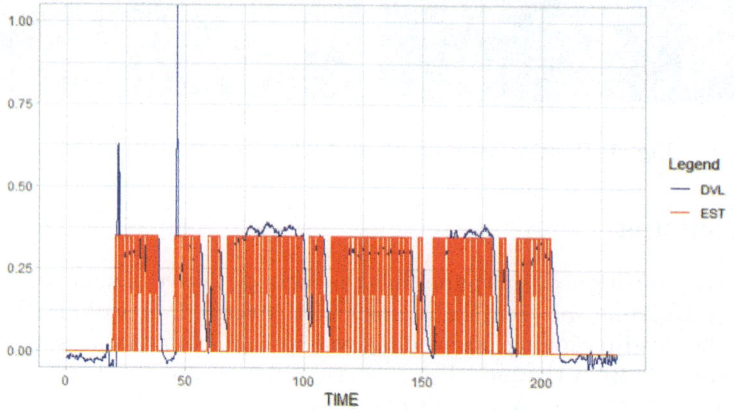

Fig. 2. Velocity estimation comparison: DVL and ANN

In order to validate the accuracy of the proposed method, we have measured the navigation error in different scenarios. Figure 3 shows the results of a navigation test with a duration of 10 min approximately in a worst case scenario, with a severe current along the track which overestimates the velocity in one direction and underestimates it in the other direction. In order to compare the position estimate using the above mentioned ANN velocity, we compare the computed position with the GPS reading when the ROV surfaces, point which is marked as "emersión". The underwater estimated position paths are marked in pink, the GPS position paths are marked in yellow. The DVL based position presents an error of 3.5 m, and the ANN based position an error of 10 m, which is enough for the purposes of such a low cost vehicle.

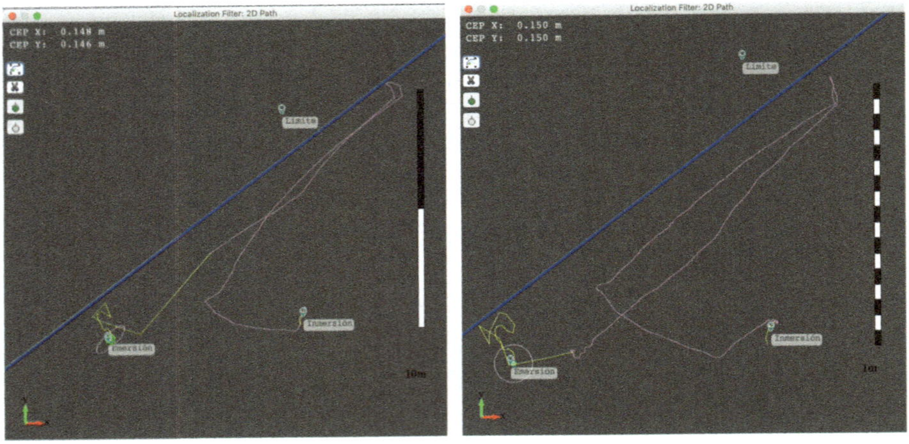

Fig. 3. Position estimation: a) ANN based b) DVL based

3 Mosaicing

As previously mentioned, this paper has the main objective of showing the development of a system for creating mosaics of the seabed, geocoding the images through the dead reckoning positioning system that obtains an estimate of the velocity variable from the model of a neural network previously trained with a dataset generated thanks to the use of a DVL device from another ROV. In this section we will see that process that is necessary for mapping the seabed. In Fig. 4 we can see an illustration of the process of mosaicing of the seabed. In the raw data stage is where the data generated during the mission stage is extracted, once obtained those data are carried out the coordinate transformations, later the corrections are made in the images provided by the sidescan sonar and finally we project the image in the corresponding position calculated, in this way we create a mosaic of the open-loop seabed.

Fig. 4. Process of mosaicing

3.1 Transformations

The position of the ROV is taken through the calculation obtained with dead reckoning, the coordinates obtained are in UTM for its ease of working with them. The transformation from UTM coordinates to pixel coordinates is represented in the Eq. 1 where k is the relation $\frac{Pixel}{Metro}$, in this way we project the information from the side scan sonar within our canvas.

$$Pixel = k \cdot metersUTM \tag{1}$$

The pose of the ROV is taken through the gyrocompass specifically the yaw which is a turn around the z axis of the ROV, it is very important to take this variable into account when generating the mosaic because with it we will be able to reconstruct with the mosaic the turns that have been made during the track (Fig. 5).

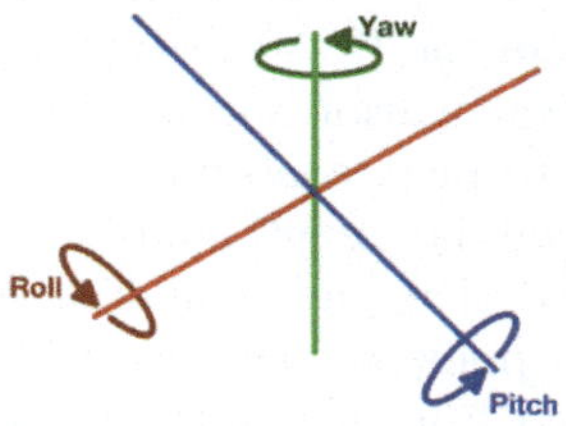

Fig. 5. Pitch, roll and yaw illustration

To be able to transform the coordinates of the sidescan sonar it is necessary to know its geometry, the geometric characteristics that make up the sidescan sonar can be seen in Fig. 6. The distance traveled by the acoustic signal between the transducer and the seabed is called the Slant Range. The Ground Range is the horizontal distance between the vertical of the emitting device and the point of incidence of the sound ray measured on the seabed. The angle formed by the beam direction and the seafloor is called Grazing Angle, while the angle formed by the beam direction and the vertical of the transducer is called elevation angle [1].

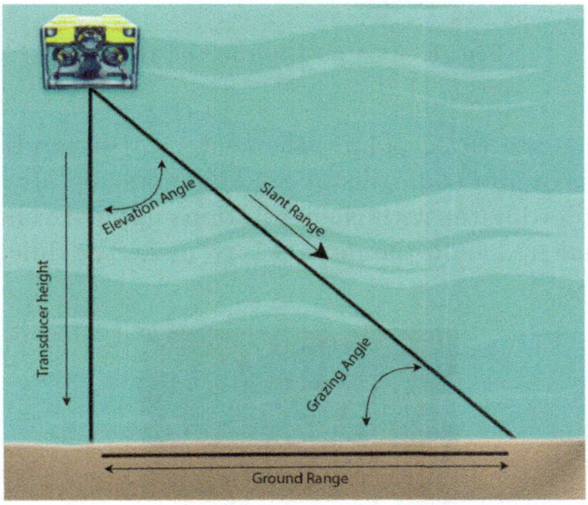

Fig. 6. Beam geometry.

Normally, sonar devices return an array of received echo strength in a range between 0 and 255. The length of the array is related to the distance at which the sonar operates. The farthest slant-range from the ROV is the one that we will use to project the coordinates of the sonar within the mosaic.

The pixel coordinates of the sidescan sonar are obtained adding and subtracting the distance of the sonar from the UTM coordinates obtained from the ROV. This calculation must only be done on the y axis of the UTM coordinates as shown in Fig. 7 the y axis of the ROV corresponds to the beams of the sidescan sonar. After doing this calculation we use Eq. 1 that we saw previously to perform the transformation to pixel coordinates. These coordinates along with the coordinates of the ROV will be essential for later projecting the sonar image.

Previously we commented that yaw would serve to apply the rotations in the images obtained, at this point is where we will apply those rotations. The coordinates we have obtained from where the sonar is located is the one that will rotate around the ROV using the yaw, we can see Eqs. 2 and 3 that are used for the application of this rotation.

$$newX_{pixel} = cos \cdot (x - ROV_x) - sin \cdot (y - ROV_y) + ROV_x \tag{2}$$

$$newY_{pixel} = sin \cdot (x - ROV_x) + cos \cdot (y - ROV_y) + ROV_y \tag{3}$$

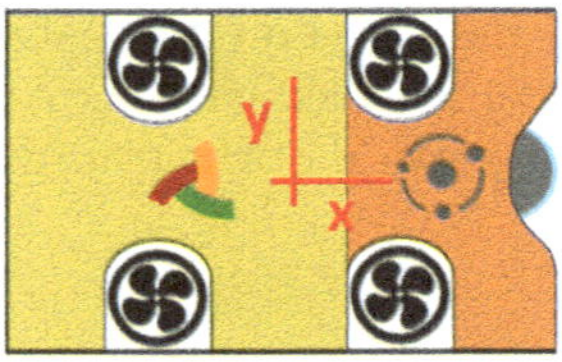

Fig. 7. Axis position in the ROV.

In Fig. 8, we can see three points, the center corresponds to the ROV and at the ends the maximum sonar distance. Therefore, we already have mapped all the necessary coordinates with their rotations applied. The next step before projecting the sonar image on the mosaic is to make the different corrections.

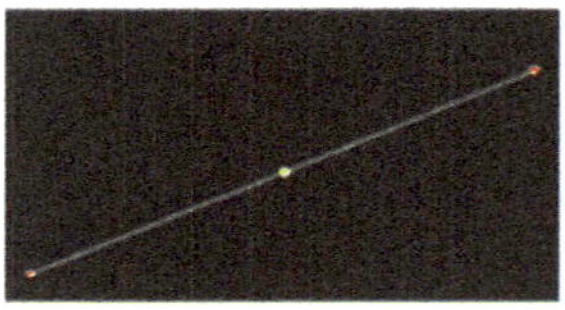

Fig. 8. Mapping of ROV and Sonar.

3.2 Corrections

Radiometric corrections aim to visually improve the images obtained with the sidescan sonar from the level transformation based on the histograms. Two radiometric corrections are proposed in this paper: Histogram equalization and Remove water column.

Histogram equalization seeks a balance between the different values of the pixels that make up the image, thus trying to improve the contrast, for this it uses the probability distribution. In this work a CLAHE filter is applied that limits the amplification of the contrast in order to avoid generating noise in the image [7,17].

Water column is a completely blind area, full of noise that does not contain any useful information, so it is usually chosen to eliminate it. The elimination of this blind zone can be done in the geometric slant-range correction or in the radiometric correction using intensity intervals [18].

Geometric corrections aim is to try to recover that information that has been affected distorted by angular and depth variations. Two geometric corrections are proposed in this paper: Slant-range corrections and anamorphosis.

Slant-range correction try to rescale the image to its original state. To achieve this correction, it is necessary to calculate the distance between the ROV with the bottom and the distance of the beam with the bottom, in this way we can find the distance between D and D_1 as we see in the Fig. 9 extracted from the paper [19].

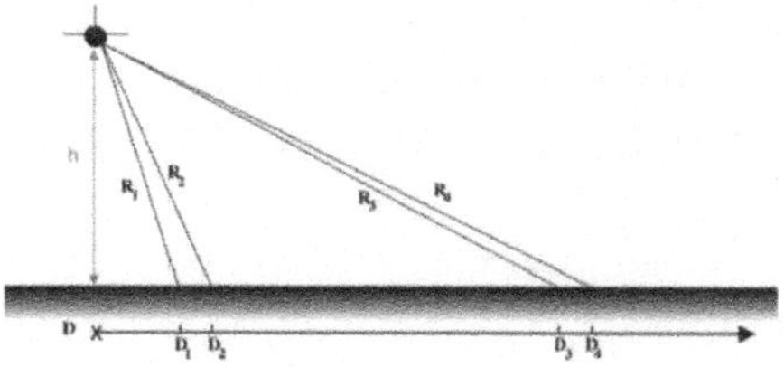

Fig. 9. Slant-range correction [4]

During the scanning, the velocity increases cause the different lines of the scanning to be very separated from each other or very close in the case of a reduced velocity, anamorphosis try to solve this, in this work, the euclidean distance is used and the given image is rescaled Eq. 4, where k represents the relationship $\frac{Pixel}{Metro}$, this equation provides us the width that the scanned line should have within our mosaic. In this work when there is a high speed of rotation it is avoided to make this correction [4].

$$height = distance \cdot k \tag{4}$$

3.3 Updating the Mosaic

At this point we are already at the end of the process and we have prepared: ROV coordinates, sonar coordinates, radiometric corrections and geometric corrections, all that remains is to update the different lines that make up the mosaic, so we will project the pixel linea obtained by sidescan sonar into the mosaic.

The Bresenham algorithm is a method for fast and efficient drawing of lines having two points as input, as we have previously commented, we have the pixel coordinates of the sidescan sonar and those of the ROV. Bresenham returns all points between points passed as input, in this case, the starting point is the ROV and the end point is the sonar. In the Fig. 10 we can see an illustration of the Bresenham algorithm. The drawback of this algorithm is aliasing, there are other algorithms that try to solve this, such as Xiaolin Wu's algorithm [20].

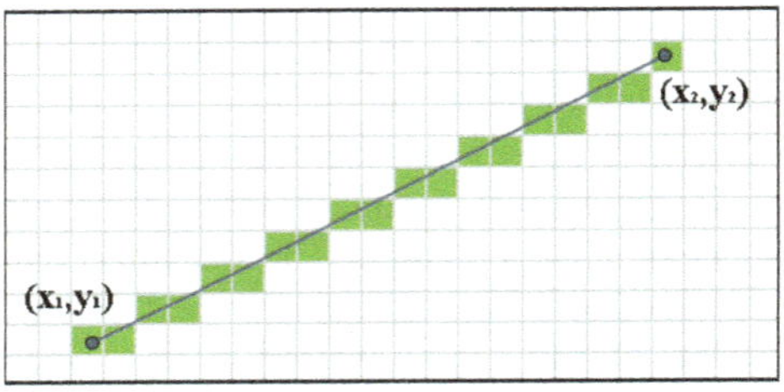

Fig. 10. Line plot

This completes the process of creating mosaics of the seabed with sidescan sonar images applying the different corrections and transformations.

4 Results

This section shows real results carried out in a mission located in the thermal power plant Central Térmica Litoral de Almería, Endesa. In this mission, the positioning system relies only on the estimation of the velocity provided by the neural network. We have selected two tracks, namely A and B, in Fig. 11 we can see the path that the ROV follows during the mission, which incorporate the detection of objects found at seabed with a minimum of quality. Figure 12 shows both tracks with relevant detected objects labelled with colored squares. Blue squares correspond to amphoras, green squares correspond to pipelines, and brown squares correspond to seabed sediments.

In order to qualitatively analyze the results, Figs. 13 and 14 show the comparison of the raw sidescan sonar images and the constructed sonar mosaics. It can be noticed the distortion in the raw images, which makes much more difficult the identification of the objects.

We are able to identify the different objects on the seabed because on this occasion for this mission, the water was clear enough so the sun's rays penetrated

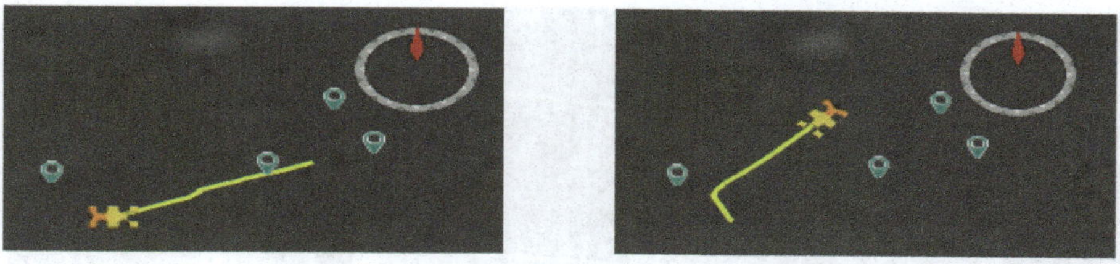

Fig. 11. Selected tracks, A and B

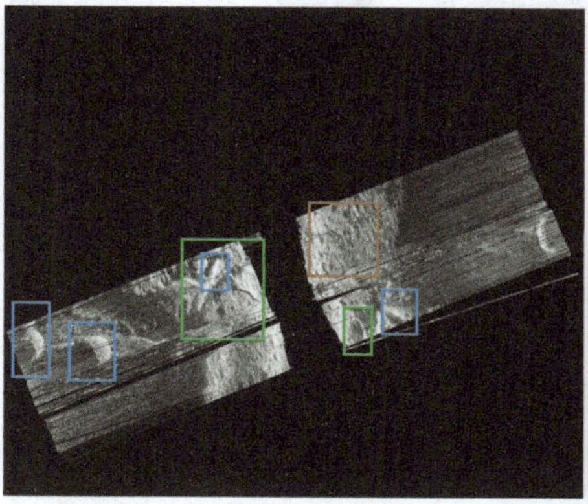

Track A mosaic.

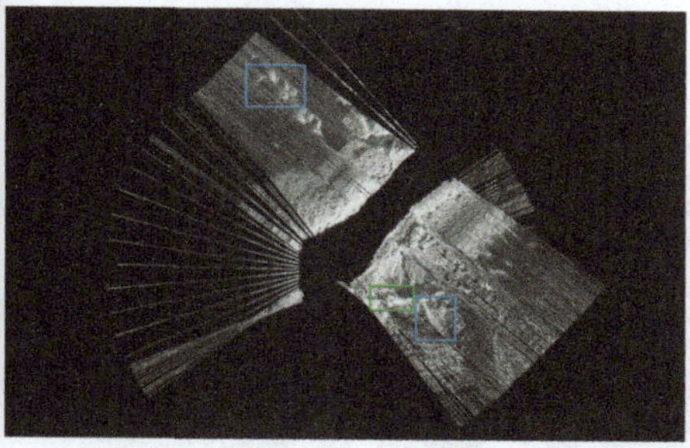

Track B mosaic.

Fig. 12. Sample mosaics with underwater objects

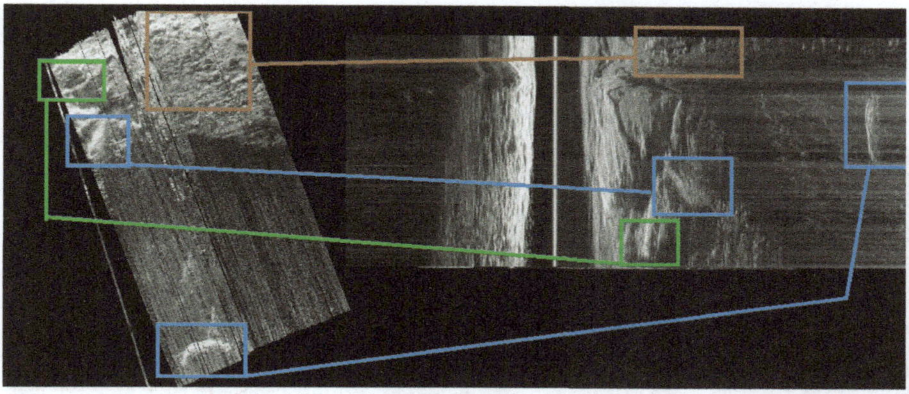

Fig. 13. Mosaic vs raw image details. Track A.

to the seabed. In the Figs. 15 and 16 we can see images that help us to check what we are seeing with the sonar images. These images are provided by the ROV camera, in this way, as we have said before, we can check if the objects we are seeing are amphoras or pipes.

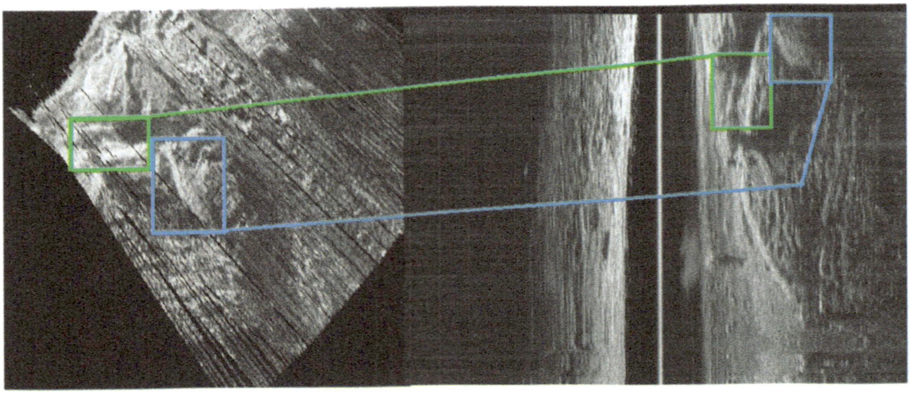

Fig. 14. Mosaic vs raw image details. Track B.

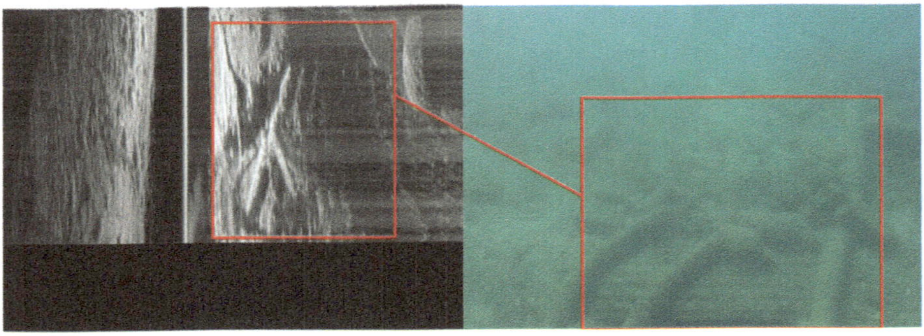

Fig. 15. Image from ROVs camera, track A.

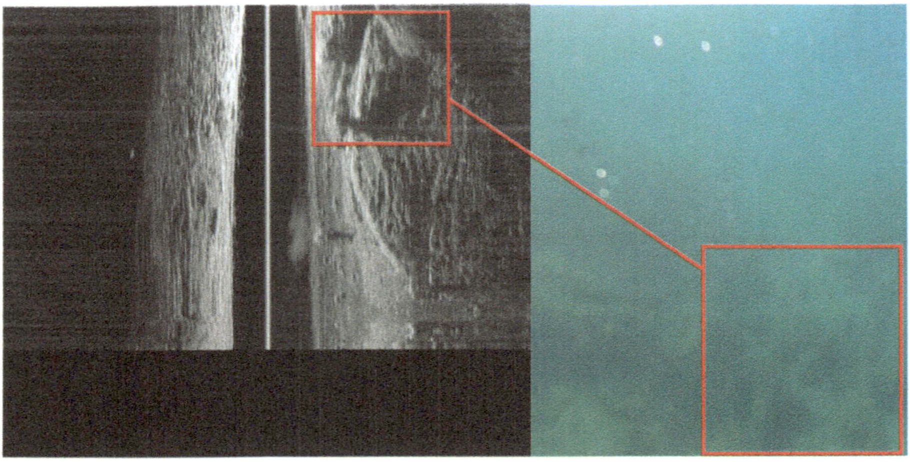

Fig. 16. Image from ROVs camera, track B.

5 Conclusions and Future Work

The development of this paper has been to show the construction of a system capable of generating mosaics of the seabed with the images obtained by a sidescan sonar, applying to these images the necessary corrections and transformations depending on the position and pose of the ROV. This mosaic system shown is an open-loop system, but as we previously discussed is the preliminary process for a closed-loop mosaic.

On the other hand, a neural network capable of estimating velocity has been presented with good results, although the track shown presented adverse conditions for the use of this ANN.

This paper leaves open the possibility of solving overlapping using non-symbolic and symbolic methods, thus achieving a closed-loop mosaic. In addition, the development of a system that is capable of detecting objects such as amphoras from the images obtained by a sidescan sonar for future research. Finally, the development of a neural network that takes the gain at an instant of time T as input, thus improving the network developed in this paper, since a good way to estimate of the velocity is very important when we are working with dead reckoning.

References

1. León Martín, J.D.: Del Escandallo a los Sondadores Multihaz. Escuela de Hidrografía, Armada Española (2004)
2. Zitová, B., Flusser, J.: Image registration methods. In: Todorovic S. (ed.) Image and Vision Computing, pp. 977–1000 (2003). https://doi.org/10.1016/S0262-8856(03)00137-9
3. Bennell, J.D.: Mosaicing of sidescan sonar images to map seabed features. In: Davies, J. (ed.) Marine Monitoring Handbook, pp. 1–5 (2001)
4. Blondel, P.: The Handbook of Sidescan Sonar (2009)
5. Palomer, A., Ridao, P., Ribas, D.: Multibeam 3D underwater SLAM with probabilistic registration. Sensors **16**, 560 (2016). https://doi.org/10.3390/s16040560
6. Rahman, S., Quattrini Li, A., Rekleitis, I.: Sonar visual inertial SLAM of underwater structures. In: IEEE International Conference on Robotics and Automation (ICRA), pp. 1–7 (2018). https://doi.org/10.1109/ICRA.2018.8460545
7. Zhao, J., Yan, J., Zhang, Z., Meng, J.: A new radiometric correction method for side-scan sonar images in consideration of seabed sediment variation. Remote Sens. **9**(6), 575 (2017). https://doi.org/10.3390/rs9060575
8. Barron, J.L., Beauchemin, S.S.: The computation of optical flow. ACM Comput. Surv. **35** (1995). https://doi.org/10.1145/212094.212141
9. Chailloux, C., Zerr, B.: Non symbolic methods to register sonar images. Oceans 276–281 (2005). https://doi.org/10.1109/OCEANSE.2005.1511725
10. Vandrish, P., Vardy, A., Walker, D., Dobre, A.O.: Side-scan sonar image registration for AUV navigation. In: Underwater Technology (UT), 2011 IEEE Symposium on and 2011 Workshop on Scientific Use of Submarine Cables and Related Technologies (SSC), pp. 1–7. IEEE Xplore (2011). https://doi.org/10.1109/UT.2011.5774096

11. Zhao, J., Wang, A., Zhang, H., Wang, X.: Mosaic method of side-scan sonar strip images using corresponding features. IET Image Process 616–623 (2013). https://doi.org/10.1049/iet-ipr.2012.0468
12. Chailloux, C., Le Caillec, J.M., Gueriot, D., Zerr, B.: Intensity-based block matching algorithm for mosaicing sonar images. IEEE J. Oceanic Eng. **36**(4), 627–645. https://doi.org/10.1109/JOE.2011.2141850
13. Fotouhi, A., Montazeri, M., Jannatipour, M.: Vehicle's velocity time series prediction using neural network. Systems Simulation and Control Laboratory, Iran (2011)
14. Thorsell, E.: Vehicle speed-profile prediction without spatial information. Chalmers University of Technology, Sweden (2013)
15. Karras, G.C., Kyriakopoulos, K.J.: Localization of an underwater vehicle using an IMU and a laser-based vision system. In: Mediterranean Conference on Control Automation, Athens, pp. 1–6 (2007). https://doi.org/10.1109/MED.2007.4433777
16. Zhao, B., Blanke, M., Skjetne, R.: Particle filter ROV navigation using hydroacoustic position and speed log measurements. In: American Control Conference (ACC), Montreal, pp. 6209–6215 (2012). https://doi.org/10.1109/ACC.2012.6315511
17. Ye, X., Yang, H., Li, C., Jia, Y., Li, P.: A gray scale correction method for side-scan sonar images based on Retinex. Remote Sens. (2019). https://doi.org/10.3390/rs11111281
18. Burguera, A., Oliver, G.: High-resolution underwater mapping using side-scan sonar. PLOS One 1–41 (2016). https://doi.org/10.1371/journal.pone.0146396
19. Gautier, S., Morissette, G.: Unsupervised extraction of underwater regions of interest in sidescan sonar imagery. J. Ocean Technol. **15**, 1–15 (2019)
20. Wu, X.: An efficient antialiasing technique. In: ACM SIGGRAPH Computer Graphics, pp. 143–152 (1991). https://doi.org/10.1145/127719.122734

360° Real-Time 3D Multi-object Detection and Tracking for Autonomous Vehicle Navigation

Javier Del Egido(✉), Carlos Gómez-Huélamo, Luis M. Bergasa, Rafael Barea, Elena López-Guillén, Javier Araluce, Rodrigo Gutiérrez, and Miguel Antunes

Electronics Department,
University of Alcalá (UAH), Alcalá de Henares, Spain
cram3r95@gmail.com,
{javier.egido,luism.bergasa,rafael.barea,elena.lopezg,
javier.araluce,miguel.antunes}@uah.es, res.rodry@gmail.com
https://www.robesafe.uah.es/

Abstract. This paper presents a real-time 3D Multi Object Detection and Tracking (DAMOT) method proposed for the UAH autonomous electric car. It allows the vehicle to recognize 360° surrounding objects and uniquely identify them to be able to follow their trajectory in scene by only receiving a 3D point cloud through ROS framework. First, we describe our proposal of 3D object detector, based on PointPillars [11], processing LiDAR data to locate objects in space obtaining their dimensions and location. Secondly, we use BEV-MOT [7], our Multi-Object Tracking technique in order to uniquely identify each object over a Bird's-Eye View (BEV) through a combination of 2D Kalman filter and Hungarian algorithm, allowing the ego-vehicle to follow surrounding objects trajectories. A comparison of the performance of our proposal with other state-of-the-art methods is carried out applying KITTI-3DMOT evaluation tool extracted from AB3DMOT [21] on KITTI [5] validation dataset. Finally, we validate our DAMOT in several traffic scenarios implemented in CARLA [4] open-source driving simulator by using AB4COGT tool, designed by authors, studying its performance in a controlled but realistic urban environment with real-time execution, providing several demonstration videos (https://cutt.ly/3rU113d).

Keywords: Real-time · CARLA · LiDAR · 3D Multi-object Tracking · ROS · DAMOT · Autonomous navigation

1 Introduction

Autonomous driving systems have to perform safe driving behaviours following conventional traffic rules to achieve a programmed destination. When travelling through this route many unforeseen objects may force the controller to react

L. M. Bergasa et al. (Eds.): WAF 2020, AISC 1285, pp. 241–255, 2021.
https://doi.org/10.1007/978-3-030-62579-5_17

to avoid fatalities. The reliability of the crash-avoidance system relies on the performance of the environment detector and its ability to predict future situations. Detection and Multi-object Tracking (DAMOT) methods analyze an object since its first detection to determine its future position, allowing the car to act in consequence to avoid critical situations. The increase in object detection performance has allowed to advance on MOT techniques, improving its accuracy but carrying a high computational cost, making prohibitive its use in real-time systems. Many different technologies have been designed to accomplish an optimal environment detector following different approaches, based on different benchmarks such as KITTI [5], which provides manual labeled data from urban scenes taken from different cameras and LiDAR mounted on a vehicle.

Our proposed method applies PointPillars [11], a state-of-the-art 3D object detector network with a striking performance, achieving accurate predictions at a high frame rate taking advantage on GPU and 2D convolutions to process a 3D point cloud.

Our tracking module is formed by our BEV-MOT proposal [7]. It works over a BEV and is composed by a 2D Kalman filter to predict vehicles position between detection frames and Hungarian algorithm in order to associate predictions and detections, running traditional tracking techniques at high speed without sacrificing precision.

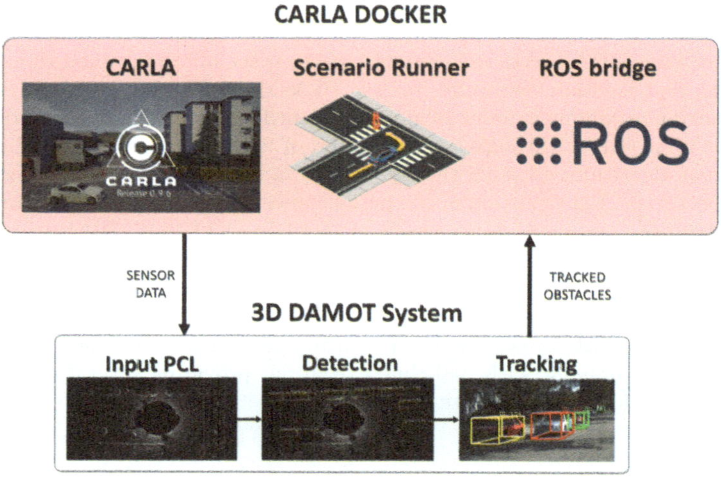

Fig. 1. Architecture of our MOT proposal simulated in CARLA.

We propose to take advantage of the good processing ratios of PointPillars 3D objects detector and traditional tracking techniques to implement a real-time 3D DAMOT able to be used in challenging urban traffic scenarios. The architecture of our proposal integrated to CARLA driving simulator is shown in Fig. 1, which receives a point cloud from ego_vehicle and returns the tracked objects around it at its current and future positions.

2 Related Works

2.1 Vision-Only 3D Object Detection

3D scene understanding through RGB or RGB-D images is possible by processing the information using Convolutional Neural Networks followed by applying complex projection transforms as in [9, 14]. GPUs may be used to reduce computational cost, getting well processing time values. Minor pixel mistakes in image detection may produce meaningful differences between estimated and reality when the object is at a relevant distance, and detecting the object orientation and depth is a difficult task.

2.2 3D LiDAR Object Detection

3D LiDAR sensors provide accurate 3D information, avoiding precision problems when positioning the objects in the three-dimensional space. In contrast to vision-only detection, point clouds provided by LiDAR sensors are independent from adverse weather or light conditions. LiDAR information must be precisely enough to recognize the surroundings in the point cloud but it may not exceed the computational limits due to a excessive detail. Different ways has been explored to develop object recognition using 3D information, such as LiDAR clustering, reducing the complexity to 2D BEV [24], dividing the scene into 3D voxels to apply 2D CNNs as PointPillars [11] or 3D CNNs [12, 25], or processing the whole point cloud [19].

2.3 2D Multi-object Tracking

2D tracking may be carried out on BEV or image frame. Despite of useful results achieved by batch MOT works like [18], future information is not available when applying tracking to real-time systems such as autonomous vehicles. Otherwise, online 2D MOT traditional techniques make data association using Hungarian Algorithmscon [3, 10], and more recently Convolutional Networks to extract object characteristics to avoid identity switching when crossing trajectories [23]. Bird's eye view perspective prevents tracking system to experience object occlusions, allowing for a better scene comprehension.

2.4 3D Multi-object Tracking

3D MOT allows the system to estimate real velocity, undistorted by bidimensional transformations, being able to perform predictions based on linear and angular velocities and precise location and size information. The simplest but efficient way to create 3D MOT is to extend 2D Kalman filter by adding a third dimension such as in AB3DMOT [21]. Other works apply simple 2D Kalman filter over BEV and recover the height using different strategies [7, 13].

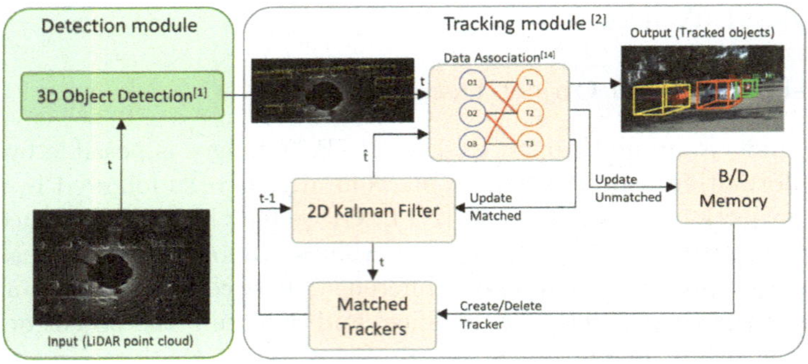

Fig. 2. 3D DAMOT proposed architecture.

3 Architecture

Our 3D DAMOT system is featured by a modular architecture where first of all a 360° object detector by using a LiDAR sensor is applied. Following, a Kalman Filter and Hungarian Algorithm based tracking approach over the BEV detected objects is carried out. The Fig. 2 outlines the pipeline of our system. The input of the 3D DAMOT consist of a LiDAR point cloud which feeds PointPillars [11] neural network, giving rise to the 3D bounding boxes of the most relevant obstacles of the scene detected in the 360° surrounding of the vehicle. These detections are projected to the road plane (BEV) and analyzed along with our BEV-MOT [7], composed by a 2D Kalman filter, which predicts positions to associate the detected items to previous trackers. Matched trackers update Kalman filter while unmatched objects are processed by birth and death (B/D) module to create new trackers for unmatched objects and delete old trackers for the unmatched ones. All trackers make up the Kalman filter, that estimates the future position of the detected objects to restart the cycle.

3.1 3D Object Detector

360° surrounding objects are detected processing LiDAR point cloud data with PointPillars [11] network. This end-to-end network analyzes the point cloud by dividing it into vertical columns to process the data as a 2D pseudo-image to take advantage of highly efficient 2D convolutions on GPU, avoiding hand tuning parameters by learning by itself instead. It is trained using KITTI [5] tracking dataset which provides 3D hand-labeled information of vehicles, pedestrians and cyclists in urban environments.

In order to increase the performance of the object detector, several network architectures have been trained for PointPillars, showing their results on the ablation study carried out in Sect. 4.2. Our 3D object detector is able to inference at 50 Hz when running on a 1080 Ti GPU.

3.2 Multi-object Tracking

Detected objects in a BEV have to be processed by our Multi-Object Tracking module (BEV-MOT) in order to assign a unique identifier to distinguish it along the covered path. Analyzing the objects contained in a scene is a difficult task and has to be accomplished by dividing it into several modules, as seen in Fig. 2.

2D Kalman Filter. By using the constant velocity model, the future position of each tracked object is predicted in the BEV. Kalman filter is updated when present frame objects are matched with current trackers or new ones are created at each step. It also corrects the detected orientation to match it to the assumed previously to avoid a 3D IoU reduction produced by flipping 180° the orientation of the object. To get 3D IoU, height is recovered from the detection module. Different future positions are predicted to calculate possible collisions.

Hungarian Algorithm for Data Association. Detected objects in current frame and predicted positions from past frame are matched applying the Hungarian algorithm according to 3D IoU, recovering height information from 3D object detector, returning matched and unmatched objects and trajectories to be managed by 2D Kalman filter and Birth and Death memory respectively.

Birth and Death Memory. B/D module handles unmatched trackers and objects. When an unmatched object is detected several times (because it newly appears on scene or it cannot be associated with a previously detected one) exceeding F_{min} threshold a new tracker is created. On the other hand, an existing tracker is deleted (due to the object has abandoned the scene or it is associated with a new one) when passed Age_{max} frames.

Matched Trackers. Trackers module manages all current trackers to correctly configure the Kalman filter according to the objects detected in the previous frame.

4 Evaluation

A previous performance evaluation has to be accomplished before to implement a system in a real prototype. To carry out this evaluation on 3D DAMOT metrics two different validation ways have been followed: KITTI real and CARLA simulated scenes.

4.1 3D DAMOT Evaluation Metrics

Traditionally DAMOT evaluation tools study the working behavior on bi-dimensional plane, designed for 2D DAMOT on images. In order to know the performance of our architecture we applied 3D DAMOT benchmark developed by AB3DMOT [21] to analyze 3D IoU from detected and tracked objects with KITTI and CARLA groundtruth. This tool applies traditional and integral metrics described following.

Traditional DAMOT Metrics. Mainstream metrics applied to multi-object tracking systems are extracted from CLEAR [2] DAMOT metrics like MOTA (accuracy), MOTP (precision), Mostly-tracked trajectories, False-Negatives (FN), False-Positives (FP), Precision, F1 score, Identity Switching (IDS), Fragmentations (FRAG, trajectory interruptions by false positives), and moreover. This metrics analyze the performance of the DAMOT system at a specific score threshold, not taking into account the confidence provided by the object detector and possibly misunderstanding the capability of the method.

MOTP measures the precision by analyzing the total error in estimated position for each object and its ground-truth.

$$MOTP = \frac{d}{num_{gt}} \tag{1}$$

MOTA considers all error made by the system, false positives, false negatives and identity switching.

$$MOTA = 1 - \frac{FP + FN + IDS}{num_{gt}} \tag{2}$$

where num_{gt} is the number of ground truth objects in all frames.

Integral DAMOT Metrics. To solve the problem presented by traditional DAMOT metrics, AB3DMOT [21] propose two integral metrics, evaluating the MOTA and MOTP of the tracking system at every score threshold.

$$AMOTA = \frac{1}{L} \sum_{\{\frac{1}{L}, \frac{2}{L}, \ldots, 1\}} (1 - \frac{FP + FN + IDS}{num_{gt}}) \tag{3}$$

where L is the number of different recall values. FP, FN and IDS in AMOTA are modified for each threshold value.

Similarly, AMOTP will be estimated by integrating MOTP along all recall values.

4.2 Object Detector Evaluation in KITTI

Firstly, we modify the 3D object detector by modifying both PointPillars [11] architecture and training steps and keeping the rest of the DAMOT architecture unchanged. All training phases are accomplished by using KITTI [5] database.

Table 1. 3D DAMOT evaluation tool default parameters

Parameter	Value
Age_{max}	3
F_{min}	1
IoU	0.1

In order to compare the results obtained with KITTI-3DMOT evaluation tool we show the performance accomplished by FANTrack [1], which is given in [21]. 3D DAMOT Evaluation parameters are shown in Table 1. To evaluate PointPillars [11] architectures we extracted them from Second. Pytorch GitHub, testing their results for given network weights, which are displayed as PP_lite and PP_onestage, only improving MOTP results.

With the aim of enhance the performance we trained the PP architecture during both 593,920 and 1,187,840 steps by using KITTI [5] tracking dataset for car object class. Table 2 shows the results obtained by the explained detectors, writing the best result in black and the second in blue, using BEV-MOT [7] as tracking module. It can be observed that PointPillars [11] architecture after 1,187,840 steps (PP Full steps) has the second best performance in 4 of 8 evaluated parameters, improving the overall performance of other PP configurations but no overcoming the [1] output. However, PP configurations run at least two times faster, which is a critical factor in a real-time implementation.

Table 2. Ablation study replacing 3D Object Detector in KITTI-3DMOT evaluation tool in KITTI tracking validation dataset with BEV-MOT [7] tracking.

Method	AMOTA	AMOTP	MOTA	MOTP	IDS	FRAG	FP	FN
[22] Detector	**31.37**	**64.29**	**62.38**	68.26	**1**	**24**	**1215**	**1894**
PP Mid steps	26.14	63.41	57.54	75.12	122	182	4746	**5353**
PP_lite	9.27	56.69	21.54	**79.19**	31	**40**	5635	13219
PP_onestage	10.25	51.51	23.68	**77.78**	**12**	992	7881	10478
PP Full steps	**27.30**	**63.45**	**59.26**	75.24	131	186	**4310**	5364

4.3 CARLA: A Realistic Urban Environment Simulator

Despite impressive efforts made by AB3DMOT [21], where a tool for evaluating 3D DAMOT systems directly in 3D space is designed, it is based on KITTI dataset [5], which provides prerecorded sequences over which the user cannot interact with the environment. Moreover, these sequences are usually based on common driving scenarios, such as a daily quiet street or a highway in which no challenging traffic scenarios as pedestrian crossing, give way, *etc.* takes place.

Regarding levels of automation, neither any research nor industry organization has demonstrated a ratified testing methodology for L4/L5 (being identified the level 5 with a fully-autonomous navigation architecture, according to J3016 SAE document [20]) autonomous vehicles. The reason is quite simple: even though some regulations have been defined for these L4/L5 levels, simulation is a critical aspect to build safe autonomous vehicles. Nevertheless, in spite of the fact that current automotive companies are very good at testing the individual components of the navigation architecture, these tests are not powerful enough to validate a fully-autonomous navigation architecture on the road, so there is a need to figure out how to test intelligent vehicles full of advanced sensors and sharing information among them [17].

In terms of 3D Multi-Object Detection and Tracking, the answer is quite similar. Since the urban environment is highly complex, the whole architecture and particularly the 3D DAMOT system must be tested in countless traffic scenarios and environments, which would escalate the development time and cost exponentially with the physical approach, either testing at the real-wold or waiting for using new sequences of KITTI (recorded by a physical system), not studying the global advantages and drawbacks of the DAMOT system. For that reason, virtual testing (simulation) and an appropriate design of the traffic scenarios are the keys to build robust and safe autonomous vehicles in the future, as shown in [6]. Since the proposed 3D DAMOT architecture of this work is open-source, we decided to validate the ability to detect and track the most relevant objects around the vehicle in CARLA [4], an open-source hyper-realistic autonomous driving simulator that offers an outstanding environment in terms of perception, flexibility, traffic situations and real-time, which are key concepts for our system.

CARLA is an open-source autonomous driving simulator implemented as a layer over Unreal Engine 4 (UE4) [16]. This simulation engine provides to CARLA an ecosystem of interoperable plugins, a realistic physics and a state-of-the-art image quality. CARLA is designed as a server-client system so as to support this functionality provided by UE4, where the simulation is rendered and run by the server. The environment is composed of 3D models of static objects, such as buildings, infrastructure or vegetation, as well as dynamic objects like pedestrians, cyclists or vehicles. These objects are designed using low-weight geometric textures and models though maintaining visual realism by making use of variable level of detail and carefully crafting the materials. Moreover, one of the main advantages when using CARLA is the possibility to modify in an easy way the vehicle on-board sensors and their features in order to obtain accurate data, the weather and even the possibility to create realistic traffic scenarios as in Fig. 3.

In order to obtain the point cloud required by our 3D object detector, we use the CARLA ROS bridge, associated to the CARLA simulator. The CARLA ROS bridge is a ROS package that aims at providing a bridge between CARLA and ROS (Robot Operating System [15]), sending the information captured by the on-board sensors and other variables of interest to the vehicle in the form of parameters and topics understood by ROS. In this paper we used the 0.9.9 version of CARLA in such a way the ROS bridge was configured according to this version. In terms of the sensors perspective, the agent sensor suite can be modified in a flexible way. Most common sensors in CARLA world are GPS, RGB cameras and LiDAR (in addition to their corresponding pseudo-sensors, such us the semantic segmentation and the ground-truth associated to the RGB information of a camera). Since in this work our object detector is only LiDAR based, we configure the sensor as shown in Table 3. Based on the bridge, the 3D point cloud captured by the LiDAR is published in ROS format as PointCloud2, with the X axe inwards, Y left and Z pointing up. Also an RGB camera sensor

Fig. 3. CARLA driving simulator

is defined backwards the car to show the performance of the 360° 3D DAMOT system to detect and track the detected objects all around the vehicle as seen in Fig. 8b, but not interfering with the operation of our system.

Table 3. LiDAR configuration in CARLA simulator

Parameter	Value
X (m)	0.0
Y (m)	0.0
Z (m)	2.5
Points per second	500.000
Upper FoV (°)	2.0
Lower FoV (°)	−15.0
Rotation frequency (Hz)	20

As shown, CARLA provides a straightforward way to add or remove sensors from the vehicle or even modify their parameters, to adjust the simulation to the real-world as best as possible.

One of the best advantages of CARLA is the possibility to create ad-hoc urban layouts, helpful to validate the 3D DAMOT system under different traffic and weather conditions. CARLA Scenario Runner module can be downloaded from CARLA GitHub, obtaining an execution engine for CARLA and traffic scenario definition. These scenarios can be modified by editing an OpenSCENARIO [8] script definition where town, vehicles, climate conditions and also driving behaviours are defined. Figure 3 depicts an scenario on Town10 with the ego_vehicle and a predefined route, showing waypoints on a curved street to achieve the destination point.

As previously explained, DAMOT evaluation in KITTI is mainly carried out along daily streets where many cars are parked on the road, so it mostly evaluate the system performance to track static vehicles when the main difficult is found in dynamic obstacles, such as pedestrians, vans, trucks or cars. To achieve more challenging situations we validated our DAMOT method on different CARLA simulated scenarios.

4.4 DAMOT Evaluation in CARLA

Figure 4 illustrates the followed process, validating both DAMOT architectures based on our BEV-MOT [7] and AB3DMOT [21] (our baseline) on CARLA simulator using AB4COGT, a specific tool designed by authors to extract CARLA objects information used as ground-truth on KITTI-3DMOT evaluation tool developed by AB3DMOT.

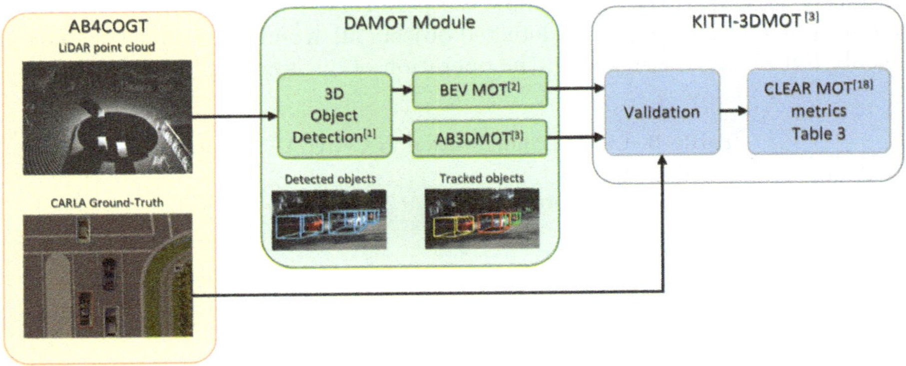

Fig. 4. DAMOT validation through AB4COGT and KITTI-3DMOT

To demonstrate the potential of validation on simulated environments by using AB4COGT, several scenarios are defined in CARLA driving simulator. The ego_vehicle used in all scenarios mounts a top LiDAR configured as described in Table 3. The scenes are saved using AB4COGT tool, recording the object when it is around the ego-vehicle and within a 50 m radius, storing both ground-truth labels and LiDAR point clouds to ensure synchronization. Then, point clouds are reproduced as ROS message to stimulate PointPillars [11] input, detecting the object in the scene. Tracking is then applied to obtain CLEAR MOT [2] metrics.

Scenario 1. Parked Aside Vehicles. In this scenario we replicated a KITTI-like scene to demonstrate the similarity between simulated and real scenes. The autonomous ego_vehicle travels along a street where several vehicles are parked aside, reproducing an environment similar to KITTI traffic scenes, without difficulties to complete the planned path.

Figure 5 shows the disposal of the objects in scene, displaying a BEV of CARLA simulator scene, a BEV of detected and tracked vehicles, and a perspective view of LiDAR point cloud and tracked objects to verify correspondence.

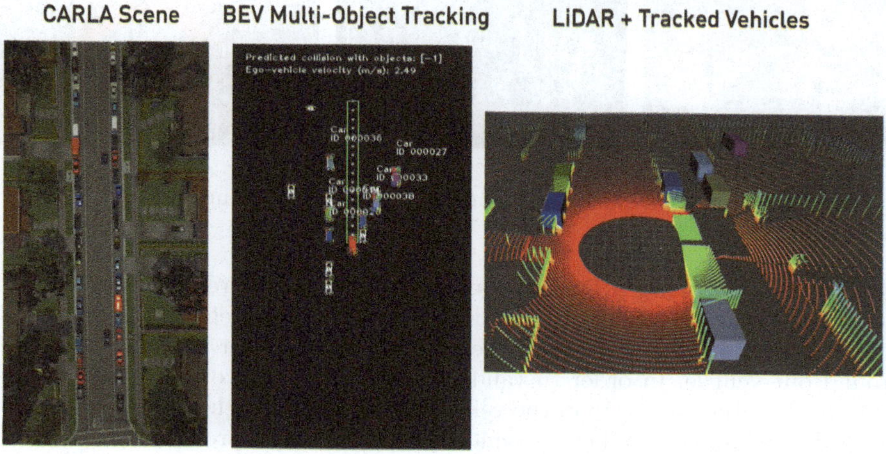

Fig. 5. Scenario 1. Parked aside vehicles.

It can be observed in Fig. 5 that CARLA provided LiDAR point cloud is not accurate enough to vehicles shape, decreasing the ability of PointPillars to detect cars due to its training on real scenarios, with accurate vehicles shape. A similar scene provided by KITTI is displayed in Fig. 6. In this case, a real LiDAR point cloud is given to PointPillars input, being able to detect much more vehicles and consequently producing a better and more stable objects tracking.

(a) KITTI point cloud. (b) Detected and tracked vehicles.

Fig. 6. KITTI scene DAMOT and point cloud.

Scenario 2. Stopped at Junction. The second scenario takes place at a simulated traffic-light regulated junction, where the ego_vehicle detects and tracks moving vehicles, introducing detection and tracking on dynamic objects. It can be observed on Fig. 7 that, despite of time delays shown on tracked vehicle markers, the vehicles are correctly detected and uniquely identified, showing a fuzzy trail on predicted future positions.

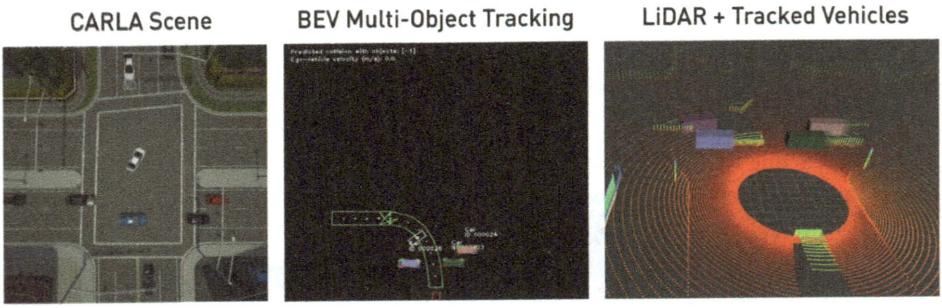

Fig. 7. Scenario 2. Stopped at traffic-regulated junction.

Scenario 3. Entering a Roundabout. Last scenario recreate a situation considering the opposite condition of KITTI, that is, our ego-vehicle, represented by a Tesla Model 3, is static aside of the road whilst the other vehicles are moving around our vehicle. In order to validate the performance of our DAMOT system, we store the odometry of the vehicles obtained through the odometry ROS topics defined for each vehicle (ground-truth), while their predicted position and associated identifier are obtained from the MOT system.

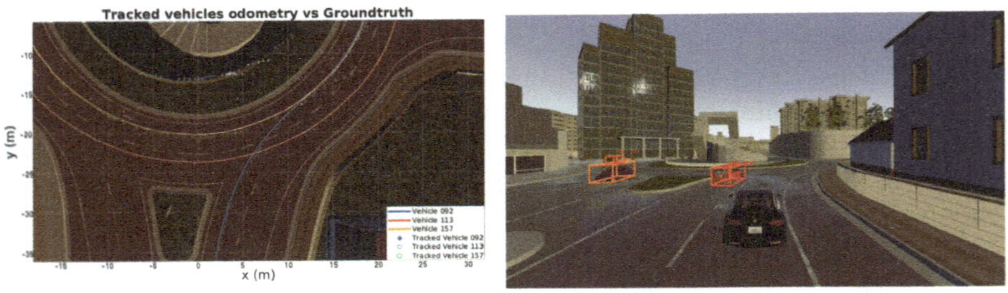

(a) Tracked odometry vs ground-truth. (b) RGB view.

Fig. 8. Scenario 3. Entering a roundabout.

In Fig. 8a vehicles ground-truth trajectory are drawn as a continuous line while tracked vehicles are shown as asterisks, squares and circumferences respectively and other detected objects appears as crosses. It can be appreciated how close are predicted and ground-truth positions for each vehicle, and the consistency of the tracker, which keeps a correct association to each vehicle for almost all the scene. Also it can be noted that most false positives are due to roundabout urban decoration elements.

Figure 8b shows main ego_vehicle centered on image and projected 3D bounding boxes overlapping the tracked vehicles. RGB camera is not needed for tracking and is only used for demonstrative use.

DAMOT Validation Results. Table 4 shows the results of the validation tests for the above scenarios using the KITTI-3DMOT tool based on validation code developed by [21], comparing two tracking methods, providing a methodology to test this kind of systems in every traffic condition without taking any risk.

Table 4. DAMOT validation on CARLA scenarios.

Method	AMOTA	AMOTP	MOTA	MOTP	IDS	FRAG	FP	FN
AB3DMOT [21]	21.92	**63.67**	32.60	**69.27**	0	**26**	68	**178**
BEV MOT [7]	**26.45**	12.04	**48.77**	22.09	0	28	**0**	187

It can be seen that our BEV-MOT [7] tracking method achieves better AMOTA and MOTA results than AB3DMOT [21], which are intrinsically related with tracking metrics, obtaining a lower result on AMOTP and MOTP metrics, associated with a worst detection. According to these results, BEV-MOT tracking method is more efficient and demonstrates a better performance than AB3DMOT, increasing AMOTP and MOTP values when obtaining a better detection in a more realistic LiDAR point cloud.

Further information can be found in attached videos[1], showing the performance of the proposed method in different environments and situations.

5 Conclusions

This work presents the architecture and validation of our 360° Real-Time 3D Multi-Object Detection and Tracking system on KITTI dataset with 3D DAMOT evaluation tool and CARLA simulator, a flexible and hyper-realistic open-source simulator for autonomous driving. The implemented method allows to track static and dynamic objects around an autonomous car in real-time to enhance its safety system by only using a 3D LiDAR point cloud as input. The validation has consisted of firstly analyze the results on KITTI validation set modifying the 3D object detector architecture and the trained weights. After that, different traffic scenarios has been implemented in CARLA to study the performance of our DAMOT architecture in a realistic urban environment, comparing our BEV-MOT proposal with AB3DMOT (our baseline). As future works, the system will be mounted on our real electric autonomous car to fully test its ability to detect and track vehicles and pedestrians in traffic urban scenarios, improving the robustness of our current tracking system. Also different sensors such as radars and cameras will be added in order to strengthen the DAMOT module behaviour. Furthermore, we hope CARLA will improve LiDAR point cloud sensor in order to produce more realistic vehicles shapes to approximate PointPillars ability to detect object to real world datasets.

[1] https://cutt.ly/3rU113d.

Acknowledgment. This work has been funded in part from the Spanish MICINN/FEDER through the Techs4AgeCar project (RTI2018-099263-B-C21)and from the RoboCity2030-DIH-CM project (P2018/NMT- 4331)), funded by Programas de actividades I+D (CAM) and cofunded by EU Structural Funds.

References

1. Baser, E., Balasubramanian, V., Bhattacharyya, P., Czarnecki, K.: FANTrack: 3D multi-object tracking with feature association network (2019)
2. Bernardin, K., Stiefelhagen, R.: Evaluating multiple object tracking performance: the CLEAR MOT metrics. EURASIP J. Image Video Process. **2008** (2008). https://doi.org/10.1155/2008/246309
3. Bewley, A., Ge, Z., Ott, L., Ramos, F., Upcroft, B.: Simple online and realtime tracking (2016)
4. Dosovitskiy, A., Ros, G., Codevilla, F., Lopez, A., Koltun, V.: CARLA: an open urban driving simulator (2017)
5. Geiger, A., Lenz, P., Urtasun, R.: Are we ready for autonomous driving? The KITTI vision benchmark suite. In: 2012 IEEE Conference on Computer Vision and Pattern Recognition, pp. 3354–3361. IEEE (2012)
6. Gómez-Huelamo, C., Bergasa, L.M., Barea, R., López-Guillén, E., Arango, F., Sánchez, P.: Simulating use cases for the UAH autonomous electric car. In: 2019 IEEE Intelligent Transportation Systems Conference (ITSC), pp. 2305–2311. IEEE (2019)
7. Gómez-Huelamo, C., del Egido, J., Bergasa, L.M., Barea, R., Ocaña, M., Arango, F., Gútierrez, R.: Real-time bird's eye view multi-object tracking system based on fast encoders for object detection. In: 2020 IEEE Intelligent Transportation Systems Conference (ITSC). IEEE (2020)
8. Jullien, J.M., Martel, C., Vignollet, L., Wentland, M.: OpenScenario: a flexible integrated environment to develop educational activities based on pedagogical scenarios. In: 2009 Ninth IEEE International Conference on Advanced Learning Technologies, pp. 509–513. IEEE (2009)
9. Konigshof, H., Salscheider, N., Stiller, C.: Realtime 3D object detection for automated driving using stereo vision and semantic information, pp. 1405–1410 (2019). https://doi.org/10.1109/ITSC.2019.8917330
10. Kuhn, H.W., Yaw, B.: The Hungarian method for the assignment problem. Naval Res. Logist. Q. 83–97 (1955)
11. Lang, A.H., Vora, S., Caesar, H., Zhou, L., Yang, J., Beijbom, O.: PointPillars: fast encoders for object detection from point clouds. In: 2019 IEEE/CVF Conference on Computer Vision and Pattern Recognition (CVPR) (2019). https://doi.org/10.1109/CVPR.2019.01298
12. Luo, W., Yang, B., Urtasun, R.: Fast and furious: real time end-to-end 3D detection, tracking and motion forecasting with a single convolutional net, pp. 3569–3577 (2018). https://doi.org/10.1109/CVPR.2018.00376
13. Patil, A., Malla, S., Gang, H., Chen, Y.T.: The H3D dataset for full-surround 3D multi-object detection and tracking in crowded urban scenes (2019)
14. Qin, Z., Wang, J., Lu, Y.: Triangulation learning network: from monocular to stereo 3D object detection. In: The IEEE Conference on Computer Vision and Pattern Recognition (CVPR) (2019)
15. Quigley, M., Conley, K., Gerkey, B., Faust, J., Foote, T., Leibs, J., Wheeler, R., Ng, A.: ROS: an open-source robot operating system (2009)

16. Sanders, A.: An Introduction to Unreal Engine 4. AK Peters/CRC Press, Natick (2016)
17. Schöner, H.: The role of simulation in development and testing of autonomous vehicles. In: Driving Simulation Conference, Stuttgart (2017)
18. Schulter, S., Vernaza, P., Choi, W., Chandraker, M.: Deep network flow for multi-object tracking (2017)
19. Shi, S., Wang, X., Li, H.: PointRCNN: 3D object proposal generation and detection from point cloud (2018)
20. SAE Taxonomy: Definitions for terms related to driving automation systems for on-road motor vehicles (J3016). Technical report, Society for Automotive Engineering (2016)
21. Weng, X., Kitani, K.: A baseline for 3D multi-object tracking (2019)
22. Weng, X., Kitani, K.: Monocular 3D object detection with pseudo-lidar point cloud (2019)
23. Wojke, N., Bewley, A., Paulus, D.: Simple online and realtime tracking with a deep association metric (2017)
24. Yang, B., Liang, M., Urtasun, R.: HDNET: exploiting HD maps for 3D object detection. In: Billard, A., Dragan, A., Peters, J., Morimoto, J. (eds.) Proceedings of The 2nd Conference on Robot Learning, Proceedings of Machine Learning Research, vol. 87, pp. 146–155. PMLR (2018). http://proceedings.mlr.press/v87/yang18b.html
25. Zhou, Y., Tuzel, O.: VoxelNet: end-to-end learning for point cloud based 3D object detection, pp. 4490–4499 (2018). https://doi.org/10.1109/CVPR.2018.00472

Towards Fine-Tuning of VQA Models in Public Datasets

Miguel E. Ortiz[1(✉)], Luis M. Bergasa[1], Roberto Arroyo[2], Sergio Álvarez[2], and Aitor Aller[2]

[1] Electronics Department, University of Alcalá (UAH), Alcalá de Henares, Spain
eduardo.ortiz@edu.uah.es, luism.bergasa@uah.es
[2] R&D AI Department, Nielsen Connect, Madrid, Spain
{roberto.arroyo,sergio.alvarezpardo,aitor.allerbeascoechea}@nielsen.com

Abstract. This paper studies the Visual Question Answering (VQA) topic, which combines Computer Vision (CV), Natural Language Processing (NLP) and Knowledge Representation & Reasoning (KR&R) in order to automatically provide natural language responses to questions asked by users over images. A review of the state of the art for this technology is initially carried out. Among the different approaches, we select the model known as Pythia to build upon it, because this approach is one of the most popularized and successful methods in the public VQA Challenge. Recently, an exhaustive breakdown was done to the Pythia code by Facebook AI Research (FAIR). We choose to use this updated framework after confirming that the two implementations had analog characteristics. We introduce the different modules of the FAIR implementation and how to train our model, proposing some improvements regarding the baseline. Different fine-tuned models are trained, obtaining an accuracy of 66.22% in the best case for the test set of the public VQA-v2 dataset. A comparison of the quantitative results for the most important experiments jointly some qualitative results are discussed. This experimentation is performed with the aim of finally applying it to eCommerce and store observation use cases for VQA in further research.

Keywords: Computer Vision · Natural Language Processing · Knowledge Representation & Reasoning · Visual Question Answering · Artificial Intelligence

1 Introduction

Recently, there is a renewed enthusiasm in multi-discipline Artificial Intelligence (AI) research problems. Part of this excitement stems from a belief that multi-discipline tasks are a step towards solving AI. What makes for a compelling "AI-complete" task? In [1], it is stated that in order to spawn the next generation of AI algorithms, an ideal task should require multi-modal knowledge beyond a single sub-domain and should have a well-defined quantitative evaluation metric to track progress. Within this context, Visual Question Answering (VQA) is

L. M. Bergasa et al. (Eds.): WAF 2020, AISC 1285, pp. 256–273, 2021.
https://doi.org/10.1007/978-3-030-62579-5_18

a recently popularized topic, where different types of knowledge are combined. Mainly, VQA is based on Computer Vision (CV), Natural Language Processing (NLP) and Knowledge Representation & Reasoning (KR&R). VQA can be defined as a system that takes as input an image and a free-form, open-ended, natural-language question about the image and produces a natural-language output answer, as depicted in the examples presented in Fig. 1.

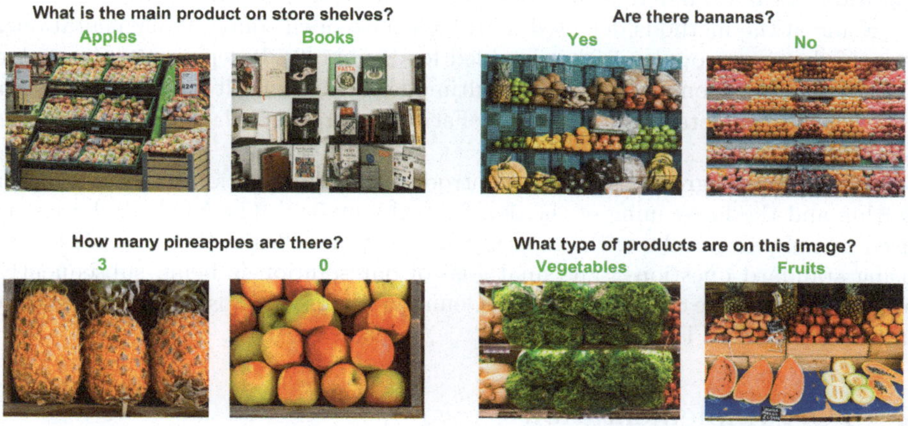

Fig. 1. Examples of inputs and outputs for a standard VQA approach.

There are different methods in the state of the art able to provide solutions to the different use cases related to the generation of automatic answers to questions performed over images. The most important ones are summarized in Table 1. For a deeper explanation, we refer the readers to the original paper of each method.

Table 1. Summary of VQA state-of-the-art methods.

Method	Reference	Datasets and accuracies
Vanilla VQA	[1]	54.06 (VQA [1])
Stacked Attention Networks	[2]	58.90 (DAQUAR [3])
Faster R-CNN in VQA	[4]	63.15 (VQA-v2 [1])
Neural-Symbolic VQA	[5]	99.80 (CLEVR [6])
FVTA	[7]	66.90 (VQA-v2 [1])
Bottom-up and Top-Down	[8]	65.32 (VQA-v2 [1])
Pythia	[9]	72.27 (VQA-v2 [1])
Differential Networks	[10]	68.59 (VQA-v2 [1])
LoRRA	[11]	69.21 (VQA-v2 [1]), 27.63 (TextVQA [11])

Most of the methods have been validated in the VQA-v2 dataset. This dataset contains 204,721 images from the MS COCO dataset [12] and other 50,000 images from some abstract scenes. The MS COCO dataset has images depicting diverse and complex scenes that are effective at eliciting compelling and diverse questions. The dataset collects as well some realistic abstract scenes to enable research focused only on the high-level reasoning required for VQA by removing the need to parse real images. The VQA-v2 dataset contains more than 1M questions with around 10M answers and it can be accessed from [13].

Some of the methods depicted in Table 1 have open-source implementations. One of the most popular is the provided for Pythia [9]. It represents a modular framework for vision and language multimodal research built on top of PyTorch [14]. The implementation developed by Facebook AI Research (FAIR) is available in GitHub [15].

With this background, this paper introduces a VQA solution that builds upon Pythia and the fine-tuning of the derived VQA models. The VQA-v2 dataset is used to compare results with respect to other approaches of the state of the art using standard questions. The final goal of our solution is being subsequently applied in VQA use cases focused on eCommerce and store observation (as shown in the examples of Fig. 1).

2 VQA Implementation

In this section, we describe the VQA approach implemented for this paper. Firstly, we introduce an overview of the Pythia's method, which is the baseline of our proposal. Then, we present the implementation details of the algorithm, explaining the different modules of its architecture. We finally highlight our contributions regarding the original Pythia solution.

2.1 Pythia's Overview

Pythia was the winner architecture in the VQA Challenge 2018 [9]. This model is based on the Bottom-up and Top-Down attention model [8], which is a network that processes the image and text separately, adding attention to image processing to get better features for the final classification. Pythia provides some changes over this model to increase its accuracy in the VQA-v2 dataset. The main improvements that Pythia introduces in the Bottom-up and Top-Down attention model are:

- **Weight normalization with ReLU.** This new method replaces the gated hyperbolic tangent activation in the image attention module for a weight normalization followed by ReLU to reduce computation charge. Features concatenation is also replaced by element-wise multiplication. With these improvements, the accuracy is increased to 66.91% according to the original paper of Pythia.

- **Learning Schedule.** Pythia researchers decided to reduce the batch size, typically fixed to 512, and make use of the learning rate scheduler to first linearly increase the initial learning rate from 0.002 to 0.01, a process known as warm-up, and then reduce it every some thousand iterations. With this improvement they obtained an accuracy of 68.05%.
- **Detectron Fine-Tuning.** In this phase, a fine-tuning of the Detectron output (a detector based on feature pyramid network) [16] is made. The architecture uses ResNeXt [17] as backbone together with two fully connected layers (FC6 and FC7) for region classification. In this case, fine-tuning occurs at layer FC7. This boosted accuracy to 68.49%.
- **Data Augmentation.** Additional data is added from external datasets and mirroring is done to the images in the VQA-v2 dataset. The number of training iterations is also increased. With all this, the accuracy is increased to 69.24%.
- **Grid feature and 100 objects.** Grid level features are added to the bottom-up features, because they do not seem to represent all the required information. These new features are obtained from ResNet152 [18]. Object-level features and grid-level features are separately fused with features from questions and then are concatenated to fed to classification. In addition, adaptive object proposals (10 to 100) are no longer used and a fixed number of 100 object proposals for all images is used instead. With these two upgrades, the accuracy increased to 70.01%.
- **Ensemble.** Two types of model ensembles are developed. The first one uses the best single model and trains the same network with different seeds, and finally averages the predictions from each model to obtain an accuracy of 70.96%. In the second case, several trained models are used. These models are trained with different Detectron models and with/without data augmentation. This approach gives 72.18% of accuracy.

Table 2. Original Pythia improvements on VQA-v2.

Model	test-dev
Bottom-up and Top-down	65.32
Weight normalization with RELU	66.91
Learning Schedule	68.05
Detectron & Fine-tuning	68.49
Data Augmentation	69.24
Grid feature & 100 bboxes	69.81
Ensemble	72.18

A summary of the original Pythia improvements on VQA-v2 is shown in Table 2. However, these improvements also present in the Pythia's framework (except for the ensembles) report a lower accuracy of 66.7%, according to the current FAIR reference [15]. In this framework, all the upgrades mentioned before are codified and validation is carried out over the same test set. This accuracy difference can be reported because the implementation of Pythia or the features are not exactly the same, or because the numbers presented in the original implementation are overvalued. As our numbers are also closer to the provided by FAIR and these numbers are more up to date, we will take them as baseline for comparing the performance provided by our VQA models.

2.2 Implementation Review

As previously described, the implementation of Pythia includes modules classified in three main sections: CV, NLP and KR&R. Each of these sections can be seen working together in Fig. 2. CV section is represented by the Features and ImageEncoder modules. Questions are processed in the NLP modules, which are WordEmbedding and TextEmbedding. KR&R section receives the CV and NLP outputs in the ImageEmbedding module to generate new highlighted features. Finally, text and image attention features are fused in the ModalCombineLayer module, which loads the data in the ClassifierLayer to achieve the final scores of answers.

As it can be seen, the implementation of Pythia does not perform the extraction of features, so it is necessary to obtain them from some external models in a previous stage. This schema just processes as inputs the features and the questions of the subject.

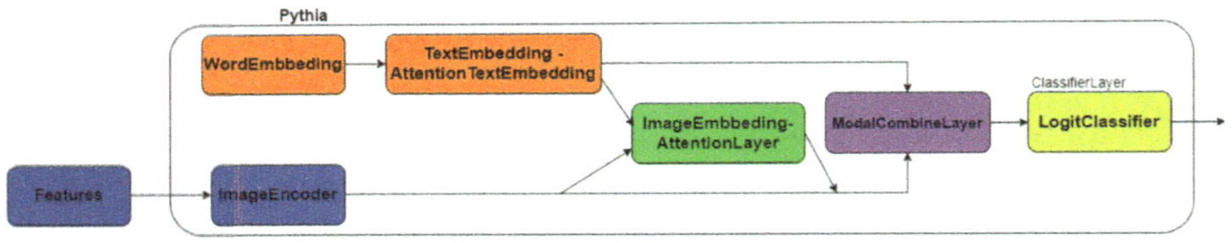

Fig. 2. Full Pythia implementation diagram.

Hereafter, the main modules that make up the Pythia model will be explained.

The Pythia implementation receives as inputs the features of the MS COCO images from two different networks: Detectron and ResNet152. These features are the inputs of the first module of the Pythia implementation (ImageEncoder).

ImageEncoder (CV) is a module that processes the last layer of the Detectron network, FC7 (see Fig. 3). Here, the approach loads the features of the FC6 layer of Detectron and trains FC7 with the rest of modules. The FinetuneFasterRcnnFpFc7 module is a fully connected net that is in charge with the FC7 training process. For the grid features (coming from ResNet152), Identity is used. This module just sends the input to the output without any arithmetic operation. In case any other operation is needed, the Identity module can be changed by the corresponding one. The same strategy can be done with FinetuneFasterRcnnFpnFC7. This procedure gives the possibility of choosing the feature extractor model that best fits to the specific application. In general, the ImageEncoder works as a bridge between the features of the image (CV section) and Pythia, where we can add layers to the features that we receive if it is needed.

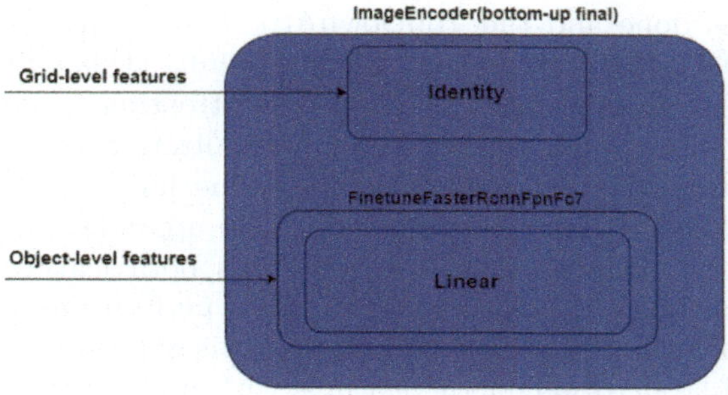

Fig. 3. Computer Vision module.

To process the questions (NLP section) two modules are used: WordEmbedding and TextEmbedding-AttentionTextEmbedding, as we depict in Fig. 4. WordEmbedding uses GloVe [19] to represent the features of the input text words. Words are represented as a vector of indexes, which are obtained from a predefined vocabulary of 100k words. That means that it is necessary to transform each word of the text into a one-hot vector. The second module is an architecture that highlights the most relevant features of the whole text, and is where the Attention process is applied. As in most text processors, the first step is to use a LSTM (Long Short-Term Memory) net to extract features of the text. It must be an RNN (Recurrent Neural Network) like LSTM or GRU (Gated Recurrent Unit), because they keep information of the previous inputs, that is, the previous words of the current question. Then, the outputs of the LSTM are forwarded to a CNN with the purpose of getting a better understanding of the features. In vanilla NLP, a global max pool is applied to the output of the RNN. In VQA, it is necessary to increase the complexity of the model to acquire better features. To obtain these improved features the model uses Dropout, Convolutional, ReLu and Softmax layers. Dropout is a layer to avoid overfitting by disabling a percentage of the features, which are randomly selected in each iteration of the training. The role of the Convolutional layers (+ReLu) is similar to the seen in the CV section, that is, to extract features with more relevance that can better represent the question. At the end of this model, a Softmax layer is used to obtain the probabilities of the features, where the ones with higher values have more impact in the current question.

Fig. 4. Natural Language Processing module.

The outputs of the ImageEncoder and AttentionTextEmbedding are processed by the ImageEmbedding - AttentionLayer. It is in this layer where the KR&R process is done and the TopDownAttention module is in charge with this process (see Fig. 5). This is the module where Pythia's first improvement with respect to the Bottom-up and Top-Down attention model is carried out. Hyperbolic tangent is replaced by weight normalization followed by ReLu to reduce computation and the concatenation of the image and text features is replaced by an element-wise multiplication. These upgrades can be found in the NonLinearElementMultiply module that receives both features. New features are received by the LinearTransform layer, which performs a linear with weight normalization operation. Finally, a Softmax layer is applied to the output of the LinearTransform layer to weight the features with probabilities from 0 to 1.

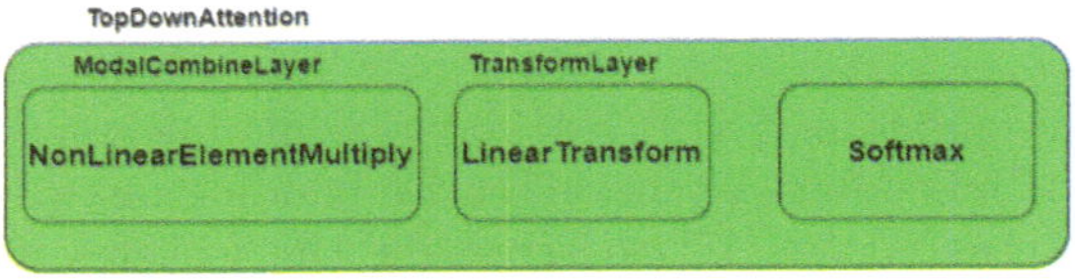

Fig. 5. Knowledge Representation & Reasoning module.

After obtaining the features with attention from images and text, they are merged using the NonLinearElementMultiply module depicted in Fig. 6. In this module, the ReLuWithWeightNormFC performs a Linear with weight normalization followed by a ReLu layer, where the Linear module is just a fully connected

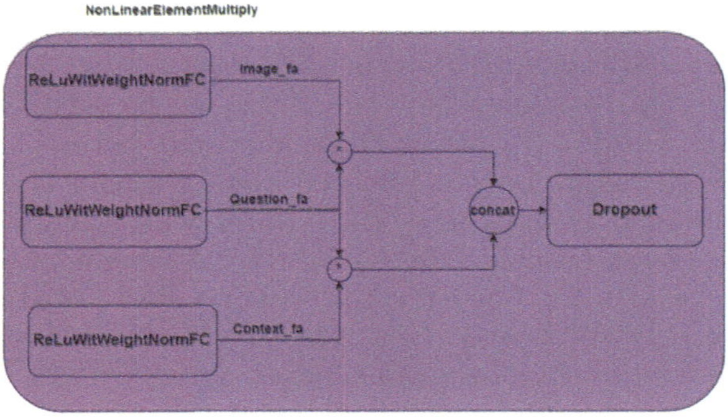

Fig. 6. Fusion module.

layer. This operation is applied to image, text and context features. Context features are only used for the LoRRa architecture [11]. The output is element-wise multiplied in order to get image-question and question-context features, and finally these features are concatenated. At the end of the module there is a Dropout layer to avoid overfitting.

The output of the fusion module is introduced as input in the last general module of the Pythia implementation, the LogitClassifier, as we can see in Fig. 7. The scores of the answers are obtained in this module. Inside the LogitClassifier, the input is connected to two different branches, one for the text features and other for the image ones. Each branch is formed by modules that we already know: ReLuWithWeightNormFC plus Linear. The difference between these two layers is the hidden dimension that returns ReLuWithWeightNorm. For text the dimension is smaller than for image. This is done to get the best possible features related to image and text. After this, the Linear layer returns the same two feature vectors of same size. Finally, both vectors are added to get the final output of the model.

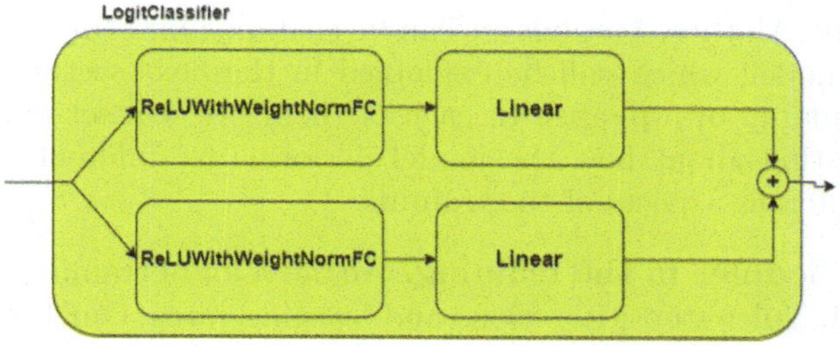

Fig. 7. Classifier module.

2.3 Contributions

Pythia's framework provides a configuration file to create new models and run it in an easy way. It offers a lot of modules that can be used for established applications or for new research. In this section, we highlight the modifications introduced on the Pythia's baseline code in order to reach the best performance results.

- Pythia's implementation does not provide a user friendly representation of the performance metrics like accuracy or loss. Instead, the framework offers a simple log file with all the performance information. We have implemented a module that parses the configuration file and gets different performance graphs to easier data understanding. This new module works offline, once the training process is over.
- Changing the training strategies provided in the baseline. Multiple fine-tuning approaches have been carried out according to the following ideas:

- Change the general rule followed by Pythia in the training scheduler.
- Put into practice long training with and without regularization.
- To combine scheduler and regularization.
- To use pre-trained models for new experiments.

3 Fine-Tuning a VQA Model

In this section, we describe the training process for fine-tuning VQA models. Firstly, we explain the different training and fine-tuning strategies we have followed. After that, we present the hyperparameters values we have assigned for an optimum performance. We finish explaining the used training loss functions.

3.1 Training Strategies

To train the Pythia models, we used the configuration file that the framework offers. In this file we found all the needed hyperparameters for the training: optimizer parameters, size of the layers, path of the dataset that the model will use to train, etc. All these parameters can be classified into three sets: optimizer, training and model; which will be explained in the next section. Our training strategy is made up of 3 phases. In each of them, we will set hyperparameters to get information about how the model performs to different configurations, focusing on different aspects of the training.

- **Using a scheduler in the training.** We did a lot of training processes with different scheduler steps, iterations and learning rates. Our first thought was that the change of learning rate should be later as the learning rate decreases, but following the results obtained from the training, this idea is not fulfilled. In general, accuracy improvements with this strategy were less than expected.
- **Using regularization in the training.** Weight decay and dropout were tested. The criteria to get the optimum value was mainly that penalization to the training accuracy was not too huge. After doing several tests, we concluded that any value for weight decay less than 10^{-6} will drastically drop the accuracy of the model. In the case of dropout, we noticed that the accuracy decreases if we deactivate more than the 40% of the features. In general, accuracy with this strategy is worse than for the first approach.
- **Using pre-trained models in the training.** This strategy turned out to be the best. The pre-trained model used in this phase was the best one that uses a scheduler without regularization. This is because those models had better performance. In the post training the scheduler is no longer used and we add both regularization methods. With this setup the model was trained for 100,000 iterations reaching to increase the accuracy obtained from the previous strategies.

3.2 Hyperparameters

In this section, we explain the values of the hyperparameters that have been used for fine-tuning, with the aim of easily understanding the training processes. These hyperparameters are depicted in Table 3 and they can be grouped into three sets: optimizer, training and model.

Table 3. Hyperparameters.

Set	Hyperparameter	Values
Optimizer	Learning Rate	0.0045 to 0.03
Optimizer	Weight Decay	10^{-10} to 10^{-05}
Training	Iterations	20000 to 100000
Training	Batch Size	128
Training	Learning Rate Scheduler	True or False
Training	Learning Rate Steps	10000 to 30000
Training	Learning Rate Ratio	0.05 to 0.1
Training	Device	CPU or CUDA
Model	Dropout	0 to 0.4

3.3 Training Loss

The problem that VQA tackles can be explained like a multi-label classification, because the model must choose between different labels, in this case answers. This means that the model must provide probabilities for a one-hot vector, where each index represents an answer. For this kind of task, a training loss based on Binary Cross Entropy with Logits (BCELogit) is the perfect solution. This loss function uses a sigmoid activation plus a Cross-Entropy loss. For each output of the model, the sigmoid layer is applied. This function returns the probabilities of each output of the model, where the output is in the range 0 to 1. The mathematical expression of the sigmoid function is shown in Eq. 1.

$$Sigmoid(x) = \sigma(x) = \frac{1}{1 + exp(-x)} \tag{1}$$

The sigmoid function is presented in the equation of BCELogit depicted in Eq. 2 and Eq. 3. There, we can see the output model represented as x and the target represented as y. As this is a multi-label classification problem, BCE is applied to each of the labels. The c is the class/answer number and n is the number of the sample in the batch. After the calculations are done, the loss function obtains a list of values for each sample. It is in this moment when Eq. 4 is calculated. PyTorch offers two options for the final stage of the BCELogit: *sum* or *mean*. *Sum* just adds all the results obtained before to get the final error, meanwhile *mean* adds all the components and divide them by the number of samples. In PyTorch, the election between *mean* and *sum* is called reduction. For Pythia's implementation, *mean* is used for the training loss as it is more representative than the other option for a loss function.

$$L_c = \{l_{1,c}, ..., l_{N,c}\}^T, l_{n,c} = -w_{n,c}[p_c y_{n,c} log \sigma x_{n,c} + (1 - y_{n,c}) log(1 - \sigma x_{n,c})] \quad (2)$$

$$l_c(x, y) = L_c = \{l_{1,c}, ..., l_{N,c}\} \quad (3)$$

$$l(x, y) = \begin{cases} \text{mean(L) if reduction='mean'} \\ \text{sum(L) \ if reduction='sum'} \end{cases} \quad (4)$$

4 Experiments and Results

In this section, we describe the experimental results we carried out to validate our VQA proposal. Firstly, we explain the used VQA dataset. After that, we study the used evaluation metrics and the experimentation we have followed. Finally, quantitative and qualitative results are presented.

4.1 Public VQA Dataset

Pythia's baseline model was trained in the VQA-v2 dataset. This dataset was used in the challenge that the VQA group runs annually. In this challenge, multiple models are presented to try to get the best accuracy, showing the best architectures that the participants could think of. Pythia was one of these models presented and the winner in 2018, with an accuracy of 72.18%. The main components of the VQA-v2 dataset are the following:

- **Images:** there are 265,016 images in the dataset. These images are the combination of MS COCO dataset, which contains 204,721 images, and the rest are abstract images done by the authors of VQA-v2 with the purpose of enriching the information and trying to cover all the possible scenarios of the real world.
- **Questions:** Each of the previous images has at least 3 questions related to the information that contains the picture. This makes more than 1 million of questions for the whole dataset.
- **Answers:** Answers are provided for the questions. These answers were obtained from a process done with human labelers, which ensures that the pool of answers is diverse. Each question has 10 answers, this makes a total of more than 11 million of answers.

All the information related to the process of the dataset creation can be read in [1]. The data is stored in two main files. The format of these files are JSON, that means that we can add all the needed information in key-value pairs. This allows us to manage the information in a very efficient way. The main files are:

- **InputQuestionFormat** is the file that contains relevant information about the images and the questions. In this file we can find the link between the two components mentioned before. The fields that this JSON file contain are:
  - **Info:** this is an object that contains information like year, version, description or a timestamp.
  - **Data type:** string type value that contains the name of the dataset used for the images.
  - **Data subtype:** string type value that specifies if the data is for train, test or validation.
  - **License:** contains the url and the license name.
  - **Questions:** it is a list formed by objects that contain the information of the images and the questions. Images and questions are represented as numbers, which help to map them when the data is loaded, increasing the searching speed.
- **AnnotationFormat** is the file that contains the link between questions and answers. This file has the same fields than the previous one, except for "Questions" and "Tasks" type. Instead, this JSON has a new key named "Annotations" that stores new information related to images, questions and answers. The field annotations is a list of objects that stores the main information of this file. Each annotation includes the following fields:
  - **Question type:** VQA questions are clustered by the first words of each question. This field gives the cluster this question belongs.
  - **Multiple choice answer:** this field is related to the most popular answers given.
  - **Image id:** this field contains the image identificator related to the question and the answers.
  - **Answer type:** determines the nature of the answer.
  - **Question id:** with this field, we obtain the correct question from the previous file.
  - **Answers:** it is a list that stores 10 answers for each question. The answers have their own properties, which hereafter will be briefly explained:
    - **Answer:** this is a literal string that contains the response to the question.
    - **Answer confidence:** this field measures the accuracy of the answers.
    - **Answer id:** it gives a number that identifies the answers.

4.2 Evaluation Metrics

To evaluate the performance of the Pythia model, the metrics that the framework uses are based on Accuracy and Loss.

Accuracy is a metric used to evaluate classifications problems. This metric measures the ratio between the number of correct predictions and the total number of input samples. The process to know if the prediction is correct or not is really simple. The expression that calculates this metric is shown in Eq. 5 The target is a one-hot vector that represents all the possible responses that the model can give, where the values of this vector are zero except for the index with the correct answer. Then, it is necessary to compare this vector with the output of the model. The output of the Pythia model is a vector where the correct answer is the index with the biggest value, where this index must be the same as the target. To get this information, it is necessary to apply a Softmax layer to the output. This layer will return a vector with real values where the total sum is one. With this information, we have a better understanding of the classification that the model has done.

$$L_c = \frac{\text{Number of correct predictions}}{\text{Total number of predictions}} \tag{5}$$

Pythia obtains two accuracy metrics: median and global avg.

- Median accuracy: apply the median operation over the accuracy returned from the batches
- Global avg accuracy: calculate the average from all the accuracy values returned for the samples.

Loss metric is just the value returned by the loss function. This function calculates how much the output model fits the target. Interpreting these metrics, we can know if the model correctly generalizes in the test data.

4.3 Experimentation

In addition to the Pythia implementation, FAIR also offers the VQA-v2 dataset ready to directly load into the model. We have used this implementation for our experimentation. It is composed of three Python packages named training, validation and testing. These packages have the same format:

- *image_name*: it stores the name of the image with the following format: *COCO_[type]_id*, where "type" refers to train, val or test.
- *image_id*: this is the number that represents the image. This number is used in the previously described *image_name*.
- *question_id*: number that refers to a question.
- *feature_path*: it is a path to the features of the image.
- *question_str*: field that contains the question.
- *question_tokens*: it is a list where its elements are each word of the question.
- *answers*: list of size 10 that contains the answers.

The sum of the three packages is 1,105,904, which is the total number of questions that the VQA-v2 dataset offers. The whole data are divided in three different sets to carry out a correct training process, according to the numbers shown in Table 4. As we can see, we will use 40% of the samples for the training, about 20% for the validation of the model and the rest 40% for the testing of the proposals.

Table 4. Dataset division.

Data type	Number of samples	Percentage
Training	443,757	40.12
Validation	214,354	19.38
Test	447,794	40.5

4.4 Quantitative Results

As mentioned in the previous sections, our experiments are focused on fine-tuning different groups of hyperparameters depending on the training strategy that we are applying. However, there are some common hyperparameters that do not change in any experiment, as the following ones:

- Batch Size: 128, due to hardware restrictions.
- Optimizer: Adamax, typical optimizer, which is explained in [20].

The total number of experiments processed was 44. From all these experiments, we show the most representative for each training phase, as we depict in Table 5, where we add the test accuracy for an easy comparison.

Table 5. Experiments grouped by training phases.

Training strategy phase	Experiment id	Accuracy
Scheduler in training	Experiment 1	65.89
Regularization in training	Experiment 4	63.61
Pre-trained model in training	Experiment 11	66.22

In the *Experiment* 1, we tried to replicate the hyperparameters used in original Pythia's trainings, where they obtained 66.7% of accuracy in [15] and 69.81% in [9], excluding ensembles because we can not reproduce them. The reason to perform this experiment is to establish a baseline model to compare with. This training does not use regularization methods, thus, these hyperparameters are set to zero. The scheduler follows the same steps mentioned in [9], with the same learning rate and the same number of iterations. The accuracy obtained in this training is less than the reported in the above original evaluations, as we can see in the first row of Table 5. In the accuracy and loss graphs presented in Fig. 8, there are depicted curves that represent the train and test values. Taking a quick look to this graphs, we can observe how the model is overfitting and the scheduler impact.

We test several modifications in the training scheduler, reaching only some slight improvements. After that, we checked some regularization methods, but we did not obtain any improvement (*Experiment* 4). However, we confirmed the

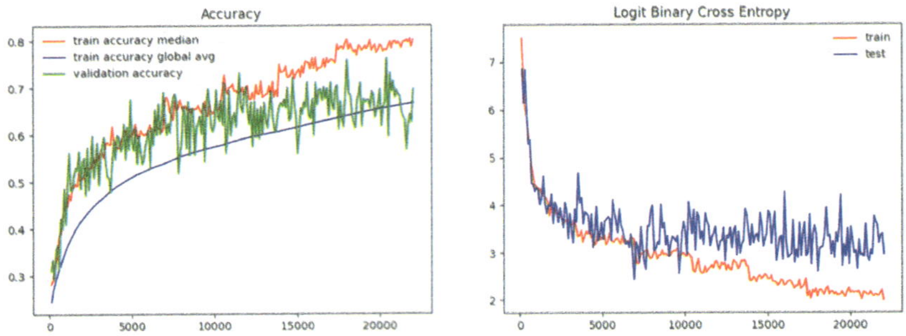

Fig. 8. Accuracy and loss graphs of Experiment 1.

fact that regularization over a pre-trained model increases the accuracy of the model. To confirm this improvement hypothesis, during the *Experiment* 11 we add 70k iterations more to the model obtained from *Experiment* 1, which was the best one from the first training strategy phase. The final accuracy increased to 66.22%, reaching this model its best performance with this configuration.

4.5 Qualitative Results

In this section, we qualitatively assess our best model with some specific images and questions to check if the answers provided by the model are the correct ones, obtaining in this way some visual results about the performance of our proposal.

Prediction	Score
yes	0.998794436454773
no	0.0010550966253504157
vegetables	3.590611959225498e-05
9	9.134103493124712e-06
6	8.617353159934287e-06

Fig. 9. Qualitative results for a generic question.

The first step for carrying out these evaluations is to load the pre-trained model, in this case corresponding to *Experiment* 11. The second step is to select the images and questions that the model will try to predict their answers for.

The first question for the example image is "is there some food?". As we can see in the Fig. 9, the most representative answer provided by the model is positive with a confidence of 99%. This shows that the performance of the model with general questions is satisfactory.

For a second question test, we decided to focus more on one of the products presented in the image (fruits, vegetables, products from animals and dairy products). Figure 10 shows these experiments. The answers are still positive, but the confidence scores decreased with respect to the initial qualitative evaluation. In the case of the products of animals and dairy products, the confidence of the most representative answer drops to 74% but still keeps the correct answer. This shows that the model can still recognize different products but it suffers if these products come from other sources.

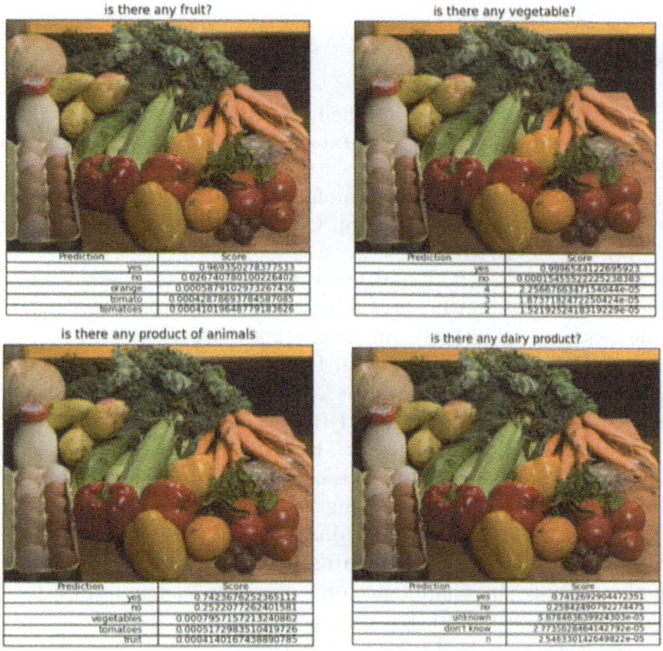

Fig. 10. Qualitative results for specific questions.

5 Conclusions and Future Works

Along this paper, we have reviewed the state of the art of the VQA technology and the different modules of the Pythia architecture. We confirmed that the framework that [15] offers is analog to the described in [9]. After that, we successfully tested the Pythia implementation following the framework proposed in [15].

In the experiments, we can see that the accuracy that we first obtained for our baseline was close to the reported results in [15]. Therefore, we fine-tune the hyperparameters and the training strategy of the model with the intention of obtaining better results regarding our baseline. We managed to slightly improve our first numbers, but the most important was to obtain the necessary knowledge to face the next stage of our project, where we will apply the described learnings for further research in eCommerce and store observation use cases for VQA, similar to the examples shown in Fig. 10.

Acknowledgments. Authors want to thank Nielsen Connect for its support and funding in the development of this project. This work has been also funded in part from the Spanish MICINN/FEDER through the Techs4AgeCar project (RTI2018-099263-B-C21) and from the RoboCity2030-DIH-CM project (P2018/NMT- 4331), funded by Programas de actividades I+D (CAM) and cofunded by EU Structural Funds.

References

1. Antol, S., Agrawal, A., Lu, J., Mitchell, M., Batra, D., Zitnick, C.L., Parikh, D.: VQA: visual question answering. In: International Conference on Computer Vision (ICCV), pp. 2425–2433 (2015)
2. Yang, Z., He, X., Gao, J., Deng, L., Smola, A.: Stacked attention networks for image question answering. In: Conference on Computer Vision and Pattern Recognition (CVPR), pp. 21–29 (2016)
3. Malinowski, M., Fritz, M.: A multi-world approach to question answering about real-world scenes based on uncertain input. In: Conference on Neural Information Processing Systems (NeurIPS), pp. 1682–1690 (2014)
4. Teney, D., Anderson, P., He, X., Van Den Hengel, A.: Tips and tricks for visual question answering: learnings from the 2017 challenge. In: Conference on Computer Vision and Pattern Recognition (CVPR), pp. 4223–4232 (2018)
5. Yi, K., Wu, J., Gan, C., Torralba, A., Kohli, P., Tenenbaum, J.: Neural-symbolic VQA: disentangling reasoning from vision and language understanding. In: Conference on Neural Information Processing Systems (NeurIPS), pp. 1031–1042 (2018)
6. Johnson, J., Hariharan, B., van der Maaten, L., Fei-Fei, L., Lawrence Zitnick, C., Girshick, R.: CLEVR: a diagnostic dataset for compositional language and elementary visual reasoning. In: Conference on Computer Vision and Pattern Recognition (CVPR), pp. 2901–2910 (2017)
7. Liang, J., Jiang, L., Cao, L., Li, L.-J., Hauptmann, A.G.: Focal visual-text attention for visual question answering. In: Conference on Computer Vision and Pattern Recognition (CVPR), pp. 6135–6143 (2018)
8. Anderson, P., He, X., Buehler, C., Teney, D., Johnson, M., Gould, S., Zhang, L.: Bottom-up and top-down attention for image captioning and visual question answering. In: Conference on Computer Vision and Pattern Recognition (CVPR), pp. 6077–6086 (2018)
9. Jiang, Y., Natarajan, V., Chen, X., Rohrbach, M., Batra, D., Parikh, D.: Pythia v0.1: The Winning Entry to the VQA Challenge 2018. arXiv preprint arXiv:1807.09956 (2018)
10. Wu, C., Liu, J., Wang, X., Li, R.: Differential networks for visual question answering. In: AAAI Conference on Artificial Intelligence (AAAI), vol. 33, pp. 8997–9004 (2019)

11. Singh, A., Natarajan, V., Shah, M., Jiang, Y., Chen, X., Batra, D., Parikh, D., Rohrbach, M.: Towards VQA models that can read. In: Conference on Computer Vision and Pattern Recognition (CVPR), pp. 8317–8326 (2019)
12. Lin, T., Maire, M., Belongie, S.J., Bourdev, L.D., Girshick, R.B., Hays, J., Perona, P., Ramanan, D., Dollár, P., Zitnick, C.L.: Microsoft COCO: Common Objects in Context. arXiv preprint arXiv:1405.0312 (2014)
13. VQA-v2 (Online) (2020). https://visualqa.org/download.html
14. PyTorch (Online) (2020). https://pytorch.org/
15. FAIR Framework (Online) (2020). https://github.com/facebookresearch/mmf
16. Detectron (Online) (2020). https://github.com/facebookresearch/Detectron
17. Xie, S., Girshick, R., Dollár, P., Tu, Z., He, K.: Aggregated residual transformations for deep neural networks. In: Conference on Computer Vision and Pattern Recognition (CVPR), pp. 1492–1500 (2017)
18. He, K., Zhang, X., Ren, S., Sun, J.: Deep residual learning for image recognition. In: Conference on Computer Vision and Pattern Recognition (CVPR), pp. 770–778 (2016)
19. Pennington, J., Socher, R., Manning, C.D.: GloVe: global vectors for word representation. In: Conference on Empirical Methods in Natural Language Processing (EMNLP), vol. 14, pp. 1532–1543 (2014)
20. Kingma, D.P., Ba, J.: Adam: A Method for Stochastic Optimization. arXiv preprint arXiv:1412.6980 (2014)

Integrating OpenFace 2.0 Toolkit for Driver Attention Estimation in Challenging Accidental Scenarios

Javier Araluce(✉), Luis M. Bergasa, Carlos Gómez-Huélamo, Rafael Barea, Elena López-Guillén, Felipe Arango, and Óscar Pérez-Gil

Electronics Department, University of Alcalá (UAH), Alcal de Henares, Spain
{javier.araluce,luism.bergasa,rafael.barea,elena.lopezg}@uah.es
{carlos.gomezh,juanfelipe.arango,oscar.perez}@edu.uah.es

Abstract. Gaze estimation is a typical approach to monitor the driver attention on the road scene. This indicator is of great importance in safe driving and in the design of the takeover control strategy for a Level 3 and Level 4 automation system. Nowadays, most of eye gaze tracking techniques are intrusive and costly, which limits their applicability over real vehicles. On the other hand, current databases used for gaze validation face the driver attention task focused on critical situations in simulation but they do not encounter actual accidents. This paper presents a low-cost and non-intrusive camera-based gaze mapping system integrating the open-source state-of-the art OpenFace 2.0 Toolkit [3] to visualise the driver attention simulation on prerecorded real traffic scenes through a heat map. The proposal has been validated by using the recent and challenging public dataset DADA2000 [9] which has 2000 video sequences with annotated driving scenarios based on real accidents. We compare our system with an expensive desktop-mounted eye-tracker, obtaining on par results and showing it is a good tool for driver attention monitoring able to be used in the design of take over systems and driving scenarios awareness systems for automated vehicles.

Keywords: Driver attention · Accidental scenarios · Gaze estimation · Heat map · Computer vision.

1 Introduction

In last years, important advances have been made in autonomous driving field from an academic and industrial point of view. According to SAE (J3016), five Levels of Automation can be applied, achieving the full automation in the Level 5. Nowadays, we have technologies that can be used to take over the functions normally reserved for the driver. In Level 1 and Level 2 the driver is still supposed to be fully engaged in supervising the actions of the vehicle under all circumstances. In Level 3 and Level 4, the driver is freed from supervision, either in limited situations or during the entire trip. Problems arise at the above levels, because the driver is in-the-loop and not always aware of what is happening (Level 1 and Level 2) or is out of the loop and needs quickly to be brought back into the loop for some unexpected reason (Level 3 and Level 4) [11].

L. M. Bergasa et al. (Eds.): WAF 2020, AISC 1285, pp. 274–288, 2021.
https://doi.org/10.1007/978-3-030-62579-5_19

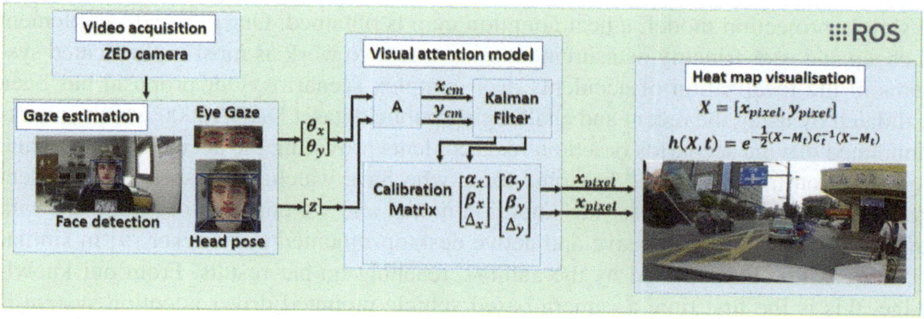

Fig. 1. General Architecture.

In this context of shared control between human and machine, driving in a safety way requires being ensured that the state of driver is suitable for driving. This is particularly important when the vehicle requests the driver intervenes in case the driving scenario is complex. In addition, it is crucial to assess driver' awareness of driving scenario (e.g. surrounding vehicles or pedestrians) right before the take-over. In conclusion, evaluating driver's visual attention is a key task in the development of automated vehicles. Gaze tracking estimation is the common way for evaluating the driver's visual attention [19]. In the literature, there are different approaches mainly based on head-mounted eye trackers [6,7,16], active desktop-mounted eye trackers [18] and passive desktop-mounted cameras [12–14,17,19]. The former provide accurate gaze information but they are intrusive and costly. The latter are low-cost and non-intrusive but less accurate. Most of the existing visual attention systems have been validated in simulation with simple scenes. However, the simulation of driver attention in complex driving scenarios is rather challenging and highly subjective [9]. The goal of these systems is to automatically estimate the regions of interest where the driver is looking. In this sense, acquisition systems and computer vision techniques have gained importance regarding the traditional human factor approaches focused on driving behavior understanding using manual tools.

There are different challenging databases to validate driver visual attention in the state of the art. Recently, the project DR(eye)VE [15] collected 555,000 frames with driver attention maps. However, only few critical scenes were captured for only one driver. Berkeley DeepDrive Attention (BDDA) dataset [18] includes 1232 videos with critical situations in-lab designed. However, the critical situations do not cause true accidents and in consequence the attention simulation does not explore the dynamic process from critical situation to actual accidents. DADA2000 dataset [9] is a larger and more diverse video benchmark with driver attention and driving accident annotation simultaneously. It contains 2000 video clips with fairly complex accidental scenarios in diverse weather and lighting conditions.

This paper presents a low-cost and non-intrusive camera-based gaze mapping system able to estimate the driver attention simulation on challenging prerecorded traffic scenes through a heat map. Gaze direction is calculated by using the open-source state-of-the-art OpenFace 2.0 Toolkit [3]. After a slight calibration process and using a simple projection model, a heat attention map is obtained. Our goal is to implement a cheap and user-friendly measurement system able to work as most sophisticated systems in the localization of accidents over complex scenarios. Our proposal has been validated by using the recent and challenging public dataset DADA2000 which collects annotated driving scenarios based on real accidents providing the accident position and the attention map generated by some users who have watched the scene. We present some performance results for our attention model and we compare our numbers with the obtained with an expensive and active desktop-mounted eye-tracker [9] in similar conditions that the reported by the authors, reaching on par results. From our knowledge, this is the first time a camera-based vehicle-mounted driver attention system is evaluated in the challenging DADA2000 dataset.

2 System Architecture

The system architecture is divided in four steps to estimate where the driver is looking at. These steps are: video acquisition, gaze estimation, visual attention model and heat map visualization. A general overview of this architecture is depicted in Fig. 1.

The complete system is connected using Robot Operating System (ROS). ROS is a framework that gives communication, standardisation and modularity facilities to the developed modules, offering an easy connectivity of each module in the general architecture of our autonomous vehicle.

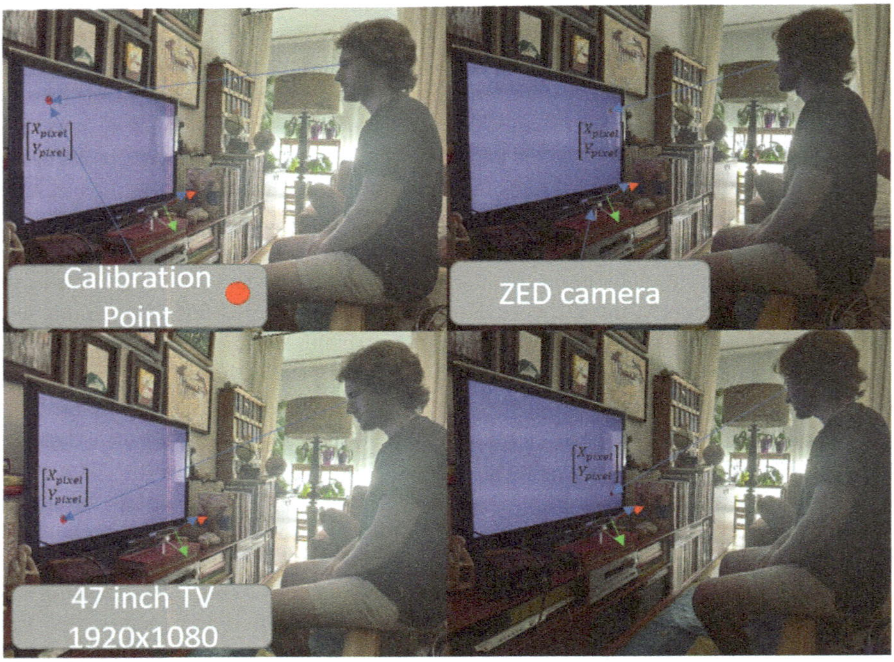

Fig. 2. Calibration step. The user under test has to watch during 8 s 4 points displayed on the screen

2.1 Video Acquisition

The measurement system consists on a ZED camera. This is an stereo camera, but for this application only the left lens is used. After a positioning study, the camera is located on the middle-bottom of the screen in front of the user where the videos are shown, as we depict in Fig. 2. This position provides an optimum working with OpenFace because the face is frontal and all the facial landmarks, including the eyes ones that are vital for this application, are correctly acquired.

ZED camera was chosen due to its ROS compatibility. The manufacturer provides a wrapper to communicate both systems. Another reason was its high working frame rate (100 fps at VGA), higher that the obtained with an standard web-cam.

In order to emulate the perspective that a driver would have in the real vehicle and to have a good gaze dynamic behavior, an study about the user's position with respect to the screen, as well as a study about the optimal screen resolution and system frame rate have been carried out in next sections.

2.2 Gaze Estimation

We are interested in non-intrusive systems that do not generate rejection nor fatigue in users and that are able to be used on a vehicle. Therefore, head-mounted eye trackers and desktop-mounted eye trackers are refused. Among vehicle-mounted cameras approaches, several gaze tracking techniques have been proposed in the literature [12–14, 17, 19]. OpenFace is one of the most popular open-source facial analysis tools due to its fine performance and robustness. We propose to use OpenFace 2.0 toolkit [3] because it provides facial behavior analysis algorithms including eye gaze. The tool begins detecting the face. Then, it extracts 68 facial landmarks and estimates the eyes gaze direction from these landmarks, as shown on Fig. 1.

The process uses a Conditional Local Neural Field (CLNF) for detecting the facial landmark, including eyelids, iris and pupil. Once the eye and the pupils are located, eye-region data are used to compute the gaze vector for each eye. OpenFace provides three different eye vectors, one for each eye and a third one for the fusion of both eyes. In this project the fusion vector is applied. Eye gaze estimation was evaluated on the challenging MPIIGaze dataset [20] obtaining a mean absolute error of 9.1° per frame, which can be enough for our application assuming this measurement uncertainty.

2.3 Visual Attention Model

In our application, gaze vector is used to know where the person under the test is looking. The vector is given as two angles, $[\theta_x, \theta_y]$, regarding the (X,Y) axis in the camera reference system. This vector is projected on the screen in order to get an attention pixel through a visual model. To minimize gaze vector uncertainty, a Kalman Filter experimentally adjusted, is implemented over the projection. To emulate driver position in the real vehicle, and following the same strategy that the authors in [9], the user is placed in front of a 47" screen, where traffic videos are shown, at a distance of 125 centimetres. In this way, his visual field varies between ±24° on the X axis and 27° on the Y axis. In these ranges the projection model is quite linear. This is the reason because we propose to use a simple projection model learned in a previous calibration step in the same way that the done in a touch screen calibration process [10]. We apply a slight calibration method at the beginning of each experiment that consist on looking at four points of the screen to get the projection matrix and the limits of the screen where the test is run.

Calibration method translates the gaze vector into coordinates that accurately represent the projection of this vector on the screen in pixels. For each calibration point, the gaze vector ($[\theta_x, \theta_y]$) is projected on the screen ($[X_k', Y_k']$) using the trigonometric equations shown in Eq. 2. z is the average distance of the user's eyes regarding the camera reference. This parameter is also provided by OpenFace.

$$z = \frac{z_{right\_eye} + z_{left\_eye}}{2} \tag{1}$$

$$\begin{pmatrix} X_k' \\ Y_k' \end{pmatrix} = \begin{pmatrix} -z \cdot sin(\theta_x) \\ z \cdot sin(\theta_y) \end{pmatrix} \tag{2}$$

The matrix A is formed by the projected points (in cm) associated to the 4 screen points that have been projected during the calibration step (red points in Fig. 2) on camera reference. In this step, these points have to be matched with the known positions of the calibration points on the screen in pixels ($[X_k, Y_k]$) regarding the screen reference (top-left corner), which are: $[X_1, Y_1] = [200, 200]$, $[X_2, Y_2] = [200, 900]$, $[X_3, Y_3] = [1700, 900]$, $[X_1, Y_1] = [1700, 200]$.

$$A = \begin{pmatrix} X_1' & Y_1' & 1 \\ X_2' & Y_2' & 1 \\ X_3' & Y_3' & 1 \\ X_4' & Y_4' & 1 \end{pmatrix} \tag{3}$$

To match these points we use a parametric model composed of the coefficients $[\alpha_X, \beta_x, \Delta_x]$ and $[\alpha_y, \beta_y, \Delta_y]$, which transform the spatial space from centimetres to pixel and change the reference from the camera to the left upper corner of the screen, according to the Eq. 4.

$$\begin{pmatrix} X_1 \\ X_2 \\ X_3 \\ X_4 \end{pmatrix} = A \begin{pmatrix} \alpha_X \\ \beta_X \\ \Delta_X \end{pmatrix} \quad and \quad \begin{pmatrix} Y_1 \\ Y_2 \\ Y_3 \\ Y_4 \end{pmatrix} = A \begin{pmatrix} \alpha_Y \\ \beta_Y \\ \Delta_Y \end{pmatrix} \tag{4}$$

Eq. 5 shows how the coefficients are calculated from Eq. 4.

$$\begin{pmatrix} \alpha_X \\ \beta_X \\ \Delta X \end{pmatrix} = A^{-1} \begin{pmatrix} X_1 \\ X_2 \\ X_3 \\ X_4 \end{pmatrix} \quad and \quad \begin{pmatrix} \alpha_Y \\ \beta_Y \\ \Delta Y \end{pmatrix} = A^{-1} \begin{pmatrix} Y_1 \\ Y_2 \\ Y_3 \\ Y_4 \end{pmatrix} \tag{5}$$

After calibration, each attention point is obtained applying the trigonometric projection of its corresponding vector gaze provided by OpenFace, where $X_{cm} = -z \cdot sin(\theta_x)$ and $Y_{cm} = z \cdot sin(\theta_y)$, and the parametric transformation carried out by Eq. 6.

$$x_{pixel} = \alpha_X X_{cm} + \beta_X Y_{cm} + \Delta_X$$
$$y_{pixel} = \alpha_Y X_{cm} + \beta_Y Y_{cm} + \Delta_Y \tag{6}$$

2.4 Heat Map Visualisation

To represent the spatial distribution of the gaze fixations on the screen, we use the common visualization approach known as heat map. To take into account the trajectory of the gaze, we integrate the spatial distribution of the fixations within a temporal window [t-w,t], where t is the current frame and w the number of last frames. This temporal window has been set to 1 s as done by [2]. In this way, we have 60 samples for frame in order to generate the heat map. With these samples a Gaussian model is built to represent the different areas of colour of the heat map, where ($M_t = [m_x(t), m_y(t)]$) is the centre of the Gaussian, C_t is the covariance matrix and $h(X,t)$ is the intensity of the heat map for the pixel (X=[x,y]). In a practical way, we draw ellipses in the heat map until two times the standard deviation of the Gaussian model.

$$h(X,t) = e^{-\frac{1}{2}(X-M_t)^T C_t^{-1}(X-M_t)} \tag{7}$$

3 Experimental Results

This section presents experimental results obtained for our architecture proposal. Evaluation is divided in two different subsections in order to compare our results with the obtained with other state-of-the-art systems.

3.1 Visual Attention Model Evaluation

To evaluate the precision of our driver attention model we have applied the following testing procedure. After the calibration step, the user has to look at eight points, placed on different positions on the screen, during eight seconds for each one. To get a more stable view, the first two second of each point are despised. This procedure has been done to test the camera parameters, the camera position, the loss of precision by using glasses and finally the performance with different users. The metric used to evaluate the accuracy of the tool is the root mean square error (RMSE) of the gaze projection on the screen and its ground-truth. Accuracy is shown in total and regarding the (X,Y) axis and is evaluated in percentage, pixels and millimeters.

Camera Parameters. To reach optimum results, a previous study about camera parameters was carried out. In this aspect, resolution and camera frame rate were changed to test precision for the different options. ZED camera allows four different resolutions with different frame rates, as we show on Table 1. Performance shows the ability of OpenFace to process frames. This test was done only by one user, but to achieve more robust conclusions each test was repeated 5 times.

Results show that, on the X axis, best performance is get by VGA resolution at 100 frames per seconds. On the Y axis, best performance is get by HD720. In total, best accuracy is obtained for HD720. In conclusion, this last configuration is taken for the following tests. Also performance shows that OpenFace is not able to work at 100 Hz with an VGA resolution.

Table 1. Testing accuracy vs camera parameters on a 1920 × 1080 screen. Test done by one person looking at eight points on a screen. Test was done 5 times.

Resolution	Frame rate	Acc_x			Acc_y			Acc_total			Performance (Hz)
		%	pix	mm	%	pix	mm	%	pix	mm	
HD2K	15	1,9	36,5	11,4	5,4	57,9	18,1	4,0	77,2	24,1	14,761
HD1080	30	1,4	27,5	8,6	5,7	61,0	19,1	4,1	79,2	24,7	29,378
HD720	60	1,9	36,1	11,3	**4,0**	43,7	13,6	**3,2**	60,5	18,9	50,74
VGA	100	**0,6**	12,3	3,8	4,7	50,2	15,7	3,3	63,7	19,9	74,63

Camera Position. Camera position is vital for the system working. Three different positions compatible with a real car were tested, all of them aligned with the middle of the screen but at different heights (10, 48 and 78 centimetres over the table where the screen is placed). Results are shown on Table 2 for the same conditions that the explained in the before section. The optimal position for the camera is just in front of the user, but this is not possible in a real car because hides the road scene. Top position is not allowed because face landmarks are occluded most of the time and the model does not work properly. Then, bottom position was chosen.

Table 2. Testing accuracy vs camera position on a 1920 × 1080 screen. Test done by one person looking at eight points on a screen. Test was done 5 times.

Position	Acc_x			Acc_y			Acc_total		
	%	pix	mm	%	pix	mm	%	pix	mm
Top	-	-	-	-	-	-	-	-	-
Mid	**1,9**	36,1	11,3	**4,0**	43,7	13,6	**3,2**	60,5	18,9
Bot	**2,5**	48,0	15,0	**10,6**	114,5	35,8	**7,7**	147,9	46,2

Glasses. Some drivers wear glasses while driving and some other drivers prefer to drive with sun-glasses. Performance of every gaze monitoring system must be analyzed under these common situations. Commercial applications like Tobii Pro Glasses [1] do not work properly because their sensors are on some special glasses and two different glasses cannot be used at the same time.

In our case, precision will get worse for sure due to the glasses hide some face landmarks. To evaluate impaired performance three different test were done: no glasses, with glasses and with sun glasses, in the same conditions that in the before case. Results are depicted on Table 3. For the sunglasses case, our system does not work because eyes landmarks are hidden and gaze can not be calculated. For the glasses case, accurate numbers decrease a little but can be used without problems.

Table 3. Testing accuracy vs glasses on a 1920 × 1080 screen. Test done by one person looking at eight points on a screen. Test was done 5 times

Eyes-glasses	Acc_x			Acc_y			Acc_total		
	%	pix	mm	%	pix	mm	%	pix	mm
Free	**1,9**	36,1	11,3	**4,0**	43,7	13,6	**3,2**	60,5	18,9
Glasses	**2,5**	48,7	15,2	**4,6**	50,0	15,6	**3,7**	71,7	22,4
Sunglasses	-	-	-	-	-	-	-	-	-

Precision Test. Once the best camera parameters and camera position have been found, it is time to test the accuracy of our model with different users. 25 users were requested for the tests. Firstly, they were asked to calibrate the tool looking at 4 points as shown on Fig. 2. Age of the users were between 17 and 55 years old and 3 of them wore glasses during the test.

Results for all of the users can be seen on Fig. 3. Red circles indicate the goal points where the person has to look at. The heat map represents the spatial distribution of the data, divided in this case in eight different clusters, modelled by a Gaussian Mixture Model (GMM) of eight Gaussian. The heat map for each Gaussian indicate the probability of a pixel to belong to each cluster. In theory, the centre of each Gaussian should be placed over the corresponding calibration point and its covariance indicates the measurement uncertainty.

Our method achieves about 1.9% error on the X axis and 4% error on Y axis, being bellow the reported by OpenFace for the MPIIGaze dataset (9.1° equivalent in our case to 19% on X axis and 34% on Y axis) [20]. This comparison should be taken with caution since in our case gaze movements are limited and the number of users who undergo the test is less.

Table 4. Testing accuracy on a 1920 × 1080 screen. Test done by one person looking at eight points on a screen. Test done by 25 users looking at eight points on a screen.

	Acc_x			Acc_y			Acc_total		
	%	pix	mm	%	pix	mm	%	pix	mm
25 people	**1,9**	36,1	11,3	**4,0**	43,7	13,6	**3,2**	60,5	18,9

Results on Y axis are worse due to the number of pixels for the vertical field of view range is smaller that the corresponding to the horizontal one. Despite these errors are considerable, they can be enough for our application. Figure 5 shows the error overprinted on an image with a Crash Object in order to compare scales. As we can see, in spite of the measurement uncertainty the object is perfectly detected in the frame, which demonstrate our method can be used to localize accidents in complex scenarios.

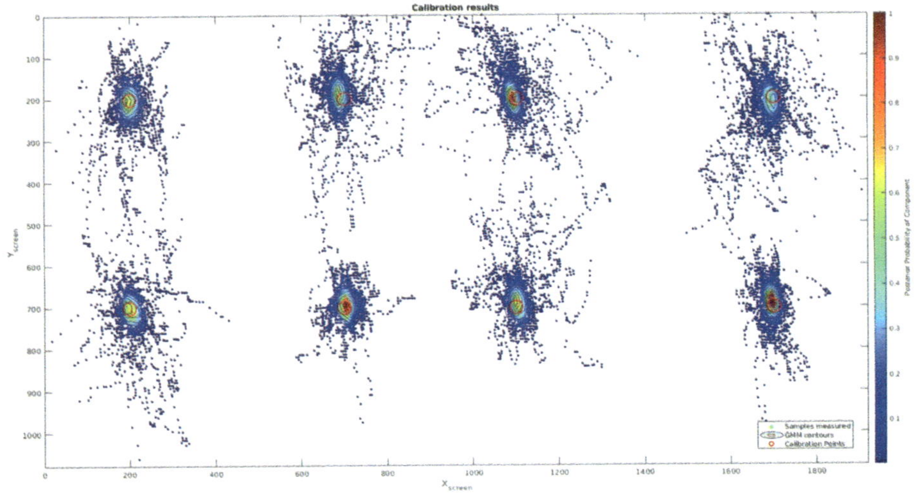

Fig. 3. Calibration results. 8 points are displayed on the screen (Red circle). The results from our output creates 8 Gaussian around the testing points.

3.2 DADA2000 Evaluation

Once shown the potential of our visual attention method to localize objects of a certain size in an image, we are ready to test it on a challenging video benchmark with driver attention and driving accidents annotated to evaluate the performance of our proposal to estimate accidents in video sequences. To have a clear idea of the quality of our method we will compare results with the obtained using an expensive and active desktop-mounted eye-tracker watching the same videos in similar conditions. Different options were raised, such as MIT300 [5], CAT2000 [4], DR(eye)VE [2] and DADA2000 [9]. The two first ones were discarded because they are not focused on drivers. Between the two lasts, authors decided to use DADA2000 because it is focused on real traffic accidents, which are the critical moments to evaluate the driving attention.

DADA2000 was done collecting accident videos from different websites, and it is composed of 658,476 frames in 2000 videos with a resolution of 1584 × 660 pixels. At the time of this work, only half of data is public. Then, we use 1000 videos for the experiments, which are classified in three groups: training (598 videos), validation (198 videos) and testing (222 videos). The videos are divided into 54 accident categories classified into two large groups, ego-car involved and ego-car uninvolved. Videos are recorded in different environment such as illumination conditions (day and night), weather (sunny, rainy and snowy) and occasion (urban, rural, highway and tunnel). The dataset provides the following information: annotated crash-objects position in pixels per frame, attention map for each frame and temporal window for each accident indicating the involved frames. Videos are partitioned in three main clips: before the accident, accident and after the accident, as it is depicted in Fig. 4. Also, the accident clip is divided in three sub-sections for a better analysis (Start, Mid and End).

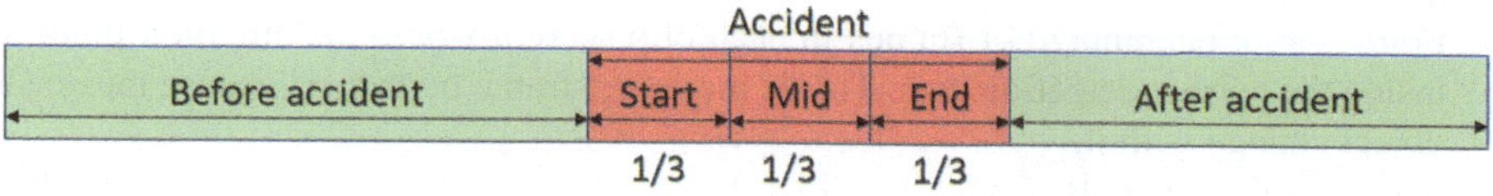

Fig. 4. Temporal partition of the video. Dividing the accident part in three pieces of same length to evaluate the results by sections.

Fig. 5. DADA2000 frame. Crash object circle for 60 and 300 pixels (blue). Error area (red).

Crash Object Detection in DADA2000 Benchmark. To test Crash Object detection, we employed 4 volunteers who were invited to watch the testing section of the dataset, composed by 222 videos. For each visualization, the output of our visual attention model was recorded in a synchronized way with the video. In order to wide the working field of view, dataset was resized from 1584 × 660 to 1920 × 1080 to match it with the screen resolution used in the experiment. The clips were played at 30 fps to compare our results with the obtained by the authors of DADA2000 [9] in the same terms. We recorded attention map without temporal aggregation, obtaining two measures per frame, because our system run at 60 Hz.

The goal of this test is to localize driver attention on a screen. We use the same metrics as the authors of the dataset to evaluate this parameter. We measure the distance between the Crash Object center (ground-truth manually annotated) and the attention points obtained for each frame. If this distance is below a certain threshold (Th) in pixels the detection is consider as true and false if it is higher. Figure 5 shows a frame with the Crash Object an its true detection area, represented by a circle, for a Th equals to 60 and 300 pixels. Performance detection is solved as a binary problem: True, if the attention point falls inside the circle, and False, if it falls outside. For this work, we only have tested experiments with the priori information option (with-priori), that means users are told that they have to find crash-objects in the sequences. With these premises we obtain two indicators:

- Frame ratio: percentage of frames in each clip section where the attention point is inside the correct detection area. This is measured frame by frame because the crash object change with it.
- Success rate: percentage of clips which frame ratio for the entire clip is over 50%. Clips refers different video sub-sections defined for the accident.

Table 5. Success rate: percentage of number of clips that frame ratio for the entire clip is more than half of the frames of the entire accident scene.

System	Ours (Based on OpenFace)			DADA2000		
Th (pixel)	Start	Mid	End	Start	Mid	End
60	9,4%	18,8%	10,9%	10,8%	10,7%	5,1%
100	39,1%	48,4%	29,7%	34,4%	30,8%	22,6%
160	76,6%	78,1%	57,8%	59,0%	57,2%	49,1%
200	81,3%	87,5%	65,6%	68,0%	67,2%	60,1%
260	90,6%	90,6%	71,9%	76,6%	76,6%	71,5%
300	90,6%	93,8%	78,1%	80,0%	81,3%	76,6%

Results for different detection area sizes and a comparison with the numbers presented by the DADA2000 authors using a Senso Motoric Instrument (SMI) RED250 desktop-mounted infrared eye tracker are shown on Table 5. Our success rate is higher than the obtained by the DADA200 authors for all the partitions except for the start one with a threshold of 60 pixels. However, our numbers are obtained for 4 users per video instead of the 5 they argue to use, even though they also argue every video was watched by at least two people, indicating that not all of were watched by the 5 users. For a 300 pixels threshold we have obtained more than 90 % of success rate for start and mid sections of the accident event. The end section always get less success rate because, at this part of the video, users normally fix their attention in other part of the scene. Results are or par or even better than the claimed by the authors of DADA2000 but it must be remarked that our experiment has been recorded on a 47 in. screen instead on a 24 in. screen. Results shows that early accident detection is worse than for the middle partition because users take some time to react. In the last part of the accident some users under test keep their eyes out the accident and the late accident detection gets worse.

Figure 6(a) shows the success rate and frame ratio curves, obtained for a threshold between 60 and 1000 pixels and for the three sub-sections presented in the accident partition, in the same way that the shown by the dataset authors in [9]. Our results outperform the obtained for our baseline for all the analyzed sub-sections.

Heat Attention Map on DADA2000. Heat attention map is a good visualization tool to represent where a user is looking when he is driving. To represent the trajectory of the gaze, a temporal window is used. For the presented experiments the temporal window has been set to 1 s, like in the DR(eye)VE project.

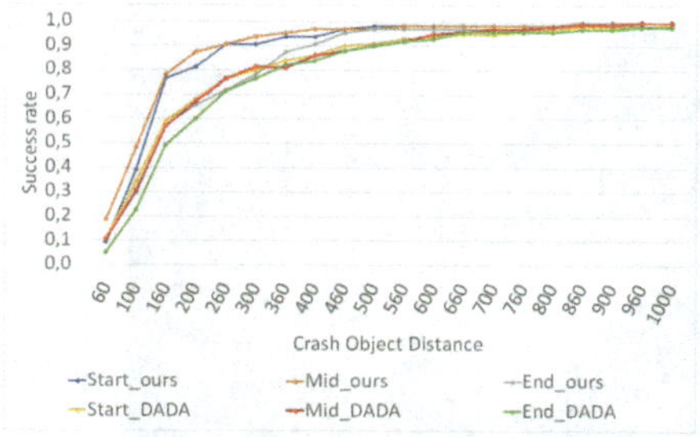

(a) Success rate obtained by the system proposed.

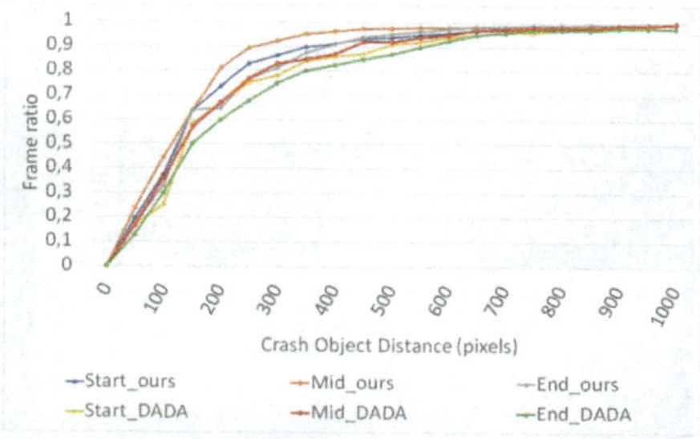

(b) Frame ratio obtained by the system proposed.

Fig. 6. Results obtained on the testing

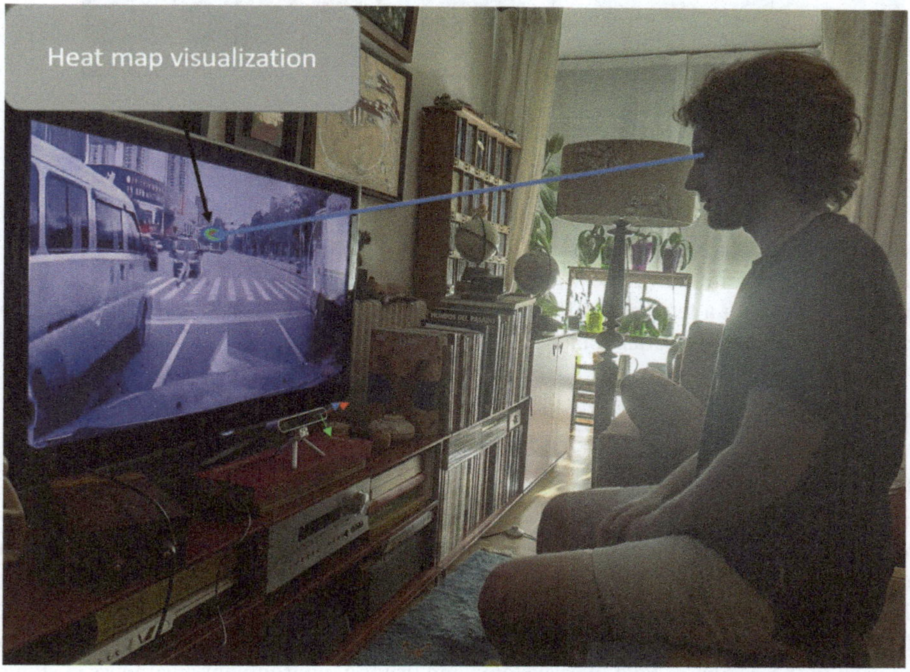

Fig. 7. Heat attention map captured on real time during a DADA test.

Figure 7 shows the way this experiment has been done. The user is looking at the accident while the camera is recording him from a lower position. The attention map generated by the samples is shown. As we can see, our attention map is included in the baseline map, showing the usability of our tool but using a cheaper and non-intrusive sensor.

Fig. 8. Attention map with temporal aggregation for the frame t, t-1 s and t-2 s.

Figure 8 shows how temporal aggregation on heat attention map can help to predict where the accident will happen. The figure shows 2 accident situations with their corresponding maps in the image where the accident takes place as a heat map. Images displayed are with a temporal difference of 1 s. As we can see, the heat map is able to correctly predict all the crash objects.

4 Conclusions and Future Works

This paper presented a new low-cost and not intrusive method to get visual attention maps, based on a passive vehicle-mounted camera, as an alternative to the head-mounted eye-tracker and the active desktop-mounted eye trackers (our baseline), which are intrusive and costly. Our proposal has been validated in the challenging public dataset DADA2000, showing on par results with our base line in similar conditions. This fact confirms our technique is a good tool for driver attention monitoring able to be used in the design of take over systems and driving environment awareness systems for automated vehicles.

As future works we plan to implement a more sophisticated attention model, to obtain saliency maps with temporal windows on challenging driving scenarios using deep-learning techniques and to test our proposals in our simulator based on CARLA [8].

Acknowledgements. This work has been funded in part from the Spanish MICINN/FEDER through the Techs4AgeCar project (RTI2018-099263-B-C21) and from the RoboCity2030-DIH-CM project (P2018/NMT- 4331), funded by Programas de actividades I+D (CAM) and cofunded by EU Structural Funds.

References

1. Tobii pro glasses 2 (2020). https://www.tobiipro.com/product-listing/tobii-pro-glasses-2/
2. Alletto, S., Palazzi, A., Solera, F., Calderara, S., Cucchiara, R.: Dr (eye) ve: a dataset for attention-based tasks with applications to autonomous and assisted driving. In: Proceedings of the IEEE Conference on Computer Vision and Pattern Recognition Workshops, pp. 54–60 (2016)
3. Baltrusaitis, T., Zadeh, A., Lim, Y.C., Morency, L.P.: Openface 2.0: facial behavior analysis toolkit. In: 2018 13th IEEE International Conference on Automatic Face & Gesture Recognition (FG 2018), pp. 59–66. IEEE (2018)
4. Borji, A., Itti, L.: Cat2000: a large scale fixation dataset for boosting saliency research. arXiv preprint arXiv:1505.03581 (2015)
5. Bylinskii, Z., Judd, T., Borji, A., Itti, L., Durand, F., Oliva, A., Torralba, A.: MIT Saliency Benchmark (2019)
6. Cognolato, M., Atzori, M., Müller, H.: Head-mounted eye gaze tracking devices: an overview of modern devices and recent advances. J. Rehabil. Assistive Technol. Eng. **5**, 2055668318773,991 (2018)
7. Dalmaijer, E., Mathôt, S., Stigchel, S.: Pygaze: An open-source, cross-platform toolbox for minimal-effort programming of eyetracking experiments. Behav. Res. Methods **46** (2013). https://doi.org/10.3758/s13428-013-0422-2
8. Dosovitskiy, A., Ros, G., Codevilla, F., Lopez, A., Koltun, V.: CARLA: an open urban driving simulator. In: Proceedings of the 1st Annual Conference on Robot Learning, pp. 1–16 (2017)
9. Fang, J., Yan, D., Qiao, J., Xue, J., Wang, H., Li, S.: Dada-2000: can driving accident be predicted by driver attention*f* analyzed by a benchmark. In: 2019 IEEE Intelligent Transportation Systems Conference (ITSC), pp. 4303–4309. IEEE (2019)
10. Fang, W., Chang, T.: Calibration in touch-screen systems. Texas Instruments Incorporated **10** (2007)
11. Jimenez, F.: Intelligent Vehicles: Enabling Technologies and Future Developments. Butterworth-Heinemann, Oxford (2017)
12. Jiménez, P., Bergasa, L.M., Nuevo, J., Hernández, N., Daza, I.G.: Gaze fixation system for the evaluation of driver distractions induced by ivis. IEEE Trans. Intell. Transport. Syst. **13**(3), 1167–1178 (2012)
13. Mizuno, N., Yoshizawa, A., Hayashi, A., Ishikawa, T.: Detecting driver's visual attention area by using vehicle-mounted device. In: 2017 IEEE 16th International Conference on Cognitive Informatics & Cognitive Computing (ICCI* CC), pp. 346–352. IEEE (2017)
14. Naqvi, R.A., Arsalan, M., Batchuluun, G., Yoon, H.S., Park, K.R.: Deep learning-based gaze detection system for automobile drivers using a nir camera sensor. Sensors **18**(2), 456 (2018)
15. Palazzi, A., Abati, D., Solera, F., Cucchiara, R., et al.: Predicting the driver's focus of attention: the dr (eye) ve project. IEEE Trans. Pattern Anal. Machine Intell. **41**(7), 1720–1733 (2018)
16. Shen, J., Zafeiriou, S., Chrysos, G.G., Kossaifi, J., Tzimiropoulos, G., Pantic, M.: The first facial landmark tracking in-the-wild challenge: benchmark and results. In: 2015 IEEE International Conference on Computer Vision Workshop (ICCVW), pp. 1003–1011 (2015). https://doi.org/10.1109/ICCVW.2015.132
17. Vicente, F., Huang, Z., Xiong, X., De la Torre, F., Zhang, W., Levi, D.: Driver gaze tracking and eyes off the road detection system. IEEE Trans. Intell. Transport. Syst. **16**(4), 2014–2027 (2015)
18. Xia, Y., Zhang, D., Kim, J., Nakayama, K., Zipser, K., Whitney, D.: Predicting driver attention in critical situations. In: Asian Conference on Computer Vision, pp. 658–674. Springer, Cham (2018)

19. Yang, L., Dong, K., Dmitruk, A.J., Brighton, J., Zhao, Y.: A dual-cameras-based driver gaze mapping system with an application on non-driving activities monitoring. IEEE Trans. Intell. Transport. Syst. **21**, 4318–4327 (2019)
20. Zhang, X., Sugano, Y., Fritz, M., Bulling, A.: Mpiigaze: real-world dataset and deep appearance-based gaze estimation. IEEE Trans. Pattern Anal. Machine Intell. **41**(1), 162–175 (2017)

Deep-Learning and Robotics

Embedded Deep Learning Solutionbreak for Person Identification and Followingbreak with a Robot

Ignacio Condés[1(✉)], José-María Cañas[2], and Eduardo Perdices[2]

[1] Universidad Carlos III de Madrid, Leganés, Madrid, Spain
ignacio.condes.m@gmail.com
[2] Universidad Rey Juan Carlos, Fuenlabrada, Madrid, Spain

Abstract. This research presents a robust embedded system for following a person, making use of a pipeline of convolutional neural networks. Besides, it features an optical tracking system for supporting the inferences of the neural networks, allowing to determine the position of a person using an RGBD camera. The system is deployed using ROS, and runs in a NVIDIA Jetson TX2, an embedded SoM (*System-on-Module*), capable of performing computationally demanding tasks onboard, and coping with the complexity required to run a robust tracking and following algorithm. The board is attached to a robotic mobile base, which receives velocity commands to move the system towards the target person.

Keywords: Deep learning · Computer vision · Robotics · Person following

1 Introduction

Last decades, the production prices of digital cameras and high-resolution sensors have been greatly reduced, bringing these devices into the consumer market segment. This, beside an increase in the hardware performance, has resulted in a strong drive for the computer vision research.

In particular, *deep learning* has been massively employed for addressing high complexity tasks, such as language understanding, speech recognition and computer vision problems.

Moreover, robotics applications can be really useful at daily tasks, specially when a robot behavior tends to emulate the human one. This requires a polished (and somehow complex) behavior, triggered by a certain input. Depending on the source of the behavior, two main branches emerge in robotics: *teleoperated robots*, useful for hazardousness or high-precision environments, and *autonomous robots*, which aim to act by themselves. The decision-making and actuation capabilities rely on the robot, due to a compromise in the design, or for purpose-specific systems, such as social robots [1,2].

L. M. Bergasa et al. (Eds.): WAF 2020, AISC 1285, pp. 291–304, 2021.
https://doi.org/10.1007/978-3-030-62579-5_20

The important advances on computer vision and robotics allow to explore outstanding synergies, as this work aims to take advantage of: obtaining a robot capable of following a certain person, navigating towards them on a robust reactive behavior, and using deep-learning-based visual perception. As it will be explained later, this behavior is composed of two main components: the *perception block*, in charge of processing the images from an embedded RGBD camera, and the *actuation block*, which moves the robotic base accordingly to the relative position of the person to be followed. This application can be specially interesting on social robots, which are designed to follow a person at home or in a hospital [3].

This research improves the system proposed on [4], where a neural-network-based following system was run in a standard laptop, with a camera and a robot plugged.

2 Related Works

The person following behavior is a old problem in Human-Robot-Interaction field with many existing solutions in the literature [5,6]. Another approach [7] uses laser sensors to detect the legs and a particle filter to provide robustness and continuity to the person estimation.

The topic has been extensively explored using a single camera [6] or stereo vision. [8] combines clothes color and position information to detect and track multiple people around, using Kalman filters and fusing the plain-view map information with a face detector. [9] uses depth templates of person shape applied to a dense depth image and an SVM-based verifier for eliminating false positives, employing an Extended Kalman Filter to continuously estimate the person positions. In addition, Histograms of Oriented Gradients are a classical and effective method for human detection [10], which is a relevant ingredient of the person following behavior.

In the last years, other approaches make use low cost RGBD sensors [11,12], taking advantage of their depth information. It simplifies not only the robustness of the detection itself but also the control of the robot movements, which can be the developed as position-based controllers. In [13] deep learning techniques on color and motion cues have been used to successfully detect people on RGBD videos for surveillance applications.

The identification of a particular person is a complementary problem to general people detection. There are image-based methods [14] and video-based methods [15]. Recent papers also use deep learning to identify the detected persons [16]. Other methods use features and skeleton keypoints [17]. [18] use image and range data combining color, height and gait features (step length and speed) for a robust identification, even in severe illumination environments. [19] is an interesting work that uses neural networks to both multiple person detection and a particular person re-identification, working in real-time on moderate hardware using a single camera.

Deep learning techniques yield interesting results as well. [20] uses deep neural networks in order to extract features from an image in order to identify the person in the image. Other approaches such as [21] make use of recurrent networks and siamese architectures to reidentify a person.

3 Design

The proposed system features a *Perception* module and an *Actuation* module, as shown in Fig. 1. The *Perception* module is responsible for extracting pertinent information from the image: the position and identity of the person to be followed. This information serves as input to the *Actuation* module, which takes a decision about the action to be performed in order to move the robot towards the person. These movements have to be reactive, happening as soon as possible whenever the person position changes.

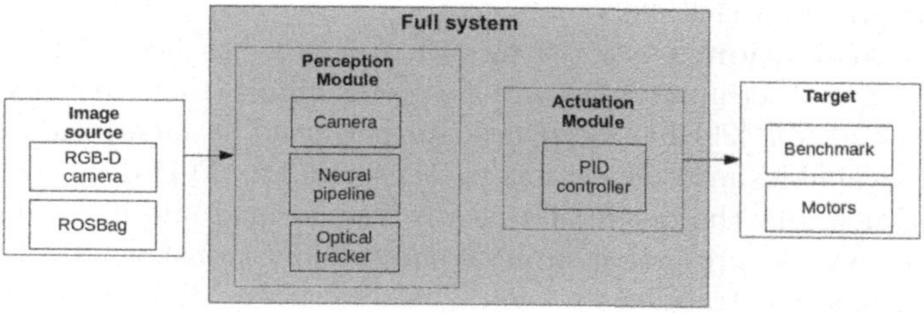

Fig. 1. Architecture of the proposed solution.

3.1 Perception Module

This module processes the camera images for determining the location of the person to be followed. It is composed of the components shown in Fig. 1, which are described below.

Camera: The robot uses an RGBD device (ASUS Xtion Pro Live) as visual sensor. The system makes use of ROS topics to retrieve the latest *(RGB, depth)* pair on an asynchronous way.

It supports both an *online* source (camera) or an *offline* source (recorded ROSBag), with a transparent adaptation to the rest of the system.

Neural Pipeline: Images are passed through a pipeline of three neural networks, which detect the persons in the scene and identify which one must be followed. The main objective of this work is to take advantage of the robustness offered by deep-learning-based techniques.

1. **Person Detection:** For addressing this task, several deep models have been tested, varying the base network architecture and its depth. Since one of the objectives of the system is to work on a portable (low-power) system, only

those architectures which yield a good performance with a sufficiently low inference time are considered. The two tested models for this purpose are the SSD-MobileNet [22,23] using a MobileNet [23] for feature extraction, and the *tiny* version[1] of YOLOv3 [24]. These models are already trained and publicly available on the TensorFlow Model Zoo [25] and on repositories hosted on GitHub[2]. In-depth tests have been conducted to compare the performance of these two models, as it will be described in Sect. 4.

2. **Face Detection:** This problem can also be addressed using a detection neural network. Since the models previously described are not suitable for detecting faces, the adopted solution is a single-class detection system. The network trained in [26] implements a two-stage neural network capable of detecting faces. This detector is based on YOLOv2 [27], which ensures a high-speed and efficient detection based on a class-specific neural network. [26] contains a video sequence comparing the accuracy of this system against a classical Haar cascade approach [28].
3. **Face Identification:** Once the face of a person has been detected, it can be used as a discriminant feature for determining their identity. For this purpose, *FaceNet* [29] has been used to perform the identification, using a publicly available implementation in TensorFlow[3]. This deep identification system transforms the image of a face is transformed into a 128-dimensional vector, known as projection or *embedding*. This transformation is learned after a triplet-loss training process, which separates different faces as much as possible, while projecting similar faces as close as possible. Ideally, it produces similar projections when two images of the same face are evaluated, despite different lighting conditions (as a channel-wise normalization step is performed before passing the image through the network).

The described neural pipeline provides *person locations, face detections and face projections* from a single image, taking advantage of the flexibility and robustness that deep learning methods offer, in order to address three different problems in an efficient way. Its functionality has been depicted in Fig. 2.

Furthermore, these neural networks have been optimized using the library TensorRT[4]. This engine modifies segments from the architecture of a given network, tuning certain parameters:

MSS (*Minimum Segment Size*): the threshold above which a segment is selected to be replaced by the TensorRT optimization. Increasing this value makes the optimizer more selective, optimizing only the heaviest segments of the network. A low value may cause an excessively high overhead, resulting in a worse performance than using the original graph.

[1] The tiny version of YOLOv3 is used to the limited memory on the Jetson TX2 board. The full YOLOv3 model requires more memory than the available size.
[2] https://github.com/mystic123/tensorflow-yolo-v3.
[3] https://github.com/davidsandberg/facenet.
[4] https://developer.nvidia.com/tensorrt.

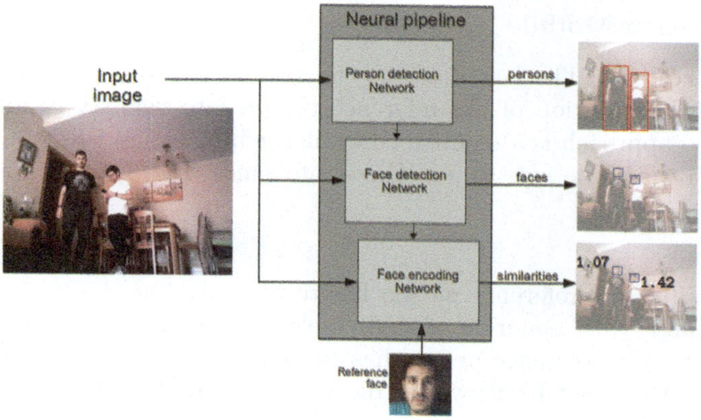

Fig. 2. Neural pipeline, showing the cascade of the three neural networks used to output persons, faces and similarities with the reference face.

MCE (*Maximum Cached Engines*): TensorRT keeps a cache of engines on runtime, with the purpose of reducing the time spent for loading them into the GPU. This parameter modulates the amount of engines cached.

Precision mode: typically, the weights and parameters of the trained neural networks are handled as 64-bit floating point numbers. A reduction in the precision to 32-bit or 16-bit achieves very similar results (as shown on Sect. 4), making the operations much lighter. An additional reduction to 8-bit numbers can be performed as well, mapping the weights values to 256 steps. This greatly reduces greatly the inference time.

Tuning these parameters led to optimized versions of the networks included in the neural pipeline, as Sect. 4 shows.

Motion Tracker: The previously depicted *Neural* component outputs reliable inferences with a certain refresh rate, namely k frames. If k is too high, the system may be affected by an important delay when the movement is performed. This may lead to unsteady movements, increasing the probability of losing the reference person. To avoid this, a *Motion Tracker* component is added to the system. Its functionality is to *estimate* the person movement along k frames, while the neural pipeline is performing the next detection.

The method chosen for this purpose is a *Lucas-Kanade* visual tracker [30]. This technique estimates the *motion field* between the images taken in two time instants, addressing the problem using a differential approach [31]. However, in this use case, the objective is not to compute the entire optical flow, but limited to the pixels inside and surrounding the persons in the scene. For this purpose, the tracking is performed on the corners inside the detected persons. These corners are found using a Shi-Tomasi corner detector [32].

3.2 Actuation Module

PID Controllers: The last block of the system is responsible of translating the location information of the reference person into velocity commands. The implemented approach is a reactive controller, which moves the robot aiming to keep the person inside *safe zones*. These safe zones (Fig. 3) can be described as follows:

Angular zone: the reference person has to be placed at the horizontal center of the image, with a margin of ±50 pixels on the sides.

Linear zone: the reference person has to be placed at a distance of 1 m with respect to the robot front, with a distance margin of ±30 cm.

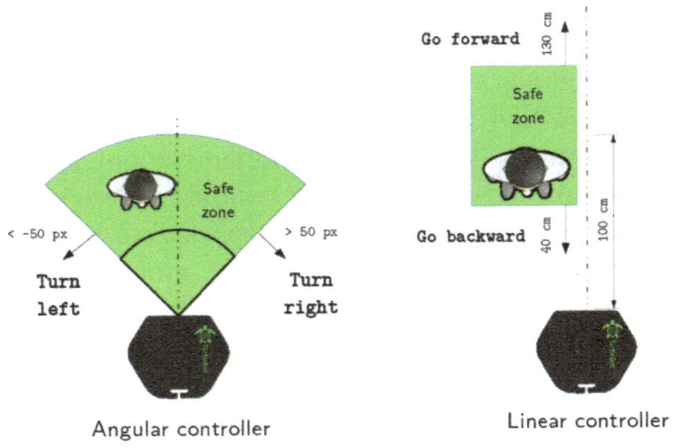

Fig. 3. Safe zones for each controller.

To place the person inside these safe zones, the robot has to move on certain directions. For determining a movement, an *error* vector (e_x, e_w) is computed (Fig. 4), using the tracked person coordinates:

e_x**:** the linear error is computed using the depth image, estimating the distance from the robot to the person. Since the Xtion sensor registers the depth image into the RGB one, the person coordinates can be used in the depth image in order to find the distance of each pixel inside the bounding box of the reference person: the *person depth map*. For avoiding accidental samples of the background of the person, the edges of the depth map are trimmed. Later, a 10×10 grid is computed to have 100 uniformly distributed samples of the depth of the person. The linear error can be computed as the median value of the samples.

e_w**:** the angular error is computed computing the difference on the horizontal coordinate between the image center and the center of the bounding box of the reference person.

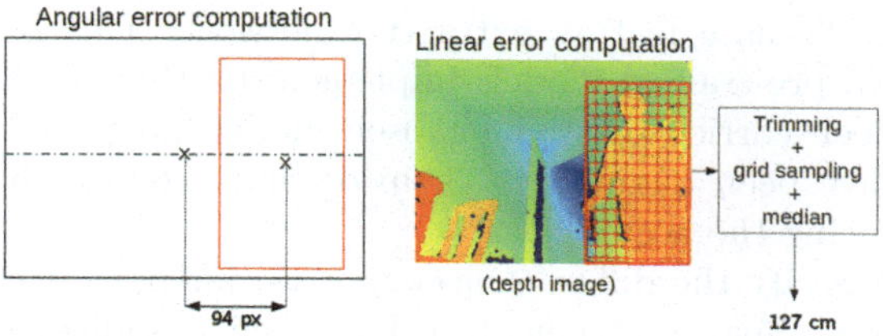

Fig. 4. Error computation on each controller.

These errors yield a reactive response, dependent on the relative distance between the person and the safe zones. In order to compute the suitable velocity commands (linear and angular responses), a PID controller is implemented for each dimension, alleviating overshooting and inertial behaviors. The PID parameters (k_p, k_i, k_d) were tuned to deliver a soft response on the robot movements. Table 1 shows the chosen PID values for following a wandering person on typical indoor conditions.

Table 1. Optimal found values for the parameters in each PID controller.

	Linear	Angular
k_p	0.4	0.005
k_d	0.04	0.0003
k_i	0.05	0.006

4 Experiments

Several experiments have been conducted in order to tune the parameters for improving the system performance, accuracy and robustness, evaluating the same video sequence under different configurations. A recurrent metric on these test is the IoU score, which assesses the overlapping quality between the predicted and the true bounding boxes. As this metric requires ground-truth labels on the videos, the annotation tool LabelMe[5] has been used to indicate the ground-truth position of the persons on each sequence.

4.1 Person Detection Experiments

This experiment compares the performance of the two object detection architectures tested on our system: YOLO [24] and SSD [22]. In the case of YOLO, the implemented version is YOLOv3, in its *tiny* version. For an SSD-based detector, on a real-time application the most convenient variants are those which

[5] https://github.com/wkentaro/labelme.

use a MobileNet [23] as a feature extraction network. The TensorFlow Model Zoo[6] offers several pre-trained models implementing this network, along which a selection has been carried out. The chosen model integrates a MobileNet v1 whose weights have been quantized [33] in order to reduce the computational cost without reducing the accuracy.

In order to quantify the different accuracy vs. inference time tradeoffs that these architectures offer, a specific test has been designed. A specific video sequence of 721 frames long has been recorded, containing a person wandering across the field of view of the camera. For every frame of the sequence, the persons are detected using YOLO and SSD respectively, and the IoU with the ground-truth labels, as well as the inference time have been measured, as it can be seen on Fig. 5. Some gaps can be noticed on the detections, corresponding to frames where the person was out of the sight of the camera.

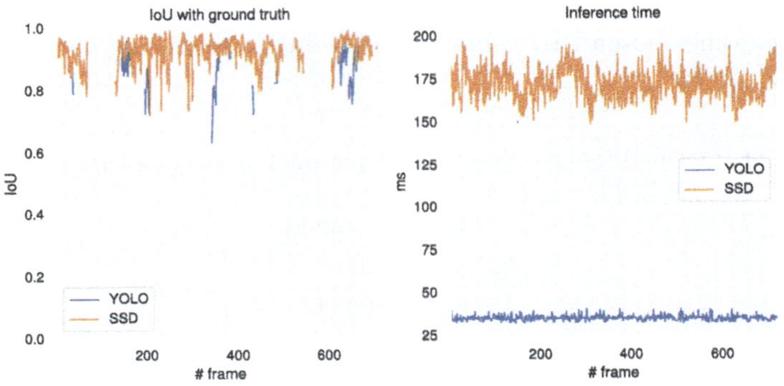

Fig. 5. Results of the person detection test: IoU score with ground truth (left) and inference time per frame (right). A discontinuity represents absence of detections.

The YOLO-based detector offers a slightly minor average IoU than the SSD-based one (0.858 and 0.926 respectively), while taking 5 times less average time to make inferences (35 ms vs. 172 ms). On these terms, the YOLO-based detector seems much more efficient. However, the YOLO detector is able to find the person only in 17% of the frames, whereas the SSD one finds the person in 74%. However, these percentages are a relative measurement for comparing both models, since not all the frames in the sequence contained a person.
This shows that the YOLO detector is too dependent on pose and lighting conditions for the detections to be successful. On the other hand, the SSD detector yields steady predictions, only cutting on the periods where the person was truly out of the field of view. Hence, this system is much more robust for our application scenario, so YOLO is discarded in favor of SSD.

[6] https://github.com/tensorflow/models/blob/master/research/object_detection/g3doc/tf2_detection_zoo.md.

4.2 Face Detection Experiments

This experiment is devoted to compare the performance of Haar cascade classifier and the neural face detection, using the same video sequence than the previous experiment. For each frame in the sequence, the faces are extracted using each one of the described methods, and the IoU score is computed with the ground truth face bounding box. The result can be visualized in Fig. 6.

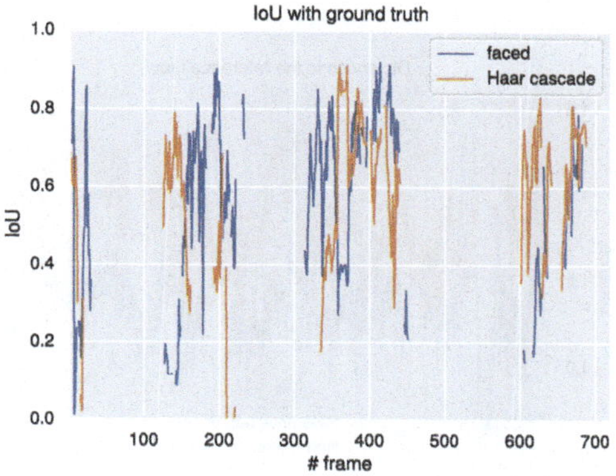

Fig. 6. IoU score with the ground truth for each one of the face detection systems.

It can be seen that both methods yield similar IoU scores and drop at the same time when the person turns their back to the camera. However, the `faced` implementation (which uses deep learning to predict the face positions) is capable of keeping a non-zero IoU at several instants where the Haar performance drops to zero. This is due to pose variances of the person, as the main drawback of the Haar cascade classifier is that it is only capable of detecting frontal faces, dropping the performance whenever the person turns the face towards a side. Both methods yield similar IoU on average, but the deep-learning approach, `faced`, detects a face in 36.89% of the frames, whereas the Haar cascade slightly drops the detection rate to 34.40%. Hence, this test validates the improvement of the face detection performance when using a specific neural network trained for that purpose.

4.3 Face Recognition Experiments

The last component of the neural pipeline is a *face recognition* neural network, devoted to confirm the identity of the reference person. This subsystem is based on a FaceNet [29] network, which projects a face into a 128-dimensional space. These projections are used by the proposed system, as their Euclidean distance to the projection of a reference face is used to determine if the input face belongs to the reference person.

This experiment is designed to assess the quality of the projection system, which should yield far points for a different face and near points for a matching face. For this proposal, a video sequence was recorded containing two persons wandering in front of the robot. The faces of each frame are labeled, separating the faces of the two persons in two different classes. For computing the distance, a reference face was set using an image, and the distance to the reference face of each one of the faces in the video was stored. The result can be observed on Fig. 7.

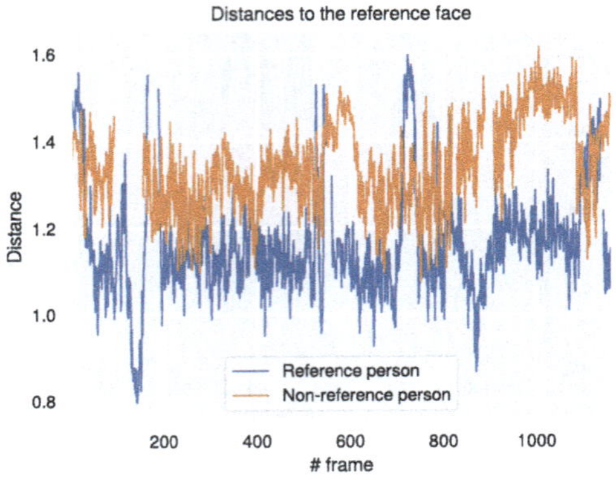

Fig. 7. Resulting distance for the two faces in the video.

The results obtained on Fig. 7 allow to extract two conclusions about the quality of the projections of the faces. First, the encodings of the reference person have an overall remarkable stability. On average, the obtained projections for every frame are located at an approximate distance of 1.16 (threshold chosen for accepting a person as the reference one). Exceptional rises in the distance are due to changes in the pose of the face and occlusions, that reduce the quality of the projection. Second, the encodings of a person different than the reference one have an overall higher distance (1.34) from the reference face. This is convenient for avoiding false positives while determining that a face is the reference one.

These results validate FaceNet as a robust approach to perform person recognition tasks, since the distance of a projection to the reference face has to be below the threshold for being labeled as the reference face.

4.4 Deep Learning Optimization with TensorRT

Among the parameters that allow to configure the TensorRT optimization, there is a reduction in the floating-point precision level of the nodes. This experiment aims to compare the loss of precision this entails with the reduction in the inference time, result of an optimization process on the SSD-based object detector used in subsection 4.1.

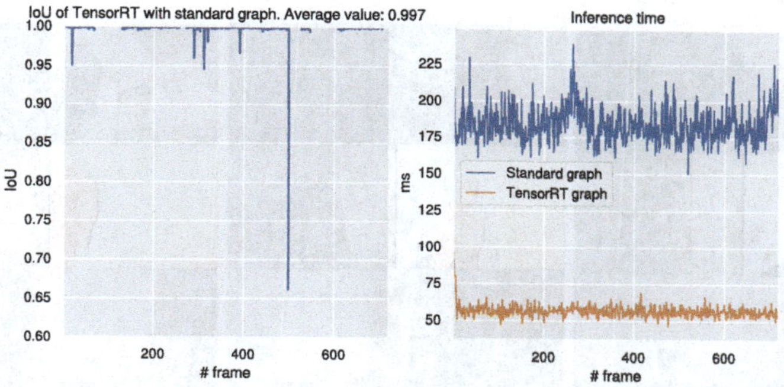

Fig. 8. IoU between the standard graph and the TensorRT graph inferences (left) and inference times for both networks (right).

Figure 8 illustrates the differences between a standard graph and an optimized one. On the one hand, this loss of precision is small, with some observable exceptions with a loss above 5% of the original performance. On the other hand, the inference time gap can be observed as well. The difference is more notorious, as the TensorRT optimized model performs the inferences 3 times faster than the original graph (56.769 ms vs. 184.477 ms). Aside of more stability on the inference time. Given these results, the TensorRT optimizations are a convenient tool to greatly increase the performance of the system, allowing the slower component (the neural pipeline) to experiment an important reduction on the inference time. This greatly improves the overall performance, as reliable neural updates are performed more often.

4.5 Typical Execution

As a result, The whole system running may seen on YouTube[7]. This video presents a sequence where the reference person enters into the field of view of the robot, shows his face to the camera. After some consecutive frames detecting the person, it is identified using the detected face. When the face projection is close enough to the reference one, the robot starts following that person. For each frame, the linear and angular errors are computed, even if the face of the person is not seen anymore, as the system has checked previously that is the right person. If the errors are outside the safe zones, a velocity command is computed and sent to the robot. This routine is executed until the person gets lost, causing the robot to stop waiting for the person to be seen again.

Figure 9 shows the behavior of the robot, expected to follow the person properly. The top region of the images shows several screen captures of the program output, with the RGBD images (left), and the movement commands and tracked persons (right). The bottom region of the plots show the scene recorded from a mobile phone, allowing to observe the performance externally.

[7] https://www.youtube.com/watch?v=WZ0riKMwJWA.

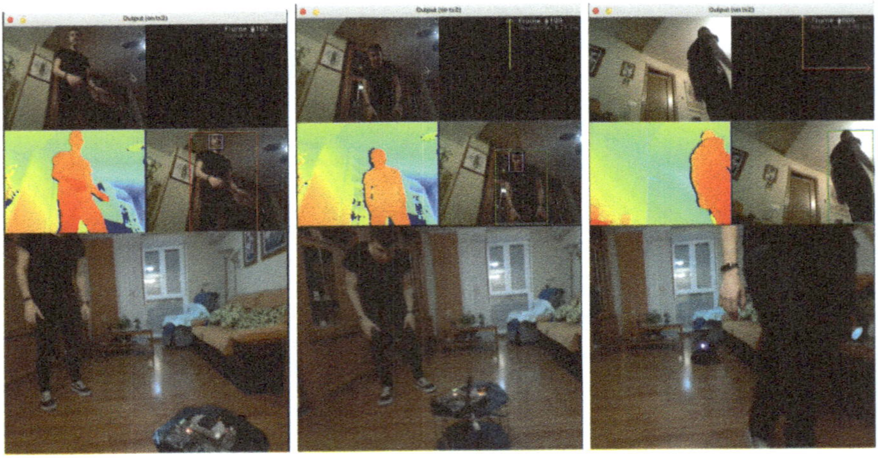

(a) Person detection. (b) Person recognition and following. (c) Following without face feedback.

Fig. 9. Typical execution.

5 Conclusions

The proposed approach presents an embedded system that follows a reference person, relying on the robustness of deep learning for being capable of working in real environments. As it has been shown, the detection and recognition pipeline has been exclusively designed using deep neural networks, ensuring a robust performance in non-controlled environments. This robustness is crucial, given that the camera is located at a very low position and has a vertical inclination in order to see the full body of the person. However, this also causes the camera to perceive an excessive brightness from ceiling lamps, dimming the persons on the image. Classical systems tend to fail due to this issue.

This neural pipeline is by a tracking component, improving the performance under certain issues, such as partial occlusions or a higher inference time if the power available makes the inferences less efficcient.

The main contributions of this research may be summarized as follows:

- The system is mounted on a battery-powered *mobile base*, which features a high-performance GPU embedded on a SoM. This assembly can operate autonomously, without requiring an external computer to perform the deep learning inferences or running algorithms in parallel. Remote monitoring is available, but not required for the system to work.
- Specific optimization engines allow the system to run efficiently 3 neural networks on a low-consumption hardware. These networks perform inferences over the images captured by the RGBD sensor attached to the system. The inferences are devoted to detect the different persons in the scene, as well

as to distinguish them by means of a discriminant feature: their face. Using neural networks for detection and identification tasks allows the system to be robust and reliable.
- The neural system is supported also an optical person tracker. It aims to guess the trajectories followed by each person while the neural network yields a new update, as this tracker predicts the person displacement considerably faster that the neural network. Since neural inferences are also sensitive to flickering caused by occlusions, trusting just on these inferences could easily result on an unsteady behavior. Thus, the introduction of the tracker improves the system robustness and softens the robot movements.

References

1. Pennisi, P., Tonacci, A., Tartarisco, G., Billeci, L., Ruta, L., Gangemi, S., Pioggia, G.: Autism and social robotics: a systematic review, *Autism Research* (2016)
2. Cheong, A., et al.: Development of a robotic waiter system,ll IFAC-PapersOnLine, In: 7th IFAC Symposium on Mechatronic Systems MECHATRONICS 2016, vol. 49, no. 21, pp. 681–686 (2016). https://doi.org/10.1016/j.ifacol.2016.10.679, http://www.sciencedirect.com/science/article/pii/S2405896316322911
3. Gockley, R., Forlizzi, J., Simmons, R.: Natural person-following behavior for social robots. Association for Computing Machinery, New York (2007)
4. Condés, I., Cañas, J.-M.: Person Following Robot Behavior Using Deep Learning: Proceedings of the 19th International Workshop of Physical Agents (WAF 2018), November 22-23, 2018, Madrid, Spain, pp. 147–161 (2019)
5. Sidenbladh, H., Kragic, D., Christensen, H.I.: A person following behaviour for a mobile robot. In: 1999 IEEE International Conference on Robotics and Automation, 1999. Proceedings. vol. 1, pp. 670–675. IEEE (1999)
6. Yoshimi, T., Nishiyama, M., Sonoura, T., Nakamoto, H., Tokura, S., Sato, H., Ozaki, F., Matsuhira, N., Mizoguchi, H.: Development of a person following robot with vision based target detection, In: 2006 IEEE/RSJ International Conference on Intelligent Robots and Systems, pp. 5286–5291. IEEE (2006)
7. Aguirre, E., García-Silvente, M., Pascual, D.: A multisensor based approach using supervised learning and particle filtering for people detection and tracking. In: Robot : Second Iberian Robotics Conference, vol. 2016, pp. 645–657 Springer, Cham (2015)
8. Muñoz-Salinas, R., Aguirre, E., García-Silvente, M.: People detection and tracking using stereo vision and color. Image Vision Comput. **25**(6), 995–1007 (2007)
9. Satake, J., Miura, J.: Robust stereo-based person detection and tracking for a person following robot. In: ICRA Workshop on People Detection and Tracking, pp. 1–10 (2009)
10. Dalal, N., Triggs, B.: Histograms of oriented gradients for human detection. In: 2005 IEEE Computer Society Conference on Computer Vision and Pattern Recognition (CVPR 2005) (2005)
11. Ilias, B., Shukor, S.A., Yaacob, S., Adom, A., Razali, M.M.: A nurse following robot with high speed Kinect sensor. ARPN J. Eng. Appl. Sci. **9**(12), 2454–2459 (2014)
12. Shimura, K., Ando, Y., Yoshimi, T., Mizukawa, M.: Research on person following system based on RGB-D features by autonomous robot with multi-kinect sensor. In: 2014 IEEE/SICE International Symposium on System Integration (SII) (2014)

13. Xue, H., Liu, Y., Cai, D., He, X.: Tracking people in RGBD videos using deep learning and motion clues. Neurocomputing **204**, 70–76 (2016)
14. Kouno, D., Shimada, K., Endo, T.: Person identification using top-view image with depth information. In: 2012 13th ACIS International Conference on Software Engineering, Artificial Intelligence, Networking and Parallel/Distributed Computing (2012)
15. You, J., Wu, A., Li, X., Zheng, W.-S.: Top-push video-based person re-identification. In: Proceedings of the IEEE Conference on Computer Vision and Pattern Recognition (CVPR) (2016)
16. Yoon, Y., Yoon, H., Kim, J.: Person reidentification in a person-following robot. In: 25th IEEE International Symposium on Robot and Human Interactive Communication (RO-MAN) (2016)
17. Munaro, M., Ghidoni, S., Dizmen, D.T., Menegatti, E.: A feature-based approach to people re-identification using skeleton keypoints. In: 2014 IEEE International Conference on Robotics and Automation (ICRA), pp. 5644–5651. IEEE (2014)
18. Koide, K., Miura, J.: Identification of a specific person using color, height, and gait features for a person following robot. Robot. Autonomous Syst. **84**, 76–87 (2016)
19. Welsh, J.B.: Real-time pose based human detection and re-identification with a single camera for robot person following, Ph.D. dissertation, University of Maryland, College Park (2017)
20. Ahmed, E., Jones, M., Marks, T.K.: An improved deep learning architecture for person re-identification. In: Proceedings of the IEEE Conference on Computer Vision and Pattern Recognition (CVPR) (2015)
21. McLaughlin, N., Martinez del Rincon, J., Miller, P.: Recurrent convolutional network for video-based person re-identification, In: Proceedings of the IEEE Conference on Computer Vision and Pattern Recognition (CVPR), June 2016
22. "SSD: Single Shot MultiBox Detector."
23. Howard, A.G., Zhu, M., Chen, B., Kalenichenko, D., Wang, W., Weyand, T., Andreetto, M., Adam, H.: Mobilenets: efficient convolutional neural networks for mobile vision applications (2017)
24. Redmon, J., Farhadi, A.: YOLOv3: An Incremental Improvement (2018)
25. ModelZoo: TensorFlow Object Detection: Model Zoo. https://modelzoo.co/model/objectdetection
26. Itzcovich, I.: Faced: CPU Real Time face detection using Deep Learning (2005)
27. Redmon, J., Farhadi, A.: Yolo9000: Better, faster, stronger (2016)
28. Viola, P., Jones, M.: Rapid object detection using a boosted cascade of simple features, vol. 1, pp. I–511 (2001)
29. "FaceNet: A unified embedding for face recognition and clustering."
30. González-Díaz, I.: Computer Vision: Dense Motion Estimation (2020)
31. Lucas, B., Kanade, T.: An Iterative Image Registration Technique with an Application to Stereo Vision (IJCAI), vol. 81, April 1981
32. Shi, J., Tomasi, U.: Good features to track. In: 1994 Proceedings of IEEE Conference on Computer Vision and Pattern Recognition (1994)
33. GoogleAIBlog, "Accelerating Training and Inference with the Tensorflow Object Detection API." https://ai.googleblog.com/2018/07/accelerated-training-and-inference-with.html

Survival Loss: A Neuron Death Regularizer

Emilio J. Almazán(✉), Javier Tovar, and Alejandro de la Calle

Nielsen Connect AI, Salvador de Madariaga 1, Madrid, Spain
{emilio.almazan,javier.tovar}@nielsen.com
alejandro.de.la.calle.negro@hp.com

Abstract. We found that combining the L2 regularizer with Adam kills up to 60% of filters in ResNet-110 trained on CIFAR-100 as opposed to combining L2 with Momentum. It does not have a significant impact in terms of accuracy though, where both reach similar values. However, we found that this can be a serious issue if the impaired model is used as a pre-trained model for another more complex dataset (e.g. larger number of categories). This situation actually happens in continual learning. In this paper we conduct a study on the impact of dead filters in continual learning when the dataset increases its difficulty over time and more power from the network is required. Furthermore, we propose a new regularization term referred to as *survival loss*, that complements L2 to avoid filters to die when combined with Adam. We show that the *survival loss* improves accuracy in a simulated continual learning set-up, with the prospect of higher improvements as more iterations arrive.

Keywords: Neural networks · Deep learning · Continual learning

1 Introduction

It is not a secret that deep neural networks are hard to train regardless of the task they try to solve. This is related to the fact that the parameter space with respect to the loss function is highly non-convex. Reaching an acceptable local minimum in this surface requires important skills from data scientists in order to find a suitable combination of initial parameter values and hyper-parameters. The optimizer is a key component for ensuring good convergence of the trainings. They define the strategy for updating the parameters of the network. In practice differences between optimizers performance are not in terms of accuracy which vary little from 94.84% with Adam to 95.23% with Momentum [17] in Cifar-10, but rather on the convergence speed and sensitivity to hyper-parameter values. We found another important difference in terms of percentage of filters with zero or almost zero norm (i.e. dead filters). In this paper we present how some optimizers are more prone to kill filters than others when combined with L2 regularization.

L. M. Bergasa et al. (Eds.): WAF 2020, AISC 1285, pp. 305–317, 2021.
https://doi.org/10.1007/978-3-030-62579-5_21

A very general definition of "dead filter" could be a filter that does not detect discriminative features in the input images. In other words, it always returns the same value regardless the input data. The most common behavior of dead filters is to return zero values. These filters can be particularly harmful during training if they are followed by a ReLU activation function [9]; they do not get updated and kill the gradient from backpropagating. We can identify two types of dead filters:

- Filters that contain non-zero parameter values but do not get excited for any training sample, reasons for this behavior could come from bad initialization to incorrect values of hyper-parameters (e.g. too high learning rate).
- Filters that have all their parameters with zero (or almost zero) values.

The former type could activate during test and even become meaningful again for other dataset. However, the latter are completely dead and will never revive again.

Dead filters with zero values present serious limitations in continual learning scenarios, in particular in those cases where the dataset becomes more complex overtime. This complexity could refer to an increasing number of categories and, also to categories more difficult to differentiate (i.e. fine-grained). In either case, we should expect to require using more power from the network. Moreover, a general practice in continual learning is to use the final model trained with one batch of data as the initial model of the next one. This raises a problem if the training of the first iteration has killed an amount of filters that does not affect the first training, but left the model in a poor condition for the next iteration where the complexity of the problem is higher.

The novelties of this paper are the following:

- We found that the combination of the L2 with Adam in ResNet architectures outputs larger percentage of dead filters than with other optimizers such as Momentum. Having dead filters in a model is a potential problem if that model is used as a pre-trained for another dataset (e.g. continual learning)
- We propose a new regularization term called *survival loss*, that addresses the problem of dead filters. This term can be used in combination with L2 to find a balance between low magnitude parameters to improve generalization and, alive filters that avoids losing full potential of the network.
- We conduct an empirical study of the effect of dead filters in continual learning scenarios, and evidence a serious limitation in term of having the power of network diminished in cases where the datasets of subsequent iterations become more complex over time.

2 Related Work

- Optimizers: There are multiple factors and decisions that impact on the training of deep neural networks, but without doubt the type of optimizer is one of the most important decisions. Over the years, the parameters update strategies defined by the optimizers have evolved, making possible to find multitude of works aiming to improve this process. Roughly, we can split the optimizers in two categories: optimizers that use the same update step in all parameters such as vanilla SGD and Momentum [11,12], and optimizers that apply different updates for each parameter and state of the training such as RMSProp, Adagrad [2] or Adam [5] and all their variants [8,13]. This latter category of optimizers is very popular among practitioners and researchers alike, mainly due to their fame of fast convergence and being less sensitive to choice of hyper-parameters. However, it has been recently discovered a fundamental problem of these optimizers when combine with L2. Ilya Loshchilov and Frank Hutter in their 2019 paper [8] identify that the implementation of Adam and L2 in every framework does not actually do what it is expected: to decay the weights. The authors study the problem from a mathematical point of view as well as empirically comparing the training error rates for different datasets. In our paper we present another limitation of the incorrect implementation of L2 with respect to the amount of dead filters that produces.

- Dead filters: Filters with all their parameters set to zero always results from a convolution of zero values regardless the input. These filters are particularly problematic when followed by the ReLU activation function, in those cases the problem is known as dying ReLU [9]. The root of the problem is that the ReLU function has an undefined derivative at zero and for negative values its derivative is zero. We can better understand this issue and its impact if we look at the two components of a neuron, which are processed in sequence: the convolutional filter and the activation function. So, if the output of the convolution is zero, the gradient of the activation function is taken as zero. This situation impairs the training since the gradient stops to be backpropatagеd at that point. We can find a significant amount of literature trying to mitigate this problem from the activation function perspective proposing variants of the function such as Leaky ReLU [10] and some others [1,4]. However, in practice researches keep using vanilla ReLU for its simplicity, efficiency and because it does not require to tune any hyper-parameter. We approach the problem from the convolutional filter perspective instead, work from this perspective have not been found. We propose a new regularization loss that penalizes filters with low norms, which are somehow responsible for dying ReLUs in the first place.
- Continual learning: In production environments it is very common to find situations where new data arrives every day. This data needs to be incorporated into the models as efficiently as possible. In those situations we could just fine-tune the previous model using the old and the new data altogether if we can afford to keep all data. Otherwise, we need to come up with strategies that adapt the previous model to the new data without forgetting the old one. This is known as the elasticity-plasticity dilemma. Most of the existing literature on continual learning aims at solving this problem also referred to as catastrophic forgetting [6,15,16]. In this paper we study continual learning from a different perspective. Concretely, we study the negative impact of dead filters as we keep fine-tuning models over time while the newly arrived data keeps making the problem harder (e.g. more number of categories). In

other words, for a particular dataset having half of the filters dead might not be a problem in terms of accuracy. However, if the same model is used as pre-trained for another more complex dataset, we could expect dead filters to limit the accuracy as opposed as if we could use the full amount of filters. A study on the generalization error with respect to model scale and dataset size is presented by Rosenfeld et al. [14].

3 Adam and L2: Dying ReLU's

In this section we empirically show that the combination of the Adam optimizer and the L2 regularizer has a negative impact on the liveliness of the filters in a CNN model. In particular, we focus on monitoring the norm of the filters during training time. Doing so allow us to study the contribution of each of the filters in the final outcome of the model, especially if the filters belong to the same layer.

We conduct several trainings for tuning the hyper-parameters of a ResNet-110 model on the CIFAR-10 dataset [7] using Adam and L2. Using an initial learning rate of 10^{-2} that decays by 10 every 40 epochs, and setting a weight decay of 5×10^{-4} with a mini-batch size of 128, the model reaches 87.8% test accuracy. Furthermore, we train the same model using SGD Nesterov with momentum 0.9 and the same weight decay reaching 88.2% test accuracy. Both models reach similar test accuracies, however if we attend to the norm of the filters we see important differences. In Fig. 1 we compare both models in terms of their filter norms. Concretely, the figure represents the percentage of filters out of the total filters in the network (y-axis) with a norm value under a threshold (x-axis). To provide with a more concrete metric for comparing models we assume that filters with norms under 10^{-15} have a meaningless contribution to the final prediction and therefore we consider them dead. In Table 1 we show the percentage of death filters for both models along with their accuracies. We see that these two models with only 0.4 points in difference in test accuracy have more than 50% difference in dead filters.

Table 1. Accuracy and percentage of dead filters (i.e. filters with norm values under 10^{-15}). ResNet-110 on CIFAR-10

	Accuracy	Dead filters (%)
Adam+L2	87.8	51.4
SGD+L2	88.2	0.0

We run several analyses to investigate further this behavior. First, we look at the progress of the filters during the course of the training. In Fig. 2 we see how the filter norms reduce over time. However, there is not a clear match with

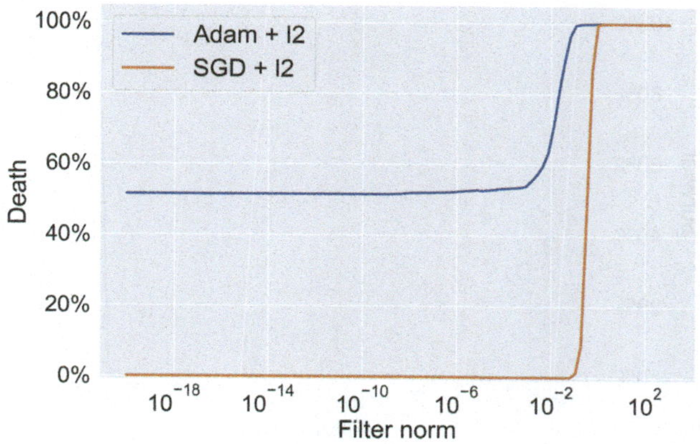

Fig. 1. Percentage of filters (y-axis) with norms under different thresholds (x-axis) for Adam with L2 and SGD Nesterov with L2.

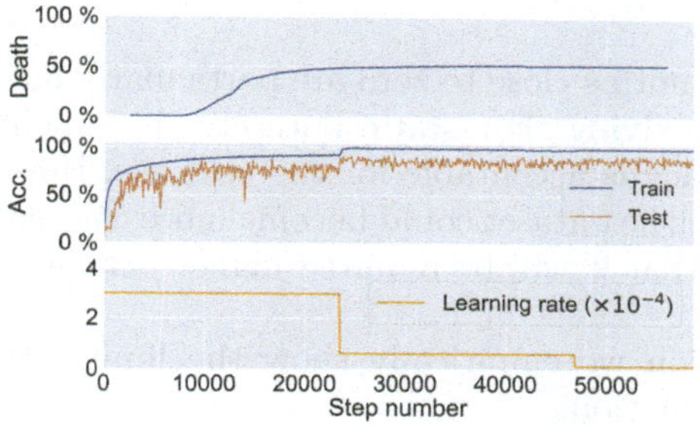

Fig. 2. (top) Progress of percentage of filter norms under 10^{-15} during training for the Adam with L2 combination. (mid) Train and test accuracy. (bottom) Learning rate schedule for the entire training

particular steps of the training such as when it reaches overfit, or with the learning rate schedule. A strategy of early stop could mitigate the problem, however many times is not trivial to decide when to stop. Second, we run another training of Adam this time without L2 and verify that this problem is not replicated as shown in Fig. 3, although the test accuracy is lower (74.8%). We hypothesize that this effect might be related to the known issue of the L2 implementation with adaptive learning rate optimizers such as Adam [5]. Finally, we analyze the status of death filters across layers in the network. In Figure 4 we show the percentage of filters with norm lower than 10^{-15} (y-axis) per layer (x-axis), where the layers are sorted from left to right from the shallowest filter to the deepest one. We can see a clear tendency for the deepest layers to have higher percentage of death filters than shallower layers.

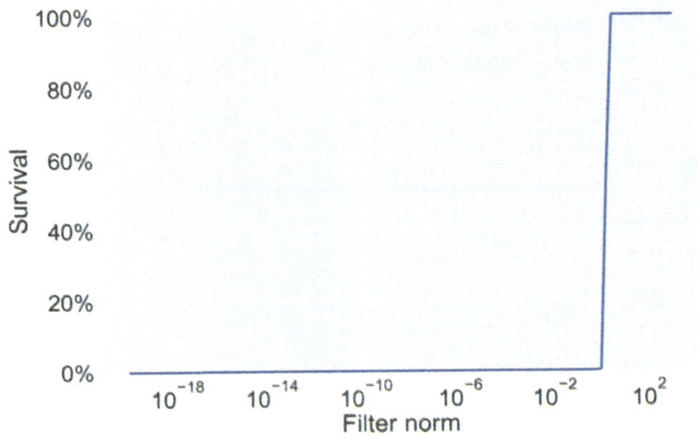

Fig. 3. Percentage of filters (y-axis) with norms below different thresholds (x-axis) for a trained Resnet-110 model on CIFAR-10 for 150 epochs using Adam without L2

Dead filters with norms close to zero are particularly damaging because they have no chance of reviving. For static datasets this might not be a problem as long as the accuracy is acceptable for the use case. However, for a continual learning set up where the dataset could become more complex over time, it seems reasonable to argue that it will be required higher percentage of the network to solve the problem.

In the next section we empirically show the limitations of dead filters in continual learning situations.

4 The Role of Dead Filters in Continual Learning

In this section we present a continual learning scenario where dead filters impact negatively on the performance as the dataset becomes more complex. To simulate a continual learning situation we split the CIFAR-100 dataset in two iterations; the first one contains only 10 classes and, the second adds the other 90 classes. We choose this naive set up with two iterations for three reasons: first, for clarity purposes when presenting the results, second we argue that is a plausible scenario in production environments, and third we believe it is sufficient to study the problem at hand. The results and conclusions should be extended to more complex configurations regardless of the number of iterations, as long as the dataset increases its complexity overtime.

Two models are trained in sequence for the two splits both with Adam+L2 to show the limitation of dead filters. The first model is trained from scratch with the first split of the dataset. For the first iteration, we start with an initial learning rate of 10^{-2} which decays by 10 every 40 epochs. We use a weight

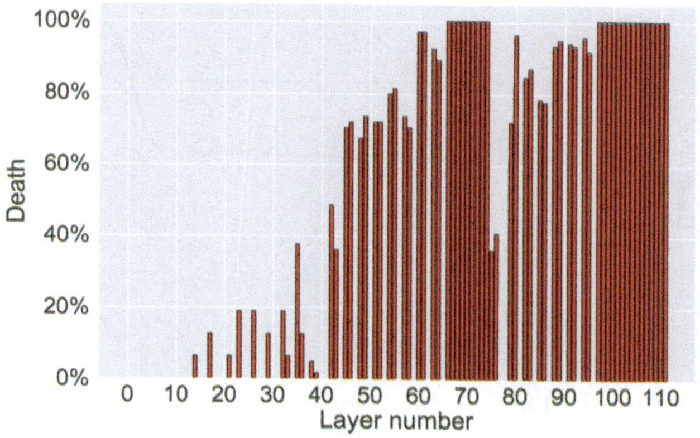

Fig. 4. Percentage of dead filters (i.e. norm lower than 10^{-15}) per layer from shallowest (left) to deepest (right)

decay of 5×10^{-4} and a batch size of 128 training over 150 epochs. For the second iteration we load the weights from the first model. Note that for the classification layer only the parameters from the 10 classes of the previous model are merged into the current one. The rest of the parameters are initialized with Xavier [3]. We continue using Adam and L2 and cross validate the weight decay and learning rate. Table 2 shows the summary of both models. In Fig. 5 we plot the comparison of both iterations in terms of dead filters,

Table 2. Comparison of the two models trained with Adam+L2 in continuous mode from the first and second split of CIFAR-100

	Accuracy	Dead filters (%)
1st Iter. (10 classes)	71.5	61.14
2nd Iter. (100 classes)	55.8	82.9

Looking at the results in the second iteration, we argue that higher accuracy could have been reached if all filters from the network were active. This argument leads to the following section where a new regularization term is proposed that results in fully alive networks.

4.1 Survival Loss Regularizer

We propose to add a regularizer in the total loss to penalize filters with small norms with the goal to keep the filters alive (i.e. non-zero), we refer to it as *survival loss*. Note that we might end up with alive filters that do not contribute

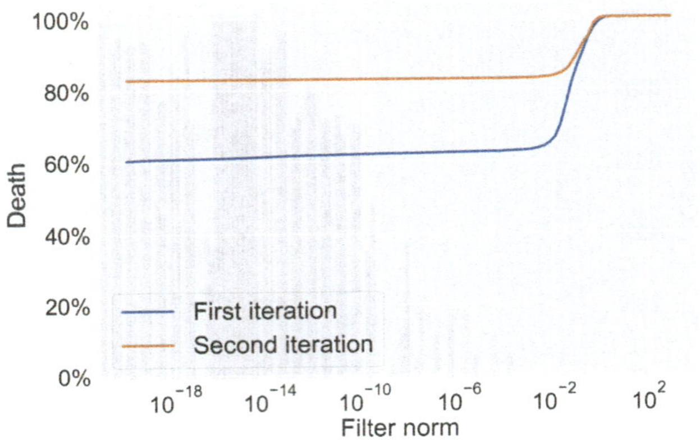

Fig. 5. Percentage of filters (y-axis) with norms below different thresholds (x-axis) for the first and second iteration of a model trained with Adam and L2 for 150 epochs

much on a particular dataset, but at least they are not dead and could be useful in future datasets. The *survival loss* has the following form:

$$\mathcal{L}_s(\theta^i) = \sum_i^N max(0, \tau - ||\theta^i||), \tag{1}$$

where θ^i is the ith filter of the network, N the total number of filters in the network, $||\cdot||$ represents the norm and τ is a hyper-parameter that defines the minimum norm of a filter to be penalized (also the maximum penalization for a filter). The penalization increases linearly from 0 if the norm is higher than τ, to τ if the norm is zero. The form of the equation is plotted in Fig. 6.

The *survival loss* is used in combination with L2 as a complimentary effect to find an adequate balanced between low weight magnitudes that improve generalization, and avoiding too low magnitudes that kill filters. The total loss is defined as follows:

$$\mathcal{L}(x, y, \Theta) = \mathcal{L}_{xe}(x, y, \Theta) + \lambda\mathcal{L}_{L2}(\Theta) + \gamma\mathcal{L}_s(\Theta), \tag{2}$$

where $\mathcal{L}_{xe}$ is the cross entropy loss, λ is the weight decay for the L2 loss, and γ is a hyper-parameter that controls the impact of the *survival loss* in the total loss.

We repeat the same set up for continual learning, but this time using the *survival loss* in combination with L2. In Fig. 7 we present the result of the first and second iterations respectively in terms of the norm of the filters. Comparing Tables 2 and 3 we can see that *survival loss* keeps the filters alive and improve accuracy not only in the second iteration but also in the first iteration.

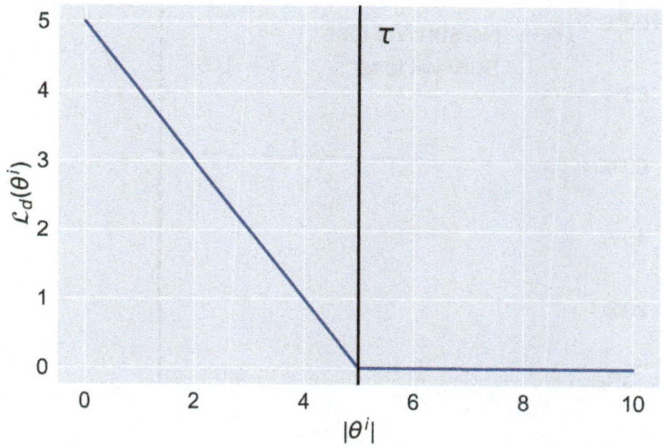

Fig. 6. Survival loss function

Table 3. Comparison of the two models trained with Adam+L2+survival loss in continuous mode from the first and second split CIFAR-100

	Accuracy	Dead filters (%)
1st Iter. (10 classes)	72.6	0.0
2nd Iter. (100 classes)	56.0	0.0

4.2 Benchmark Results

In this subsection we want to compare the results of a training that includes our *survival loss* with respect to our baseline. For that purpose we train a model on CIFAR-10 using ResNet-20. The batch size is 128. We use an Adam optimizer with momentum values $\beta_1 = 0.9$ and $\beta_2 = 0.999$, and an initial learning rate of 3×10^{-4} that decays 1 order of magnitude every 40 epochs. The model uses batch normalization layers with momentum 0.9. We train for 80 epochs. We evaluate Adam+L2 with $\lambda = 10^{-2}$ and Adam+L2+Surv. loss with a value of $\lambda = 1$, $\gamma = 0.46$ and $\tau = 0.1$.

Table 4. ResNet-20 on CIFAR-10. Test accuracy

	Accuracy (%)
ADAM+L2	80.7
ADAM+L2+Surv.Loss	**86.8**

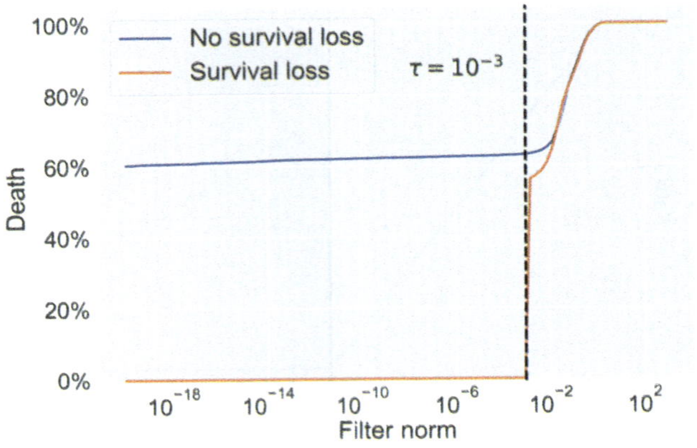

(a) First iteration

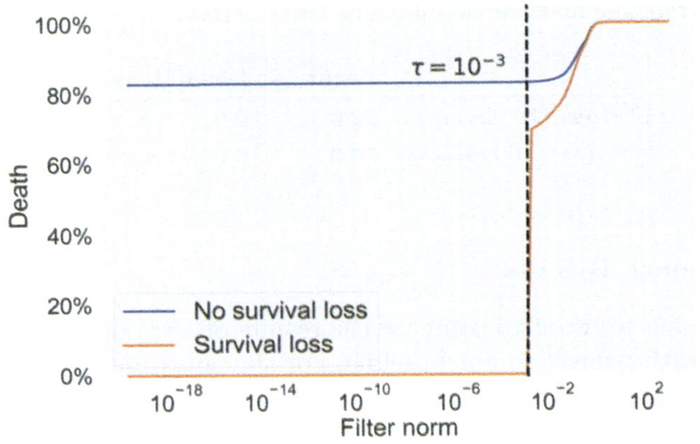

(b) Second iteration

Fig. 7. Percentage of filters (y-axis) with norms below different thresholds (x-axis) for two models of Adam+L2 with and without *survival loss*. Both models were trained for 150 epochs. (a) First iteration (b) Second iteration.

As we can see in Table 4, we have found that using *survival loss* we see an increase of 6% in accuracy with respect to a model that has not been trained with *survival loss*. A detailed validation of the hyper-parameters of the survival loss are discussed in the next section.

4.3 Hyper-parameters Discussion

The definition of the survival loss Eq. (2), requires two parameters to be set: τ that controls the norm of the filters and, γ responsible for the impact of the *survival loss* in the total loss.

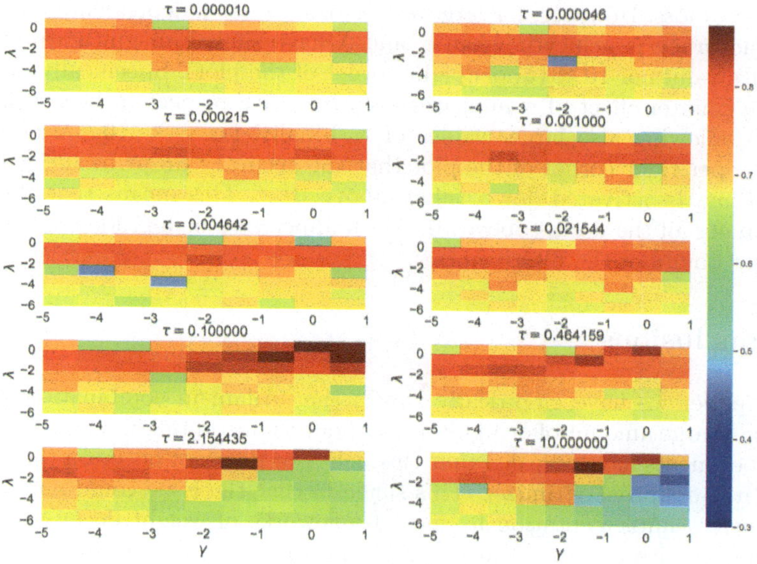

Fig. 8. Validation accuracy in function of λ, γ and τ. For better visualization, in the figure we show results fixing the value of τ. We note that both axis are in logarithmic scale.

We study the impact of λ, γ and τ on the validation accuracy for a wide range of values. As in the experiments carried out in Sect. 4.2, all the models have been trained on CIFAR-10 using ResNet-20 for 80 epochs. We use an initial learning rate of 3×10^{-4} that decays 1 order of magnitude every 40 epochs. The result is shown in Fig. 8. In order to visualize better the figure, we have plotted the results at different values of τ.

Intuitively, L2 regularizer and the *survival loss* acts in opposite ways: while L2 regularization penalizes large magnitudes for weights, the *survival loss* penalizes low filter norms. However, we must point out two main differences. Firstly, L2 regularization acts at a parameter level while the *survival loss* does it at a filter level. Secondly, L2 operates continuously on all parameters during the entire training, whereas *survival loss* only affects to certain parameters of the network (those filters whose norm are below the threshold) and at specific moments during training (particularly when the norm is below the threshold). We understand that both losses are complementary to each other. On one hand L2 aims at improving generalization by keeping magnitudes low, while on the other hand *survival loss* avoids filter death.

In Fig. 8 we can observe that below a value of $\tau = 0.1$ the validation accuracy is independent on γ, having higher accuracy for ranges of λ between 10^{-2} and 1. This can be explained if we look at role of τ in the *survival loss*: for thresholds below the actual norms of the filters, *survival loss* is zero and it has no effect on the total loss. However, for large values of τ we see a dependence on γ. Also, higher accuracy is achieved when γ and λ have similar magnitudes and both are above values of 10^{-2} (upper right corner of the plot). Specifically, we found that the counter-effect of γ improves accuracy with respect to be baseline (see Table 4), allowing also larger values of λ. Finally, for $\tau = 10$, we clearly see on the upper right corner of the plot that the accuracy drops to 40% when the *survival loss* is activated, for γ values above 10^{-2}. This means that the loss is constraining all the norms above 10, which appears to be a crude truncation for the lower bound of the filter's norms.

5 Conclusions

In this paper we have empirically shown that Adam in combination with L2 yields to a large amount of dead filters, without a substantial impact on accuracy. Our experiments have revealed that this behavior does not occur on the Momentum, a non-adaptive optimizer. This suggests that the issue could be related to the incorrect implementation by most frameworks of weight decay (i.e. L2) in combination with adaptive learning rate optimizers such as Adam [8].

We have a presented a continual learning scenario where dead filters generated from the combination of Adam and L2 leads to limitations in term of accuracy as the data becomes more complex overtime (e.g. increasing number of classes).

Finally, we have proposed a new regularization term called *survival loss* that addresses the problem of dead filters. This term works in combination with L2 and results in fully alive networks with higher generalization accuracy; around 6 points higher than without using *survival loss*. We expect further improvements of this solution as iterations keep making the data more complex. (e.g. increasing number of classes).

This research has thrown up questions in need of further investigation regarding the association between dead filters and Adam with L2. Moreover, further work must be carried out in order to prove the utility of the proposed *survival loss* in a wider range of scenarios.

References

1. Clevert, D.A., Unterthiner, T., Hochreiter, S.: Fast and accurate deep network learning by exponential linear units (elus). arXiv preprint arXiv:151107289 (2015)
2. Duchi, J., Hazan, E., Singer, Y.: Adaptive subgradient methods for online learning and stochastic optimization. J. Mach. Learn. Res. **12**(7), 2121–2159 (2011)

3. Glorot, X., Bengio, Y.: Understanding the difficulty of training deep feedforward neural networks. In: Proceedings of the 13th International Conference on Artificial Intelligence and Statistics, pp. 249–256 (2010)
4. He, K., Zhang, X., Ren, S., Sun, J.: Delving deep into rectifiers: surpassing human-level performance on imagenet classification. In: Proceedings of the IEEE International Conference on Computer Vision, pp. 1026–1034 (2015)
5. Kingma, D.P., Ba, J.: Adam: a method for stochastic optimization. arXiv preprint arXiv:14126980 (2014)
6. Kirkpatrick, J., Pascanu, R., Rabinowitz, N., Veness, J., Desjardins, G., Rusu, A.A., Milan, K., Quan, J., Ramalho, T., Grabska-Barwinska, A., et al.: Overcoming catastrophic forgetting in neural networks. Proc. Nat. Acad. Sci. **114**(13), 3521–3526 (2017)
7. Krizhevsky, A.: Learning Multiple Layers of Features from Tiny Images. University of Toronto (2012)
8. Loshchilov, I., Hutter, F.: Decoupled weight decay regularization. arXiv preprint arXiv:171105101 (2017)
9. Lu, L., Shin, Y., Su, Y., Karniadakis, G.E.: Dying relu and initialization: theory and numerical examples. arXiv preprint arXiv:190306733 (2019)
10. Nair, V., Hinton, G.E.: Rectified linear units improve restricted boltzmann machines. In: Proceedings of the 27th International Conference on Machine Learning (ICML-10), pp 807–814 (2010)
11. Nesterov, Y.: A method of solving a convex programming problem with convergence rate o(1/k2). Sov. Math. Dokl **27**, 269 (1983)
12. Polyak, B.T.: Some methods of speeding up the convergence of iteration methods. USSR Comput. Math. Math. Phys. **4**(5), 1–17 (1964)
13. Reddi, S.J., Kale, S., Kumar S.: On the convergence of adam and beyond. arXiv preprint arXiv:190409237 (2019)
14. Rosenfeld, J.S., Rosenfeld, A., Belinkov, Y., Shavit, N.: A constructive prediction of the generalization error across scales. arXiv preprint arXiv:190912673 (2019)
15. Shawn, B., Lapo, F., Thomas, M., Joel, L., Kenneth, O.S., Jeff, C., Cheney, N.: Learning to continually learn. arXiv preprint arXiv:200209571 (2020)
16. Yoon, J., Yang, E., Lee, J., Hwang, S.J.: Lifelong learning with dynamically expandable networks. arXiv preprint arXiv:170801547 (2017)
17. Zhang, M., Lucas, J., Ba, J., Hinton, G.E.: Lookahead optimizer: k steps forward, 1 step back. In: Advances in Neural Information Processing Systems, pp. 9593–9604 (2019)

Reinforcement Learning Experiments and Benchmark for Solving Robotic Reaching Tasks

Pierre Aumjaud[1(✉)], David McAuliffe[2], Francisco Javier Rodríguez-Lera[3], and Philip Cardiff[1]

[1] University College Dublin, Dublin, Ireland
{pierre.aumjaud,philip.cardiff}@ucd.ie
[2] Resero Ltd., Dublin, Ireland
david.mcauliffe@resero.io
[3] Universidad de León, León, Spain
fjrodl@unileon.es

Abstract. Reinforcement learning has shown great promise in robotics thanks to its ability to develop efficient robotic control procedures through self-training. In particular, reinforcement learning has been successfully applied to solving the reaching task with robotic arms. In this paper, we define a robust, reproducible and systematic experimental procedure to compare the performance of various model-free algorithms at solving this task. The policies are trained in simulation and are then transferred to a physical robotic manipulator. It is shown that augmenting the reward signal with the Hindsight Experience Replay exploration technique increases the average return of off-policy agents between 7 and 9 folds when the target position is initialised randomly at the beginning of each episode.

Keywords: Reinforcement learning · Robotics · Benchmark · Model-free · Sim-to-real

1 Introduction

Reinforcement learning (RL) is a paradigm in the field of machine learning that has recently gained tremendous interest [1]. Unlike supervised learning, RL is capable of learning sequential decision-making policies by interacting with an environment, while sometimes achieving super-human performance [2]. RL requires a reward function to be defined, which is why it has naturally been applied to games or toy problems [3]. However, real-world applications are still scarce and challenging. Robotics allow researchers to define controlled training environments relatively easily, thus RL has found many successful applications in this field.

L. M. Bergasa et al. (Eds.): WAF 2020, AISC 1285, pp. 318–331, 2021.
https://doi.org/10.1007/978-3-030-62579-5_22

Reinforcement learning approaches can be grouped in two categories: model-based – where the agent first learns a model of the environment and then uses it to make predictions; and model-free – where the agent directly learns a policy based only on its interactions with the environment. Historically, model-based RL has first been applied to robotics [4] due to it high sample efficiency, allowing agents to solve tasks with a limited number of policy iterations. Nevertheless, it is sometimes challenging to build accurate models for complex robotics problems, causing model-based approaches to often suffer from poor asymptotic performance. The invention of Deep Deterministic Policy Gradients [5] has paved the way for other model-free RL algorithms to successfully solve problems with continuous state and action space. Since then, considerable improvements have occurred in the field with the invention of Twin Delayed Deep Deterministic Policy Gradient [6], Soft Actor Critic [7,8] and Hindsight Experience Replay [9], which addresses the problem of sparse rewards.

One of the most common robotic tasks studied in an RL context is trajectory planning to reach a target position. A number of approaches have been adopted to solve this problem, including model-based approaches [10,11], model-free approaches [12–19], a combination of both [20], or learning acceleration via closed-loop policies [21].

In this article, we will consider solving the problem of reaching target positions both using a robotic simulator and a physical manipulator. The equipment, training environments and methods adopted to perform the performance benchmark of a number of model-free RL algorithms are described in Sect. 2. The comparative results are reported and discussed in Sect. 3, both in terms of training convergence and evaluation metrics. Finally, a general conclusion is given and directions for future works are outlined in Sect. 4.

2 Environments and Methods

2.1 Manipulator Description

The robotic arm used in this work is the WidowX MKII manipulator by Trossen Robotics [22], see Fig. 1b. It is a 6-joints robotic arm possessing an 82 cm diameter reach and equipped with Dynamixel servo actuators and parallel grippers. The robot arm is compatible with the Robot Operating System (ROS) [23] and is shipped with its associated ROS packages. The Replab project [24] has previously employed the same robot arm to solve grasping tasks and provide a reproducible benchmark platform for robotics learning. The learning environments used here are adapted from those in the Replab project. The Pybullet physics engine [25] simulates the response of the robot arm in a virtual environment, which accelerates the training process compared to training on the physical robot alone.

2.2 Reaching Task and Environment Definition

The reaching task considered here is a well-known problem in the robotic manipulation field. It consists of determining the optimal sequence of actions required

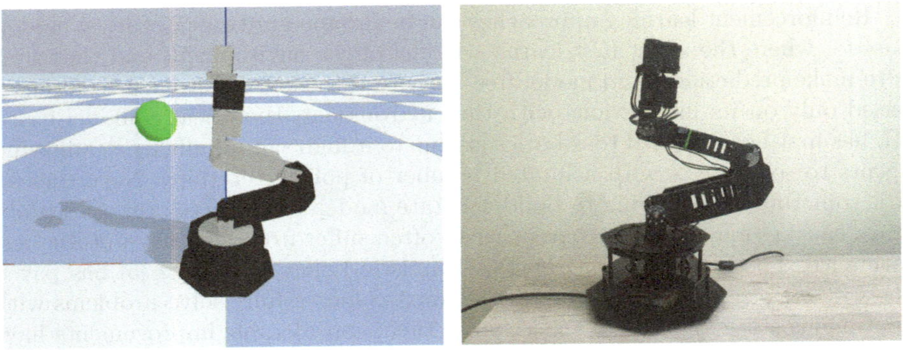

(a) Virtual environment (b) Physical environment

Fig. 1. The simulated (Pybullet) and physical (WidowX MKII manipulator) learning environments in their initial episode position. The target position is indicated by a green sphere.

to bring the robot arm's end effector as close as possible to given target coordinates within its workspace.

In order to solve the reaching task in an RL context, the problem is formalised as a Markov Decision Process. The environment's *state* s_t is defined as an array holding the current spatial coordinates of the robot's end effector and the angular position of each joint:

$$\forall s_t \in S, \qquad s_t = [x_e, y_e, z_e, r_1, r_2, r_3, r_4, r_5, r_6]_t \tag{1}$$

where (x_e, y_e, z_e) are the Cartesian coordinates of the end effector and r_i is the angular position of joint i in radians. The *actions* a_t sent to the environment are defined as an array holding the changes in joint angle from the previous position:

$$\forall a_t \in A, \qquad a_t = [\theta_1, \theta_2, \theta_3, \theta_4, \theta_5, \theta_6]_t \tag{2}$$

where θ_i is the change in angular position from the previous position of joint i. S and A denote the set of possible states and actions, respectively. These spaces are finite and continuous in this context. The *reward* r_t is defined as the negative, squared, L2 norm of the vector between the current end effector position and the goal position (dense reward setting):

$$\forall r_t \in \mathbb{R}^-, \qquad r_t = -[(x_e - x_g)^2 + (y_e - y_g)^2 + (z_e - z_g)^2]_t \tag{3}$$

where (x_g, y_g, z_g) are the Cartesian coordinates of the goal pose. Each learning iteration is composed of an action sent, a state and a reward received by the agent, which defines a *timestep t*. A sequence of 100 timesteps defines one *episode*, unless the termination condition – the episode ends if the distance between the end effector and the goal is less than 0.5 mm – is reached before this. The sum of the rewards over one episode defines the *return* R_t, see equation below.

$$R_t = \sum_{t=1}^{100} r_t \tag{4}$$

The *policy* maps the environment's current state to the next action with the objective of maximising the return. A deterministic policy is a function defined as $\pi_d : S \mapsto A$ and a stochastic policy is a probability distribution defined as $\pi_s(A|S)$, meaning that the probability of selecting an action depends on the current state.

The robot arm joints are always initialised with the same angle positions at the beginning of each episode both in the virtual and the physical environments, as shown in Fig. 1. The goal position is either constant across all training and evaluation episodes (fixed goal) or initialised randomly at the beginning of each episode (random goal). Either way, the goal position is always within the reach of the robot arm.

The learning environments are implemented with OpenAI Gym [26], a standard toolkit for developing and comparing reinforcement learning algorithms. A goal-oriented Gym environment [26] is also implemented in order to explore the observation space more efficiently using Hindsight Experience Replay. In this case, the desired goal and achieved goal coordinates are also included in the state definition. In total, four learning environments are implemented:

- **Env1**: Pybullet simulation + fixed goal
- **Env2**: Pybullet simulation + random goal
- **Env3**: Physical robot + fixed goal
- **Env4**: Physical robot + random goal

The policies are trained on the Pybullet environments only (i.e. Env1 and Env2), however they are evaluated both in simulation and on the physical robot.

2.3 Experiments and Benchmark

It is proposed to compare the performance of a number of RL algorithms at solving the reaching task using the environments described above. The RL algorithms considered in the benchmark are:

- Advantage Actor Critic (A2C) [27]
- Actor Critic using Kronecker-Factored Trust Region (ACKTR) [28]
- Deep Deterministic Policy Gradients (DDPG) [5]
- Proximal Policy Optimization (PPO) [29]
- Twin Delayed Deep Deterministic Policy Gradient (TD3) [6]
- Soft Actor Critic (SAC) [7,8]
- Trust Region Policy Optimization (TRPO) [30]
- SAC + Hindsight Experience Replay (HER) [9]
- TD3 + Hindsight Experience Replay [9]

All these algorithms are compatible with continuous state and action spaces environments, which is a requirement for the reaching task considered here. DDPG, SAC, TD3 and HER learn a deterministic policy and a stochastic policy is learnt by the rest of the algorithms. The two off-policy algorithms – SAC and TD3 – are combined with HER in an effort to reduce sample complexity. The algorithms are applied using the Stable Baselines implementation [31].

The hyperparameters of each algorithm are first optimised with the Optuna framework [32] over 100 training runs of 100 episodes each. In each run, the hyperparameters are sampled with the Tree-structured Parzen Estimator and unpromising runs are stopped early using a median pruner. Subsequently, RL agents are trained in the virtual environments Env1 and Env2 for 200,000 timesteps using each algorithm described above. The A2C, ACKTR and PPO agents are trained over 8 parallel environments in order to speed up the learning process (the other algorithm's implementations do not currently support parallelisation). The training is performed on University College Dublin's Sonic HPC cluster [33], which is equipped of two NVIDIA Tesla V100 16 GB GPUs per node. The training is repeated over 10 different initialisation seeds and the metrics are averaged in order to strengthen the robustness of the experiments. Once the optimal policies are learnt in the virtual environments, they are deployed and tested both in the virtual (Env1 and Env2) and physical environments (Env3 and Env4). All policies are evaluated with their associated agent for 100 episodes. A random policy is also evaluated to serve as a reference. In the case of physical environments, the model weights of the best-performing seed run (in terms of return) are transferred to the agent controlling the physical robot. In the case of the virtual environment, all seed runs are evaluated and the metrics are averaged. The following evaluation metrics are reported:

- **Average return**: Return averaged over the 100 evaluation episodes.
- **Train time**: Average training time over the 10 random seeds in minutes.
- **Success ratio @X mm**: number of episodes where the end effector ended within a distance threshold of X mm from the goal position, divided by the total number of evaluation episodes. The distance thresholds considered are 50 mm, 20 mm, 10 mm and 5 mm.
- **Reach time @X mm**: number of timesteps required to reach the target during a successful episode within a distance threshold of X mm.

The source code repository and an explanatory video are available at https://git.io/JJJdu and https://youtu.be/-N-6Me8UkFk.

3 Results and Discussion

3.1 Training Convergence

The training convergence curves averaged over the 10 initialisation seeds are shown in Figs. 2 and 3 for the two virtual environments considered, Env1 (fixed goal) and Env2 (random goal). The curves are smoothed with a rolling window average of 50 for clarity purposes. A comparison of the convergence curve between Env1 and Env2 is also reported for each algorithm individually, see Fig. 4. In these plots, the middle line represents the average reward over the 10 seed runs at each timestep and a shaded area is drawn at ± the standard deviation above and below the average.

It can be noted that Env2 (random goal) is a more challenging task to solve since the average return is generally lower than that of Env1 (fixed goal) at the end of the training process. All algorithms manage to learn a successful policy in Env1 after 200,000 timesteps. DDPG, TRPO and SAC achieve the highest sample efficiency as their learning curves plateau the earliest after less than 20,000 timesteps, whereas A2C exhibits the highest sample complexity. Moreover, TRPO, TD3 and ACKTR produce the most stable and repeatable training as their standard deviations are the lowest at the end of training, see Fig. 4.

When the goal is initialised randomly at the beginning of the episode, most algorithms fail to learn a successful policy. Some algorithms perform even worse than a random policy, for example DDPG in Env2. However, combining SAC and TD3 with HER yields a healthy training, even in Env2, as illustrated in Fig. 3. Accelerating the observation space exploration this way tends to reduce slightly the sample efficiency and the training stability of the off-policy algorithms SAC and TD3, see Figs. 4h and 4i.

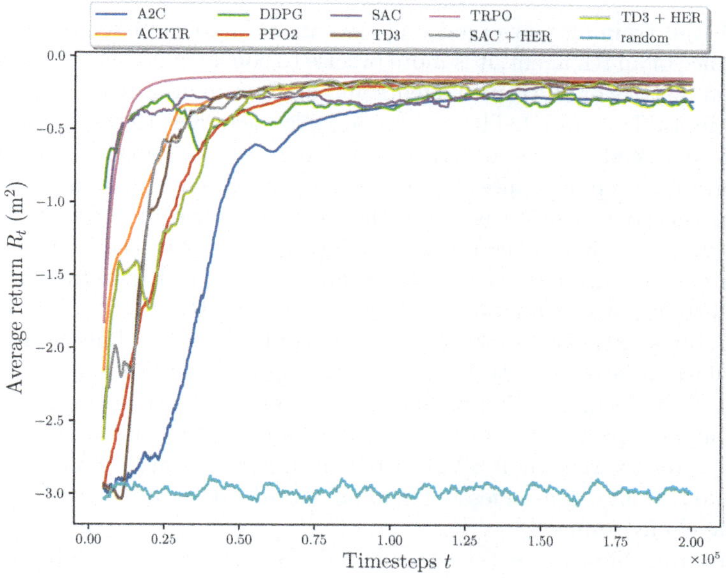

Fig. 2. Convergence curves of the algorithms solving Env1 (fixed goal)

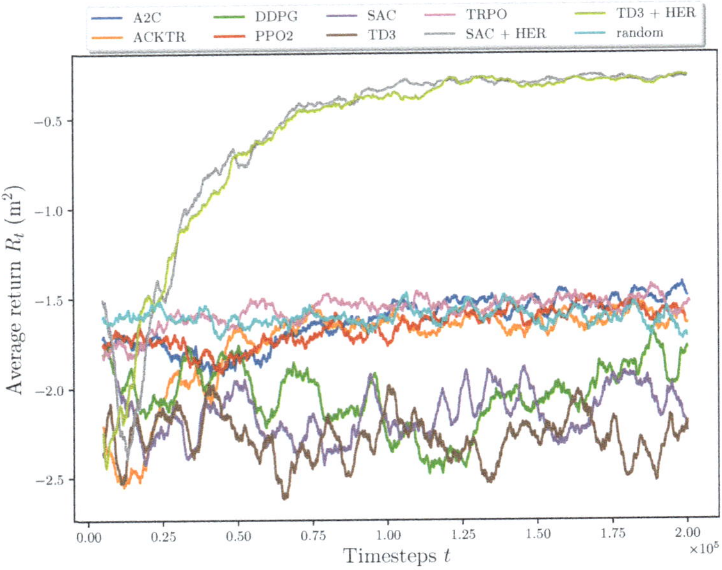

Fig. 3. Convergence curves of the algorithms solving Env2 (random goal)

3.2 Evaluation Metrics

Although the training convergence curve can be a good proxy to measure the performance of an RL agent, it is more precise to consider evaluation metrics. The evaluation metrics of all trained agents and all four environments considered are reported in Tables 1 and 2. The metrics are averaged over 10 trainings initialised with different seeds; the standard deviations are not reported here for clarity but are available upon request or in the associated source code repository.

When the goal position is fixed, the average return of successful agents is much higher than that of the random policy, which agrees with the observations from the convergence curves, see for example TRPO in Env1. In this case, the best-performing algorithms in terms of return are TRPO and TD3. When the goal position is initialised randomly however, only the algorithms combined with HER achieve an average return significantly higher than random.

The training walltime is substantially shorter for algorithms supporting parallelisation, as expected. For instance with a train walltime of 68 s, PPO is almost 19 times faster to train than SAC + HER in Env2. It is also about twice as fast as a random policy that does not exploit any deep learning architectures nor uses parallel environments.

The success ratio measures the accuracy of an agent at reaching the target and the reach time measures their speed at reaching the target during an episode (note that the maximum reach time is equal to the length of an episode, 100).

In the least challenging environment – Env1 (fixed goal), all algorithms achieve a high success ratio and a fast reach time when the distance thresh-

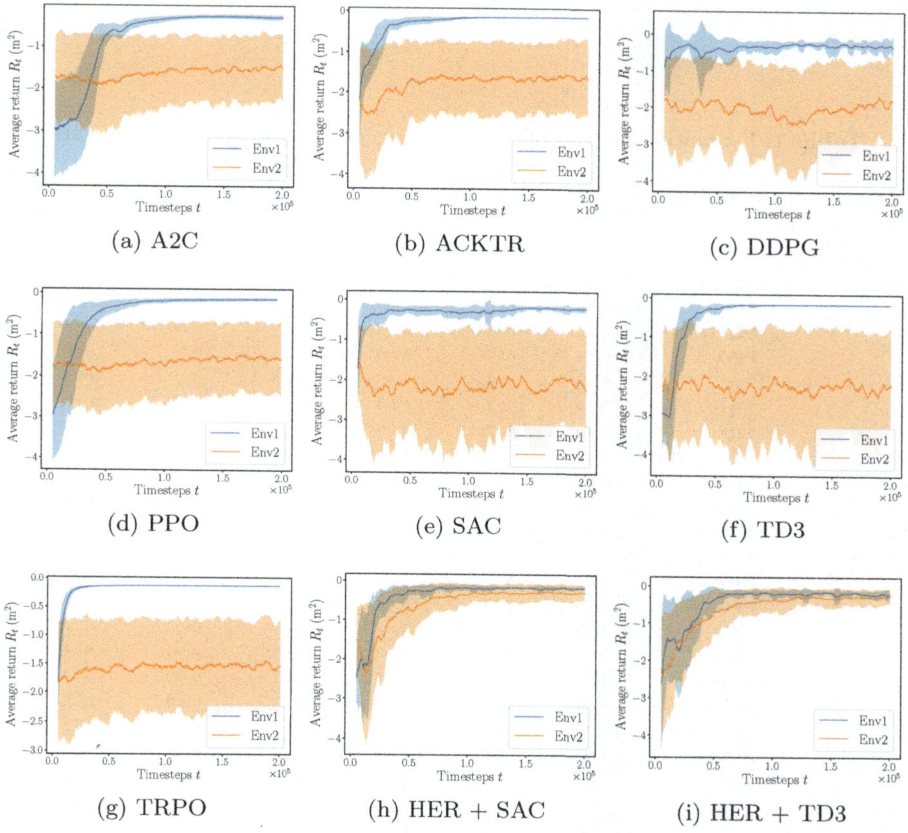

(a) A2C (b) ACKTR (c) DDPG

(d) PPO (e) SAC (f) TD3

(g) TRPO (h) HER + SAC (i) HER + TD3

Fig. 4. Convergence curve comparison between Env1 (fixed goal – in blue) and Env2 (goal initialised randomly – in orange)

old is large (i.e. 50 mm). As expected, the agent's accuracy decreases as the distance threshold is becoming tighter, and it drops dramatically beyond a distance threshold of 20 mm. In this situation, TD3 is the most accurate algorithm with a success ratio @5 mm of 0.5. The agents capable of reaching their target quickly are generally those exhibiting a high success ratio and average return. The fastest agents are TD3 and TRPO with a reach time of around 9 timesteps at the 50 mm threshold, although the other successful agents are only slightly slower.

When the goal is initialised randomly (Env2), fewer agents managed to learn a successful policy than in Env1, which is reflected by the much lower average return. The best-performing agents are those leveraging the HER exploration strategy, i.e. SAC + HER and TD3 + HER with a success ratio @50 mm of 0.67 and 0.64, respectively. This phenomenon was also observed in the training convergence, see Fig. 4. Similar to Env1, the success ratio diminishes rapidly as

Table 1. Evaluation metrics of all trained agents for the four environments

Environment	Algorithm	Average return R_t (m^2)	Average train walltime (s)	Success ratio @50 mm	Reach time @50 mm	Success ratio @20 mm	Reach time @20 mm
Env1 (simulation + fixed goal)	A2C	–0.30	55.0	0.94	18.01	0.26	37.0
	ACKTR	–0.13	55.3	1.00	9.26	0.99	13.5
	DDPG	–0.29	635.0	0.90	9.2	0.30	12.3
	PPO	–0.14	65.0	1.00	9.7	0.71	14.2
	SAC	–0.18	973.8	1.00	10.6	0.60	24.5
	TD3	–0.12	712.8	1.00	9.0	1.00	12.0
	TRPO	–0.12	267.3	1.00	9.0	1.00	12.0
	SAC + HER	–0.15	1097.4	1.00	9.2	0.80	12.1
	TD3 + HER	–0.20	984.1	0.80	9.0	0.70	12.3
	Random policy	–2.98	134.1	0.00	N/A	0.00	N/A
Env2 (simulation + random goal)	A2C	–1.49	72.3	0.03	30.8	0.00	N/A
	ACKTR	–1.64	74.2	0.04	48.1	0.00	N/A
	DDPG	–1.94	523.2	0.02	39.8	0.00	N/A
	PPO	–2.54	68.1	0.04	31.7	0.00	N/A
	SAC	–2.59	1063.9	0.02	49.7	0.00	N/A
	TD3	–2.19	783.7	0.03	34.9	0.00	N/A
	TRPO	–1.43	191.8	0.05	25.7	0.00	N/A
	SAC + HER	–0.27	1285.7	0.67	11.4	0.10	27.2
	TD3 + HER	–0.32	980.7	0.64	11.1	0.17	19.3
	Random policy	–1.68	132.6	0.05	22.0	0.00	N/A
Env3 (physical robot + fixed goal)	A2C	-1.08	N/A	0.00	N/A	0.00	N/A
	ACKTR	–0.28	N/A	0.95	13.0	0.00	N/A
	DDPG	–0.71	N/A	0.00	N/A	0.00	N/A
	PPO	–0.25	N/A	1.00	16.9	0.00	N/A
	SAC	–0.33	N/A	0.35	10.0	0.00	N/A
	TD3	–0.19	N/A	1.00	10.0	0.00	N/A
	TRPO	–0.53	N/A	0.00	N/A	0.00	N/A
	SAC + HER	–0.68	N/A	0.00	N/A	0.00	N/A
	TD3 + HER	–0.28	N/A	1.00	10.0	0.00	N/A
	Random policy	–2.19	N/A	0.00	N/A	0.00	N/A
Env4 (physical robot + random goal)	A2C	–1.29	N/A	0.10	13.0	0.00	N/A
	ACKTR	–1.82	N/A	0.05	35.0	0.00	N/A
	DDPG	–1.55	N/A	0.05	27.0	0.00	N/A
	PPO	–1.31	N/A	0.00	N/A	0.00	N/A
	SAC	–1.43	N/A	0.00	N/A	0.00	N/A
	TD3	–1.63	N/A	0.00	N/A	0.00	N/A
	TRPO	–1.76	N/A	0.00	N/A	0.00	N/A
	SAC + HER	–0.39	N/A	0.50	17.9	0.05	9.0
	TD3 + HER	–0.28	N/A	0.75	18.5	0.20	17.5
	Random policy	–1.57	N/A	0.00	N/A	0.00	N/A

Table 2. Evaluation metrics of all trained agents for the four environments (continued)

Environment	Algorithm	Success ratio @10 mm	Reach time @10 mm	Success ratio @5 mm	Reach time @5 mm
Env1 (simulation + fixed goal)	A2C	0.04	63.8	0.01	91.8
	ACKTR	0.73	22.1	0.20	43.3
	DDPG	0.20	12.5	0.00	N/A
	PPO	0.41	17.1	0.08	40.3
	SAC	0.10	14.0	0.10	15.0
	TD3	1.00	12.5	0.50	18.2
	TRPO	0.81	12.4	0.22	38.5
	SAC + HER	0.50	12.6	0.00	N/A
	TD3 + HER	0.10	13.0	0.00	N/A
	Random policy	0.00	N/A	0.00	N/A
Env2 (simulation + random goal)	A2C	0.00	N/A	0.00	N/A
	ACKTR	0.00	N/A	0.00	N/A
	DDPG	0.00	N/A	0.00	N/A
	PPO	0.00	N/A	0.00	N/A
	SAC	0.00	N/A	0.00	N/A
	TD3	0.00	N/A	0.00	N/A
	TRPO	0.00	N/A	0.00	N/A
	SAC + HER	0.01	37.1	0.00	N/A
	TD3 + HER	0.03	25.9	0.01	28.6
	Random policy	0.00	N/A	0.00	N/A
Env3 (physical robot + fixed goal)	A2C	0.00	N/A	0.00	N/A
	ACKTR	0.00	N/A	0.00	N/A
	DDPG	0.00	N/A	0.00	N/A
	PPO	0.00	N/A	0.00	N/A
	SAC	0.00	N/A	0.00	N/A
	TD3	0.00	N/A	0.00	N/A
	TRPO	0.00	N/A	0.00	N/A
	SAC + HER	0.00	N/A	0.00	N/A
	TD3 + HER	0.00	N/A	0.00	N/A
	Random policy	0.00	N/A	0.00	N/A
Env4 (physical robot + random goal)	A2C	0.00	N/A	0.00	N/A
	ACKTR	0.00	N/A	0.00	N/A
	DDPG	0.00	N/A	0.00	N/A
	PPO	0.00	N/A	0.00	N/A
	SAC	0.00	N/A	0.00	N/A
	TD3	0.00	N/A	0.00	N/A
	TRPO	0.00	N/A	0.00	N/A
	SAC + HER	0.00	N/A	0.00	N/A
	TD3 + HER	0.00	N/A	0.00	N/A
	Random policy	0.00	N/A	0.00	N/A

the distance threshold reduces. The negligible success ratios @50 mm observed in the other algorithms are comparable to that of a random policy, meaning that the agent has occasionally and fortuitously reached the target.

A sharp decrease in performance is globally observed when trained policies are transferred to the physical robot and evaluated in Env3 and Env4, both in terms of average return and success ratio. This could be further improved by continuing the training from the virtual environments in the physical environments, as the Pybullet simulation does not take into account the joint friction and other imperfections of the physical manipulator. Such transfer from a simulated to the physical world is a well-known challenge in robotics, often referred to as sim-to-real.

However, it is worth noting that PPO, TD3 and TD3 + HER achieved a relatively successful policy transfer from the simulated (Env1) to the physical environment (Env3) as exhibited by their success ratio @50 mm of 1 and a comparable reach time. Finally in the case of the physical environment with random goal (Env4), the only successful policies are those combined with HER. This was also noted in the associated virtual environment with random goal, Env2.

4 Conclusion and Future Work

A benchmark procedure is described in the present paper to compare the performance of various model-free RL algorithms at solving the reaching task with a robot manipulator. It is found that it is generally more challenging to solve a reaching task where the target coordinates are initialised randomly at the beginning of each episode. In a fixed goal setting, the highest performance in terms of average return, success ratio and reach time is achieved by TD3 and TRPO. The highest sample efficiency is achieved by SAC, TRPO and DDPG. Augmenting the intrinsic reward signal of off-policy algorithms such as SAC and TD3 using the HER exploration strategy tends to destabilise slightly the training – both in terms of repeatability and sample efficiency; however it also allows the agent to learn an efficient policy in environments where the goal is initialised randomly. Transferring policies from virtual to physical environments proved to be challenging, however similar conclusions were drawn as to which algorithms perform best. Combining TD3 with HER proved to be the most efficient strategy in physical environments.

Although the methodology described here provides a systematic and reproducible experimental procedure, the authors acknowledge that the benchmark results could be further improved by additional work.

Firstly, some algorithms failed to solve the reaching task completely when the goal is initialised randomly, despite the dense reward setting, see Fig. 2. This is likely due to failing to identify optimal hyperparameters for the configuration in question. Selecting appropriate hyperparameters for training RL agents is notoriously difficult [7]. A more exhaustive hyperparameter search may improve the performance. Additionally, scaling the environment's action space to lie in the interval [–1, 1] could help with the training convergence.

Secondly, the policy transfer from the simulated to the physical environment appears to be challenging as policies often fail to generalise to the physical system (also known as sim-to-real). By nature, there are inherent discrepancies between the dynamics of a simulator and the real world. In particular, a physical robotic arm is subject to the following noise sources and non-stationarities: a) the position and speed resolution inherent to the Dynamixel servo actuators used by the WidowX MKII; and b) overheating of the actuators and wear of the body parts. These physical imperfections tend to increase the compliance margin between the desired goal position and the actual position. Performing a rigorous calibration of the physical robot would ensure that the arm's dynamics in the simulator closely match that of the real world, thus reducing this reality gap.

Another possible solution to account for these irregularities consists in introducing noise in the simulator by altering joint friction coefficients or link masses for example, thus effectively creating a collection of imperfect training environments. The most realistic virtual environment can be identified by training RL agents on these imperfect environments, deploying the learnt policies on the physical robot and selecting the one with the smallest observed difference.

The discrepancies observed between virtual and physical environments can be further reduced by executing part of the training on the physical environment in order to adjust the policy to the real world. In addition, the robot's accuracy could be further increased by restricting the joint's amplitude motion between each timestep (i.e. reducing the range of the action space).

Lastly, the complexity of the task to solve could be further increased so that the algorithm's performance can be compared in more detail. Such increases in complexity could be brought by considering a sparse reward setting, where the reward is only incremented if the end effector is positioned close enough to the goal position. Solving such a task would greatly benefit from exploration enhancement techniques such as HER or curiosity-driven exploration [34] where the intrinsic reward signal coming from the environment is augmented. The task difficulty may also be increased by defining a more challenging reward function. For example, the orientation of the arm's end effector may be included in the reward definition. Additionally, obstacles could be included in the environment with a reward penalty applied at each collision. Finally, the environment's complexity could be increased by allowing the target to move randomly at each timestep for instance.

Acknowledgement. This Career-FIT project has received funding from the European Union's Horizon 2020 research and innovation programme under the Marie Skłodowska-Curie grant agreement No. 713654.

References

1. Sutton, R., Barto, A.: Reinforcement Learning: An Introduction, 2nd edn. MIT Press, Cambridge (2018). http://www.incompleteideas.net/book/the-book-2nd.html

2. Silver, D., et al.: Mastering the game of go with deep neural networks and tree search. Nature **529**(7587), 484–489 (2016). https://doi.org/10.1038/nature16961
3. Mnih, V., et al.: Human-level control through deep reinforcement learning. Nature **518**(7540), 529–533 (2015). https://doi.org/10.1038/nature14236
4. Deisenroth, M.P., Rasmussen, C.E.: PILCO: a model-based and data-efficient approach to policy search. In: Proceedings of the 28th International Conference on Machine Learning, pp. 465–472 (2011). https://doi.org/10.5555/3104482.3104541
5. Lillicrap, T.P., et al.: Continuous control with deep reinforcement learning. In: 4th International Conference on Learning Representation (2016). http://arxiv.org/abs/1509.02971
6. Fujimoto, S., Van Hoof, H., Meger, D.: Addressing function approximation error in actor-critic methods. In: 35th International Conference on Machine Learning, ICML 2018, vol. 4, pp. 2587–2601 (2018). http://arxiv.org/abs/1802.09477
7. Haarnoja, T., et al.: Soft actor-critic algorithms and applications. Computing Research Repository CoRR (2018). http://arxiv.org/abs/1812.05905
8. Haarnoja, T., Zhou, A., Abbeel, P., Levine, S.: Soft actor-critic: off-policy maximum entropy deep reinforcement learning with a stochastic actor. In: International Conference on Machine Learning (2018). http://arxiv.org/abs/1801.01290
9. Andrychowicz, M., et al.: Hindsight experience replay. In: Advances in Neural Information Processing Systems, vol. 30, pp. 5048–5058 (2017). http://arxiv.org/abs/1707.01495
10. Devin, C., Gupta, A., Darrell, T., Abbeel, P., Levine, S.: Learning modular neural network policies for multi-task and multi-robot transfer. In: Proceedings - IEEE International Conference on Robotics and Automation, pp. 2169–2176 (2017). https://doi.org/10.1109/ICRA.2017.7989250
11. Gupta, A., Devin, C., Liu, Y., Abbeel, P., Levine, S.: Learning invariant feature spaces to transfer skills with reinforcement learning. In: International Conference on Learning Representations (2017). http://arxiv.org/abs/1703.02949
12. Plappert, M., et al.: Multi-goal reinforcement learning: challenging robotics environments and request for research. Computing Research Repository CoRR (2018). http://arxiv.org/abs/1802.09464
13. Chen, T., Murali, A., Gupta, A.: Hardware conditioned policies for multi-robot transfer learning. In: 32nd Conference on Neural Information Processing Systems (NeurIPS) (2018). http://arxiv.org/abs/1811.09864
14. Rupam Mahmood, A., Korenkevych, D., Komer, B.J., Bergstra, J.: Setting up a reinforcement learning task with a real-world robot. In: IEEE International Conference on Intelligent Robots and Systems, pp. 4635–4640 (2018). https://doi.org/10.1109/IROS.2018.8593894
15. Tavakoli, A., Pardo, F., Kormushev, P.: Action branching architectures for deep reinforcement learning. In: 31st Conference on Neural Information Processing Systems (NIPS) (2017). http://arxiv.org/abs/1711.08946
16. Gu, S., Holly, E., Lillicrap, T., Levine, S.: Deep reinforcement learning for robotic manipulation with asynchronous off-policy updates. Computing Research Repository CoRR (2016). https://arxiv.org/abs/1610.00633
17. Luo, S., Kasaei, H., Schomaker, L.: Accelerating reinforcement learning for reaching using continuous curriculum learning. In: International Joint Conference on Neural Networks (IJCNN), pp. 1–8 (2020). https://ieeexplore.ieee.org/document/9207427
18. Pham, T.H., De Magistris, G., Tachibana, R.: OptLayer - practical constrained optimization for deep reinforcement learning in the real world. In: Proceedings - IEEE International Conference on Robotics and Automation, pp. 6236–6243 (2018). https://doi.org/10.1109/ICRA.2018.8460547

19. Lucchi, M., Zindler, F., Mühlbacher-Karrer, S., Pichler, H.: Robo-gym – an open source toolkit for distributed deep reinforcement learning on real and simulated robots. In: 2020 IEEE/RSJ International Conference on Intelligent Robots and Systems (IROS 2020) (2020). http://arxiv.org/abs/2007.02753
20. Pong, V., Gu, S., Dalal, M., Levine, S.: Temporal difference models: model-free deep RL for model-based control. In: 6th International Conference on Learning Representations (ICLR), pp. 1–14 (2018). https://arxiv.org/abs/1802.09081
21. Pinto, L., Mandalika, A., Hou, B., Srinivasa, S.: Sample-efficient learning of non-prehensile manipulation policies via physics-based informed state distributions. Computing Research Repository CoRR (2018). http://arxiv.org/abs/1810.10654
22. Trossen Robotics: https://www.trossenrobotics.com/ Accessed 07 Feb 2020
23. Quigley, M., et al.: ROS: an open-source robot operating system. In: Proceedings Open-Source Software workshop of the International Conference on Robotics and Automation (ICRA) (2009). http://ai.stanford.edu/~mquigley/papers/icra2009-ros.pdf
24. Yang, B., Zhang, J., Pong, V., Levine, S., Jayaraman, D.: REPLAB: a reproducible low-cost arm benchmark platform for robotic learning. In: International Conference on Robotics and Automation (ICRA) (2019). http://arxiv.org/abs/1905.07447
25. Coumans, E., Bai, Y.: PyBullet, a Python Module for Physics Simulation for Games, Robotics and Machine Learning (2016–2019). http://pybullet.org
26. Brockman, G., et al.: OpenAI Gym (2016). http://gym.openai.com/
27. Mnih, V., et al.: Asynchronous methods for deep reinforcement learning. In: 33rd International Conference on Machine Learning (ICML), vol. 48, pp. 1928–1937 (2016). https://arxiv.org/abs/1602.01783
28. Wu, Y., Mansimov, E., Liao, S., Grosse, R., Ba, J.: Scalable trust-region method for deep reinforcement learning using kronecker-factored approximation. In: 31st Conference on Neural Information Processing Systems (NIPS) (2017). https://arxiv.org/abs/1708.05144
29. Schulman, J., Wolski, F., Dhariwal, P., Radford, A., Klimov, O.: Proximal policy optimization algorithms. Computing Research Repository (CoRR) (2017). http://arxiv.org/abs/1707.06347
30. Schulman, J., Levine, S., Moritz, P., Jordan, M.I., Abbeel, P.: Trust region policy optimization. In: 31st International Conference on Machine Learning (ICML) (2015). http://arxiv.org/abs/1502.05477
31. Hill, A., et al.: Stable Baselines. GitHub repository (2018). https://github.com/hill-a/stable-baselines
32. Akiba, T., Sano, S., Yanase, T., Ohta, T., Koyama, M.: Optuna: A next-generation hyperparameter optimization framework. In: Proceedings of the ACM SIGKDD International Conference on Knowledge Discovery and Data Mining, pp. 2623–2631 (2019). https://doi.org/10.1145/3292500.3330701
33. University College Dublin, UCD Research Office and IT Services. ResearchIT Sonic HPC cluster
34. Pathak, D., Agrawal, P., Efros, A.A., Darrell, T.: Curiosity-driven exploration by self-supervised prediction. In: IEEE Computer Society Conference on Computer Vision and Pattern Recognition Workshops, pp. 488–489 (2017). https://doi.org/10.1109/CVPRW.2017.70

Age and Gender Recognition from Speech Using Deep Neural Networks

Héctor A. Sánchez-Hevia, Roberto Gil-Pita, Manuel Utrilla-Manso, and Manuel Rosa-Zurera(✉)

Signal Theory and Communications Department, University of Alcalá, 28805 Alcalá de Henares, Madrid, Spain
{hectoradrian.sanchez,roberto.gil,manuel.utrilla,manuel.rosa}@uah.es
https://n9.cl/rtn2

Abstract. This paper deals with joint gender identification and age group classification from speech, aimed at improving the functionalities of Interactive Voice Response Systems. Deep Neural Networks are used, because they have recently demonstrated discriminative and representation capabilities over a wide range of applications, among them, speech processing problems based on features extraction and selection. A comparative study of various neural network architectures and sizes is presented to gather knowledge about performance dependence on the network architecture and the number of free parameters. The classification framework was trained and evaluated using Mozilla's 'Common Voice' dataset, an open and crowdsourced speech corpus. The results are promising, with the best systems achieving a gender identification error lower than 2% and an age group classification error lower than 20%.

Keywords: Human-robot interaction · Speaker classification · Machine learning

1 Introduction

This paper deals with the design of an algorithm for classifying speakers into age and gender groups, to be applied in 'Interactive Voice Response' (IVR) systems. IVRs are a kind of Spoken Language Understanding (SLU) system widely used in contact centres, based on the so-called How-May-I-Help-You (HMIHY) system [11]. IVR is a technology that allows a computer to interact with humans through the use of voice, and therefore, IVR is an example of human-machine interaction. The basic scheme of an IVR system is represented in Fig. 1.

An IVR system performs automatic caller intent classification by means of continuous speech recognition, which translates speech into text, and semantic text classification, which labels speech into a predefined caller intent [5]. Modern IVR systems also extract biometrical and emotional information from speech, which complements lexical and syntactic information, so as to make human-machine interaction similar to human-human interaction.

L. M. Bergasa et al. (Eds.): WAF 2020, AISC 1285, pp. 332–344, 2021.
https://doi.org/10.1007/978-3-030-62579-5_23

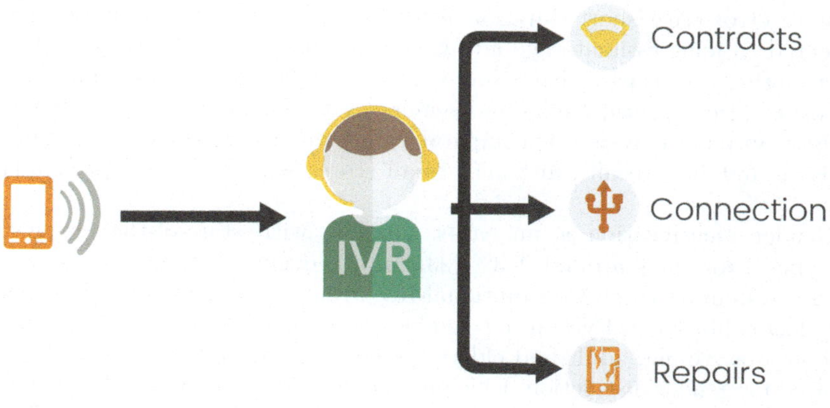

Fig. 1. Basic scheme of an IVR system

IVR systems are based on the use of artificial intelligence algorithms, to humanise the interaction with the contact centre, improving the customer experience and reducing the time dedicated to their attention. IVR systems are mainly used to reduce costs related to call duration, dropouts and repetitive calls rates, and transfers between service assistants, thanks to improved call routing. A typical addition to improve routing is anger detection, which has been analysed in [17,20]. IVR systems typically include complex algorithms:

- Speech recognition, to generate text from speech, complemented with natural language processing, to extract the meaning of speech.
- Speech processing to extract information about emotions, biometric information (gender, age, etc.) to improve routing effectiveness.
- Acoustic environment monitoring, to gather information about the environment where the user is located, and to relate that information with the meaning of the message and the speaker's emotions.

The main application of age and gender recognition in this context is to conduct automated demographic analysis, which can be used to improve market research or personalise the experience for different target audiences, among other uses. Age estimation from speech recordings (viewed as a regression problem) has been approached multiple times in the last two decades. In [16], speech signals sampled at 16 kHz and PCM coded with 16 bits, were used to calculate MFCC, ΔMFCC and ΔPower coefficients, to be applied to Linear Discriminant Analysers (LDA) and Neural Networks to estimate speakers' agedness. More recently, Support Vector Regression with i-vectors as input has been used to estimate age from telephone speech [3]. A novel age estimation system based on long shortterm memory (LSTM) recurrent neural networks (RNN) was proposed in [27]. The proposed system was tested and compared with a state-of-the-art i-vector approach, reporting a relative improvement up to 28% in terms of mean absolute error with short duration utterances. In [15] several approaches are presented to joint estimate age and gender for telephone applications. Instead of estimating age using a regression analysis, people are classified into four age groups: children, youth, adults and seniors. Authors claim that performance of the best automatic system is comparable to human performance (around 60% precision and 70% recall), although recall scores were lower by approximately 10%.

Gender identification is an easier problem with state-of-the art systems achieving error rates around 2–3% [23]. In a previous work [24], we have proposed a system based on Convolutional-Recurrent neural networks implemented using Keras library in Python, for gender distinction and age classification into three groups: young, adult and elderly people. An average error below 2% was reported for gender distinction, which is comparable to the best published in the literature. An average classification error below 20% for age group classification was obtained.

More recently, the Deep Learning paradigm has been applied to age estimation. For example, Deep Neural Networks (DNN) have been applied to predict both height and age of a speaker for short speech durations [12]. In case of age estimation, the Root Mean Squared Errors (RMSE) are 7.60 and 8.63 years for male and female respectively, when the mean duration of speech segments is around 2.5 s. Anyway, recent studies demonstrate that age estimation is a very difficult task, even for human listeners [13], who are affected by aspects that are not related to the speaker's age, such as the listeners' own age, or the speaker's speech rates.

This paper presents a framework to jointly identify speakers' gender and classify them into age groups, designed to be applied in IVRs. The age and gender recognition system is based on the Deep-Learning paradigm. The main objective is to compare different network architectures and sizes to gather knowledge about performance dependence on the architecture, and the number of model free parameters to be adjusted during training. This work is a continuation of [24].

2 Problem Statement

Artificial neural networks date back to the 1940s, thanks to the work by Pitts and McCulloch [22] who studied how to simulate how the human brain system solves general problems. The back-propagation algorithm [9] was an important milestone in this field in the 1980s, but its application brings to light very soon several problems due to the overfitting of training, lack of large-scale training data, and limited computation power, which made the application of artificial neural networks difficult in real problems. For decades, machine learning systems, of which neural networks are a subset, required careful engineering and considerable domain expertise to design a feature extractor, to transform the raw data into a suitable internal representation or feature vector from which the learning subsystem could detect or classify patterns in the input [14]. Deep learning arose to overcome these problems, and its high popularity is usually attributed to the following factors [28]:

- The collection of large-scale annotated training data, for image processing, speech, etc. The large size of these data sets helps to avoid overfitting when the number of network parameters is very high.
- The revolution of parallel computing, mainly due to the fast development of Graphical Processing Units (GPUs) clusters.
- Advances in the design of network structures and training algorithms. Convolutional and recurrent neural networks, temporal convolutional networks, and residual nets are some examples.

Deep learning has been used recently in many applications: video coding, fraud detection in financial applications, financial time-series forecasting, predictive and prescriptive analytics, medical image processing, and power systems research [25]. Most of the efforts have been dedicated to find new applications or to propose new, more powerful architectures, but not much work has been devoted to know the appropriate size of a network to solve a certain problem. Systems such as IVRs, which include complex subsystems for speech recognition, natural language processing, artificial intelligence for optimal routing, etc., must be implemented with powerful and robust algorithms, but at the same time they must not require large computational resources. This motivates the study of the performance of the different subsystems based on the number of free parameters to adjust in the training, or the complexity of the network that solves the problem, so that it is possible to determine what is the limit of computational complexity appropriate to solve a problem. This is the objective of this article.

3 Framework Description

The proposed framework takes a speech signal in the time-frequency domain and maps it to 4 outputs using various neural networks. The system performs joint age and gender classification where gender identification is viewed as a binary classification problem, while age grouping is viewed as a categorical classification problem. The following 4 types of neural networks have been tested:

- *CNN*: a Fully Convolutional Neural Network. A type of network that applies multiple sliding convolutional filters throughout multiple layers. CNNS are commonly used for many speech related tasks (e.g. [2,18,21]).
- *CRNN*: Convolutional Recurrent Neural Network. A CNN with one or more additional recurrent layers that are capable of finding patterns in sequences. CRNNs are typically used for audio event detection/classification [6,26]
- *TCN*: Temporal Convolutional Network [4]. A new type of network that is also capable of modelling sequences but using only convolutional layers. TCNs are being used for speech enhancement [19].
- *CTCN*: A CNN followed by a TCN. The same idea as the CRNN but substituting the recurrent layer by a TCN.

The different network types described above are formed by a combination of the following layers:

- *Conv*: 2D Convolutional layer with rectified linear unit (ReLU) activation function. A 2-D convolutional layer applies sliding convolutional filters to the input. The layer convolves the input by moving the filters along the input vertically and horizontally and computing the dot product of the weights and the input, and then adding a bias term. The size of the kernel (filter) is indicated by the numbers in brackets. This layer is followed by max-pooling in the feature axis with a poolsize of 2. Max-pooling is a sample-based discretisation process to down-sample the 2D Convolutional layer output matrix so as to reduce its dimensionality.
- *R*: Reshaping layer, used to change the dimensions of its input, without changing its data. It turns the 4 dimensional tensor ([batch×time×features×channels]) into 3 dimensional tensor by combining the feature and channel dimensions into a single one.
- *GRU*: Gated Recurrent Unit [7], with tanh activation and sigmoid recurent activation. The GRU aims to solve the vanishing gradient problem which comes with a standard recurrent neural network. It returns the last output in the output sequence.
- *TCN*: Temporal Convolutional Network formed by 4 increasingly dilated 1D convolution layers (dilation $\in \{1, 2, 4, 8\}$) with residual connections. 1D convolutions have a kernel size of 5 with ReLu activation. When the input size is different to the number of filters the residual connection has a 1D convolution with a kernel size of 1 and linear activation.
- *DC1D*: Depthwise Convolution 1D, a type of convolutional layer which acts on each input channel separately. ReLU activation, kernel size fixed to 5 and depth multiplier (number of filters per channel) indicated in brackets.
- *MP*: Maximum pooling performed only in the time axis, with poolsize equal to the sequence length.
- *O*: Output layer, 4 unit Fully Connected layer with sigmoid activation.

Table 1 shows the different networks tested in this work, detailing the layer structure of each network. The number of units in each layer is indicated by the number preceding the layer type. Additionally dropout regularisation with a dropout rate of 0.2 is used between every layer for all the networks.

We opted to fix the input length to streamline the classification. However, since the system needs to work with variable signal lengths, a simple time-averaging strategy is used. This way, audio signals exceeding the selected input length are divided into multiple segments to be processed independently by the network, and their outputs are later combined by averaging in a similar fashion to maximum voting.

Table 1. Tested neural networks detailing their architecture.

#	Type	Network architecture
1	CRNN	48 Conv(3,3)+24 Conv(3,3)+R+48 GRU+O
2	CRNN	48 Conv(3,3)+24 Conv(3,3)+R+64 GRU+O
3	CRNN	48 Conv(3,3)+24 Conv(3,3)+R+96 GRU+O
4	CRNN	48 Conv(3,3)+24 Conv(3,3)+R+112 GRU+O
5	CRNN	48 Conv(3,3)+24 Conv(3,3)+48 Conv(3,5)+R+48 GRU+O
6	CRNN	48 Conv(3,3)+24 Conv(3,3)+64 Conv(3,5)+R+96 GRU+O
7	CTCN	48 Conv(3,3)+24 Conv(3,3)+32 Conv(3,5)+R+32 TCN+DC1D (4)+MP+O
8	CTCN	48 Conv(3,3)+24 Conv(3,3)+48 Conv(3,5)+R+48 TCN+DC1D (4)+MP+O
9	CTCN	48 Conv(3,3)+24 Conv(3,3)+48 Conv(3,5)+R+56 TCN+DC1D (4)+MP+O
10	CTCN	48 Conv(3,3)+24 Conv(3,3)+R+32 TCN(5 × 4)+DC1D(4)+MP+O
11	CTCN	48 Conv(3,3)+24 Conv(3,3)+R+54 TCN(5 × 4)+DC1D(4)+MP+O
12	TCN	32 TCN+DC1D(4)+MP+O
13	TCN	48 TCN+DC1D(4)+MP+O
14	TCN	64 TCN+DC1D(4)+MP+O
15	CNN	48 Conv(3,3)+24 Conv(3,3)+R+DC1D (2)+MP+O
16	CNN	48 Conv(3,3)+24 Conv(3,3)+48 Conv(3,5)+R+DC1D(4)+MP+O
17	CNN	48 Conv(3,3)+24 Conv(3,3)+96 Conv(3,5)+R+DC1D(3)+MP+O
18	CNN	48 Conv(3,3)+32 Conv(3,3)+128 Conv(3,5)+R+DC1D(3)+MP+O

4 Experimental Work

4.1 Training Corpus

The Mozilla Common Voice dataset [10] has been used, an open and publicly available multi-language speech corpus. The English corpus used in this work is composed of 12 GB of labeled audio files, containing thousands of speech clips (most are short phrases between 2–5 s), recorded and annotated by anonymous speakers. Labels include speakers' ages grouped by decades (from 'teens' to 'octogenarians') and gender. This labelling precludes from implementing an age estimation system based on regression networks. In total, there are 143,170 audio files (≈98 h) with age and gender labels. Age is highly unbalanced, following the distribution depicted in Fig. 2, where only 0.3% of the total audio files (432 files) correspond to octogenarians.

The audio files have been grouped into three age groups: young people (teens and twenties), adults (thirties, forties and fifties), and seniors (sexagenarians, heptagenarians and octogenarians). The system is built to have four outputs, one for each age group and a last one for gender classification.

The original audio files are in MP3 format, with a bit rate of 64kbps. They are trans-coded to PCM with 8 kHz sampling frequency to approximate the quality of telephone lines. The losses due to MP3 coding mainly affect high frequency bands of the audio signals, and most of the information below 4 kHz remains available. It is worth commenting that some audio files contain long silences at the beginning and the end, which have been detected using a voice activity

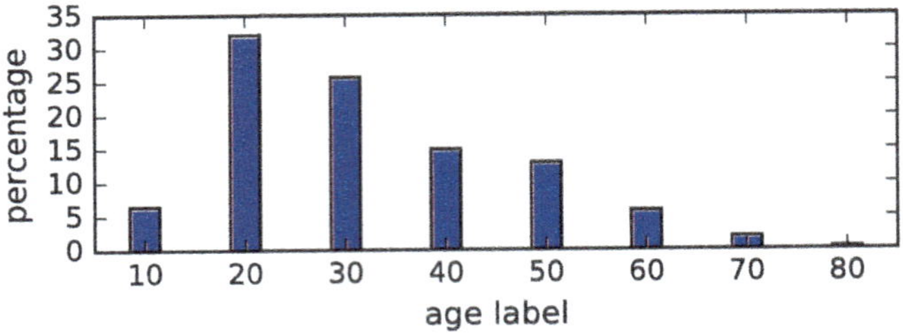

Fig. 2. Dataset age group distribution.

detector and deleted to avoid confusion. Additionally, all the audio signals have been normalised for their amplitudes to be in the range ± 1.

4.2 Experimental Settings

The dataset was divided into three subsets for the experiments: training set, which contains 70% of the available files; validation set, with 15% of the files, and evaluation set, with the remaining 15% of the files. The size of these sets was empirically determined, to have a large number of data for training, and enough data for testing. Due to privacy concerns, speakers' ID labels are not included in the corpus. For this reason, it is not possible to ensure that a particular speaker is not represented in the different sets, but it is unlikely.

Audio files are divided into one-second blocks (with 50% overlapping between adjacent blocks). The Short-time Fourier transform (STFT) is applied to each block, with 20 ms length frames (160 samples), windowed with the hamming window, and with adjacent windows overlapping 50% of their length. After STFT calculation, the energy of 20 subbands in the Mel-scale is obtained for each frame. Zero padding is used to complete the last block of each audio file. Therefore, each second of the audio file is represented by a $[100 \times 20]$ matrix. Data are normalised to have zero-mean and unit-variance according to the mean and standard deviation of the training set.

All the networks were trained for 300 epochs using Adam optimisation with 0.001 learning rate and binary cross-entropy as loss function. Mini batch size was fixed to 2048. To counter class unbalance, errors due to the different classes in the training set are weighted according to the size of the age group (0.25 for young people, 0.2 for adults and 0.55 for seniors). The best network parameters were selected in terms of the minimum validation loss. Neural networks were implemented in Python 3.5 using Keras 2.24 [8] with Tensorflow r1.12 [1] as the backend.

4.3 Results and Discussion

The models have been compared in terms of block level performance and file level performance. Block level results are obtained by evaluating the one-second audio blocks independently, while file level results are obtained by averaging the outputs of all the blocks that form each file before computing the metrics.

Table 2. Results obtained with the tested neural networks described in Fig. 1 (gender error/age error), as a function of the number of parameters of the network.

#	Params	Block error	File error
1	35884	3.22%/27.52%	2.06%/23.20%
2	47132	3.13%/25.22%	2.07%/21.14%
3	74236	2.08%/22.68%	1.87%/18.13%
4	90092	2.74%/21.72%	1.76%/16.77%
5	59260	3.21%/26.23%	2.15%/21.92%
6	94156	2.92%/24.04%	2.03%/19.64%
7	44828	3.43%/27.98%	2.17%/23.21%
8	77164	3.15%/25.53%	1.98%/19.93%
9	94756	2.99%/24.47%	1.82%/19.72%
10	51580	3.22%/26.78%	2.00%/21.23%
11	97054	2.95%/22.62%	1.79%/17.04%
12	21028	4.13%/37.30%	2.66%/31.37%
13	42484	3.64%/31.35%	2.25%/28.76%
14	84484	3.17%/25.79%	1.86%/20.12%
15	14236	4.25%/36.16%	2.81%/32.01%
16	30700	3.61%/30.14%	2.27%/24.90%
17	49372	3.40%/27.65%	2.09%/22.34%
18	81028	3.13%/25.53%	1.89%/19.57%

Table 2 shows the error for gender and age group classification of the tested networks both at block and file level and Fig. 3 depicts the age classification error at file level. The accuracy improvement obtained by averaging the outputs of all the blocks of a given speech file is quite noticeable. In relative terms, age group classification error is reduced by an average of 18% and gender classification error is reduced approximately by 35%.

Results show a clear relation between the number of free parameters to be adjusted during training and the performance of the different networks. Interestingly, the type of network does not affect performance too much. Even the CNN, which lacks long term memory, obtains very similar results to those of the other types for a comparable number of free parameters. The results obtained by the CRNN and CTCN networks show that, in this particular application, the use of 2 instead of 3 convolutional layers preceding the temporal memory layer is preferred, most likely due to the higher dimension of the reshaped data.

Age and gender classification has also been studied as a single problem, by defining a classification problem with six categories which combine gender and

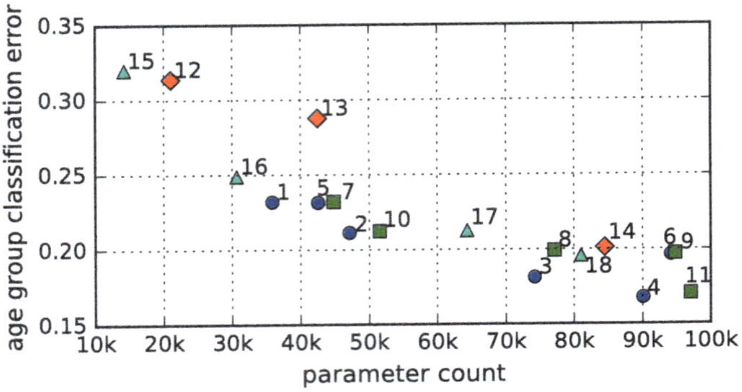

Fig. 3. Age group classification error at file level as a function of network size.

age: Young Male (YM), Young Female (YF), Adult Male (AM), Adult Female (AF), Senior Male (SF) and Senior Female (SF). Table 3 shows the results of the best performing network of each type for the 6 classes problem in terms of recall, precision and F1 score, and Fig. 4 depicts the F1 score at file level for all the tested networks. These results, compared to the age group classification error presented in Table 2, show an increase in the average error of about 2% for block level and 1.3% for file level.

Table 3. Six classes results for best performing neural network of each type (Recall/Precision/F1 score).

#	Type	Params	Block level	File level
4	CRNN	90092	0.76/0.81/0.78	0.74/0.81/0.77
11	CTCN	97054	0.72/0.79/0.76	0.74/0.83/0.78
14	TCN	84484	0.66/0.72/0.69	0.73/0.82/0.77
18	CNN	81028	0.69/0.76/0.73	0.70/0.78/0.74

Larger differences among the different types of networks are observed, with worse results for the CNN and better results for the TCN, which is interesting since it is the network type which obtains the highest improvement working at the file level compared to block level (Table 3). To better understand the meaning behind the results, Fig. 5 shows the relative confusion matrix for the overall best performing network (network number 4). The confusion matrix shows that most errors come from the confusion between adjacent age groups, and that gender misclassification errors are far more unlikely. A deeper analysis to the results shows that ages closest to the boundaries between classes (e.g. 20 s and 60 s) have larger error rates than those further away (e.g. 10 s and 80 s).

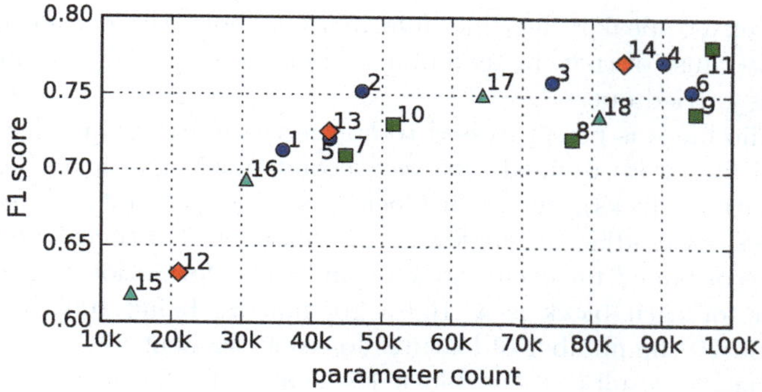

Fig. 4. F1 score at file level as a function of network size.

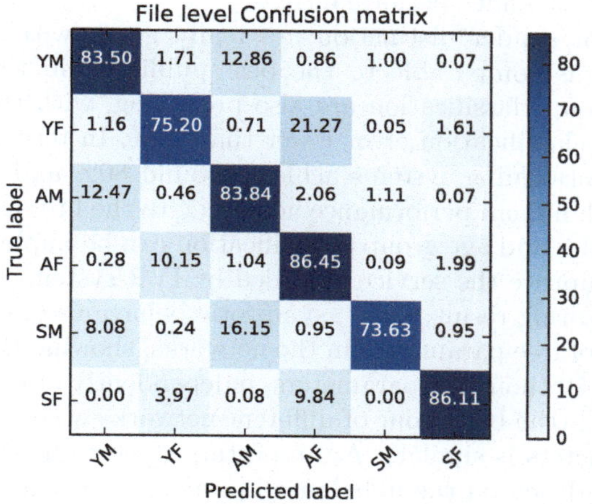

Fig. 5. Relative confusion matrix (rows sum up to 100%) at file level for network number 4.

5 Conclusions

In this paper, a study of age and gender recognition from speech using different types of neural networks is presented. The underlying application is Interactive Voice Response systems, where gender and age classification can be used for demographic analysis and to optimise the call flow, for instance, offering the immediate assistance of a human operator or trying to react with appropriate dialogue strategies.

One of the main difficulties for implementing this kind of systems is the access to a dataset that could be used for training and testing the classifiers. For this study one the few freely available datasets with information about speakers' gender and age has been used, the Mozilla Common Voice dataset, an open corpus that anyone can use to train speech-enabled applications. This dataset doesn't contain the exact speaker age, but information about the decade the speaker belongs to. Because of that, instead of age prediction, speaker classification into age groups is carried out.

Each audio file was pre-processed to obtain the matrix which feeds the classifier. Audio files were divided into one-second blocks (with 50% overlapping between adjacent blocks), and each block was segmented into 20 ms (160 samples) windows, with adjacent windows overlapping 50% of their length. Spectral features were obtained for each segment, and the information to be applied to the classifier for each block is a $[100 \times 20]$ matrix, being 100 the number of segments, and 20 the number of features for each segment.

In this paper, results at the block level, and the file level are presented, demonstrating a better performance at the file level. The tested classifiers are different models of DNNs, so as to study the influence of the number of free parameters on the classifier performance.

The results for gender distinction are really good, with errors commonly below 2%, which is comparable to the best published in the literature. The results for age group classification are also promising, with the best configurations achieving a classification error lower than 20%. In terms of precision and recall, the best performing systems achieve around 80% and 70% respectively, with is in line with human performance according to the literature. These results indicate that gender and age groups classification can be implemented with very good results to improve the service provided by IVR systems.

Another interesting result is the asymptotic behaviour of classification error with the number of free parameters in the networks, showing that it is not worth to use more than 50 thousand parameters, independently of the kind of network configuration used (the behaviour of different networks with an equivalent number of free parameters is similar). An important issue to be studied in a future work is the dependency on the unbalanced property of the available datasets.

Acknowledgement. This work has been partially funded by the Spanish Ministry of Economy, Industry and Competitiveness, with project RTC-2016-4687-7 and the Spanish Ministry of Science, Innovation and Universities, with project RTI2018-098085-B-C42 (MSIU/FEDER).

References

1. Abadi, M., Agarwal, A., Barham, P., et al.: TensorFlow: large-scale machine learning on heterogeneous systems (2015). http://tensorflow.org/. Software available from tensorflow.org
2. Badshah, A., Ahmad, J., Rahim, N., Baik, S.: Speech emotion recognition from spectrograms with deep convolutional neural network. In: 2017 International Conference on Platform Technology and Service (PlatCon), pp. 1–5 (2017)
3. Bahari, M., McLaren, M., Van Leeuwen, D., et al.: Age estimation from telephone speech using i-vectors. In: Proceedings of Interspeech 2012. Portland, USA (2012)
4. Bai, S., Kolter, J., Koltun, V.: An empirical evaluation of generic convolutional and recurrent networks for sequence modeling. arxiv 2018. arXiv preprint arXiv:1803.01271

5. Bhat, C., Mithum, B., Saxena, V., Kulkarni, V., Kopparapu, S.: Deploying usable speech enabled IVR systems for mass use. In: 2013 IEEE International Conference on Human Computer Interaction (ICHCI), pp. 1–5 (2013)
6. Cakir, E., Adavanne, S., Parascandolo, G., Drossos, K., Virtanen, T.: Convolutional recurrent neural networks for bird audio detection. In: 2017 25th European Signal Processing Conference (EUSIPCO), pp. 1744–1748 (2017)
7. Cho, K., et al.: Learning phrase representations using RNN encoder-decoder for statistical machine translation (2014). arXiv:1406.1078
8. Chollet, F., et al.: Keras (2015). https://keras.io
9. Rumelhart, R., et al.: Learning representations by back-propagating errors. Nature **521**, 533–536 (1986)
10. Foundation, M.: Common voice (2019). https://voice.mozilla.org
11. Gorin, A., Riccardi, G., Wright, J.: How may i help you? Speech Commun. **23**(1–2), 113–127 (1997)
12. Kalluri, S.B., Vijayasenan, D., Ganapathy, S.: A deep neural network based end to end model for joint height and age estimation from short duration speech. In: 2019 IEEE International Conference on Acoustics, Speech and Signal Processing (ICASSP 2007), pp. 6580–6584. IEEE (2019)
13. Kang, Y., Tsang, K.Y., Wong, K.W.Y.: The effect of speech rate on age estimation in conversational speech. Toronto Working Papers in Linguistics (TWPL) **42**, 1–10 (2020)
14. LeCun, Y., Bengio, Y., Hinton, G.: Deep learning. Nature **323**, 436–444 (2015)
15. Metze, F., Ajmera, J., Englert, R., Bub, U., et al.: Comparison of four approaches to age and gender recognition for telephone applications. In: 2007 IEEE International Conference on Acoustics, Speech and Signal Processing (ICASSP 2007), vol. 4, pp. IV–1089 (2007)
16. Minematsu, N., Sekiguchi, M., Hirose, K.: Automatic estimation of one's age with his/her speech based upon acoustic modeling techniques of speakers. In: 2002 IEEE International Conference on Acoustics, Speech, and Signal Processing (ICASSP 2002), vol. 1, pp. I–137 (2002)
17. Mohino-Herranz, I., García-Gómez, J., Utrilla-Manso, M., Rosa-Zurera, M.: Precision maximization in anger detection in interactive voice response systems. In: 145th Convention of the Audio Engineering Society, p. 10090 (2018)
18. Neumann, M., Vu, N.T.: Attentive convolutional neural network based speech emotion recognition: A study on the impact of input features, signal length, and acted speech. arXiv preprint arXiv:1706.00612 (2017)
19. Pandey, A., Wang, D.: TCNN: Temporal convolutional neural network for real-time speech enhancement in the time domain. In: 2019 IEEE International Conference on Acoustics, Speech and Signal Processing (ICASSP 2019), pp. 6875–6879 (2019)
20. Pappas, D., Androutsopoulos, I., Papageorgiou, H.: Anger detection in call center dialogues. In: 2015 6th IEEE International Conference on Cognitive Infocommunications (CogInfoCom), pp. 139–144 (2015)
21. Park, S.R., Lee, J.: A fully convolutional neural network for speech enhancement. arXiv preprint arXiv:1609.07132 (2016)
22. Pitts, W., McCulloch, W.: How we know universals the perception of auditory and visual forms. Bull. Math. Biophys. **9**(3), 127–147 (1947). https://doi.org/10.1007/BF02478291
23. Ranjan, S., Hansen, J.H.: Improved gender independent speaker recognition using convolutional neural network based bottleneck features. In: Proceedings of Interspeech, pp. 1009–1013 (2017)

24. Sanchez-Hevia, H., Gil-Pita, R., Utrilla-Manso, M., Rosa-Zurera, M.: Convolutional-recurrent neural network for age an gender prediction from speech. In: 2019 Signal Processing Symposium, Krakow (Poland), pp. 246–249. IEEE (2019)
25. Sengupta, S., et al.: A review of deep learning with special emphasis on architectures, applications and recent trends. Knowl. Based Syst. **194**(105596), 1–33 (2020)
26. Xu, Y., Kong, Q., Wang, W., Plumbley, M.: Large-scale weakly supervised audio classification using gated convolutional neural network. In: 2018 IEEE International Conference on Acoustics, Speech and Signal Processing (ICASSP 2008), pp. 121–125 (2018)
27. Zazo, R., Nidadavolu, P., Chen, N., Gonzalez-Rodriguez, J., Dehak, N.: Age estimation in short speech utterances based on LSTM recurrent neural networks. IEEE Access **6**, 22524–22530 (2018)
28. Zhao, Z.Q., Zheng, P., Xu, S.T., Wu, X.: Object detection with deep learning: a review. IEEE Trans. Neural Netw. Learn. Syst. **30**(11), 3212–3232 (2019)

Monocular 3D Hand Pose Estimation for Teleoperating Low-Cost Actuators

Francisco Gomez-Donoso(✉), Félix Escalona, Alejandro Bañuls, Daniel Abellan, and Miguel Cazorla

University Institute for Computer Research, University of Alicante, P.O. Box 99., 03080 Alicante, Spain

fgomez@ua.es

Abstract. So far, in order to interact with a virtual agent that simulates hands or to teleoperate a robotic hand it is required any kind of device like joysticks, gamepads or VR controllers. These devices map a button press to a predefined action, usually with no fine control of the entity that is being teleoperated, which is unsuitable and unpleasant. In this work, we study different setups to perform 3D hand pose estimation from color images of hands with the aim of controlling a low-cost robotic hand. The experiments we carried out demonstrated that it is possible to do so with a deviation of less than 5 mm per joint.

Keywords: Tridimensional hand pose estimation · Hand pose estimation · Joint regression · Deep learning

1 Introduction

It is well known that being able to estimate the tridimensional hand pose has many applications nowadays. This task regards to create an accurate virtual model of the hands that can be used in virtual reality worlds, for instance. The user would seamlessly interact with the virtual objects using their hands, as if it were real life. Estimating hand poses can also be used for human-computer interaction. The user would perform a set of hand gestures, that would be recognized by a computer in order to trigger different actions. Another important application for fine grain hand pose estimation is the sign language recognition. An embedded system with this capabilities would enable the sign language user to communicate with any person that is not familiar with it.

Finally, another important application, that is also tackled in this work, is the teleoperation. The teleoperation is the ability to control a remote agent and is often used to circumvent an impediment. For instance, carrying out maintenance tasks in a hazardous area by teleoperating a robot.

Tridimensional hand pose estimation becomes even more interesting if it is monocular. Namely, the estimation is performed by taking just an image as

L. M. Bergasa et al. (Eds.): WAF 2020, AISC 1285, pp. 345–359, 2021.
https://doi.org/10.1007/978-3-030-62579-5_24

input. This capability removes the requirement of having dedicated devices like gloves or tracking markers, which would harm the immersion of the user in the virtual world, or set additional constrains to the teleoperation process. It is also more appropriate and has a lower energy requirements, as it only requires a color camera to work.

Thus, in this work, we propose a tridimensional hand pose estimation pipeline that is able to state the hand pose by taking one color image as an input. The hand pose is based on the inference of each joint in a hand in a 3D coordinate frame. The pipeline is deep learning-based which provides high accuracy. Finally, the estimations are used to successfully teleoperate a low-cost robot hand. The main contributions of this work is:

- A deep learning-based pipeline for monocular 3D hand pose estimation
- A study of the best method to infer the position of the joints in a hand from images
- The teleoperation of a low-cost robotic hand actuator using the mentioned pipeline

The rest of the work is organized as follows. First, we briefly review some related works in Sect. 2. In Sect. 3, it is thoroughly explained the tridimensional hand pose estimation pipeline. In order to train the systems that compose the pipeline, we adopted the dataset described in Sect. 4. Then, the experimentation we carried out to validate our proposal is described in Sect. 5. Finally, in Sect. 6, we state the conclusions of this work and point out the directions of future research.

2 Related Works

Many works have addressed the 3D pose estimation problem since the 1990s, but it is still a hot topic in computer vision. The first studies were based on a pipeline using traditional machine learning techniques. In [20] authors used a hierarchical structure of classifiers to recognize different objects with a handcrafted descriptor. These traditional approaches experimented difficulties to estimate poses with precision, don't generalize in real world application and are computationally expensive.

With the advent of low cost sensors, such as depth cameras, new approaches tried to profit their advantages. There are two main approaches using depth images:

- **3D model tracking** [2]. A synthetic 3D hand model is usually used to generate the hypothesis to be evaluated with real 3D data. The poses must be very accurate, and the main issues are the noise impact and the changes between frames, facts that are common in real world problems and affect these methods very sensitively.
- **Discriminative techniques using depth information** [8]. Works based on this approach predict the hand pose directly from RGB-D images. As an

example, [8] and [21] use *Random Forests* for the prediction. In the case of [21], authors use *Random Forests* to classify hand masks from depth images. In this way, they are able to detect hand parts and estimate the locations of the 3D joints. This approach is computationally expensive, due to classification per pixel techniques, and requires a large dataset to resolve the discretization of the poses.

In recent years, with the popularization of the *CNN* and deep learning techniques, new approaches have appeared. Most of them use convolutional networks to extract the features from depth images [22]. Another works combine feature extraction with supervised training to predict the 3d pose of the hand [14]. These systems need a large amount of labeled data with the poses of the joints.

The approach presented in [18] is based on the use of deep learning networks trained with synthetic data to extract deep features. These features are organized according to their spatial and temporal neighborhood, and achieve a robust and fluctuation-free estimation of the 3D pose of the hand. This technique works with depth maps too with a previous crop of the detected hand. Another work based on deep learning is described in [22]. Authors present a *CNN* that calculates the 3D pose of the hand that is refined with the pose of a predefined static hand. Although most of these methods are achieving good results, all of them require the use of depth information from 3D sensors.

In [24], authors address this problem using just a RGB image and a *CNN*. 3 different *CNN* are used in this approach. The first network performs a pixel-level segmentation of the hands, the second one detects 2D joints in the image space and the third one transforms from 2D points into 3D. This technique is only training with synthetic data and therefore does not achieve good results for real data. Moreover, the use of 3 independent networks is computationally expensive.

In recent years, new publications have appeared addressing the problem of the estimation of hand poses from multiple views [15,17]. However, these methods are sensitive to calibration procedures and require a good previous depth estimation or, at least, a good 3D triangulation through multiple cameras.

As we have seen, most models require a high computational cost and do not obtain high precision. Another issue is that the most of them are training with synthetic data and do not generalized properly with real data. Along with these methods, several datasets were published to feed these systems.

In the early 2000s, two datasets appeared, one for static pose detection [11], consisting on 3000 hand images with static poses divided in 6 different classes, and another one for dynamic gesture classification [12], composed of 60 hand sequences performing gestures, divided in 4 classes. In both versions there are no information about the pose of the hand, the classes are highly imbalanced and the resolution and the number of examples are insufficient.

Later, in 2014, another dataset of hand pose classification was released [7]. This dataset, named *NUS*, consists of 10 types of poses with 24 images for each, that are captured with variations in position and hand size, but the background of the images is uniform. It does not provide joint annotations, so it is not intended for continuous estimation of hand poses. On the other hand, the uniform background affects the generalization of the learning algorithms.

The dataset introduced in [6] contains sign language Polish and American gestures. This dataset consists of 3 series of gestures with the following information: RGB images with different good quality resolutions, ground truth of the skin and the location of keypoints of the hand. Although it contains 1600 examples, they are not enough because the background is monotone and reduce its generalization capability.

Another dataset was published by [13]. It comprehends real and synthetic depth maps. Every depth map represents one segmented hand with a determined pose. The ground truth includes 2D positions of every joint and its corresponding depth value. It contains 65 real captures divided intro 6 different groups depending on their occlusion level, pose or movement. Synthetic data was generated from a seed and 200 random rotations. This dataset is very representative but it only provides depth maps, so the methods that use it must employ a 3d sensor of the same nature. Moreover, the number of samples is insufficient to feed deep learning algorithms, and synthetic data is too much different than actual examples.

Following the trend of low cost 3D sensors, NYU Hand Pose dataset was created [22]. This dataset contains 8252 test frames and 72757 train frames with RGBD data and ground truth of the hand pose. For every frame, it provides RGBD data from 3 Kinects: one frontal view and two lateral views. The training test only contains samples from one person, whilst the test set has samples from two users. Additionally, a synthetic replication of the hand pose is provided for every view. The main inconvenient of this dataset is that the RGB images are generated by projection the point clouds provided by a 3D sensor, so they are low quality and show many unknown regions beyond the scope of the sensor. For this reason, it is compulsory to make use of a 3D sensor to profit this dataset.

On the other hand, in [23] authors propose a different dataset based on stereo images. This dataset provide 18000 stereo hand pose pairs, the ground truth of their 3d joints and 3D maps. Despite the quality of the annotations, it lacks posture variability because for each scene it only contains one hand with random and naive movements from two different individuals.

As we have seen, most of these datasets, specially the older ones, use synthetic hands and uniform backgrounds with poor generalization for real data, which cause bad results off-lab.

3 Tridimensional Hand Pose Estimation Pipeline

As we mentioned before, our approach is able to take as input any color image that depict a hand, and to estimate the position of the joints in the hand in a tridimensional coordinate frame.

To do so, the pipeline is composed of two main parts. First, the Hand Area of Interest Localization (HAIL) system is in charge of detecting the hands in the input images. We do this before running the Hand Joints Estimation (HJE) step

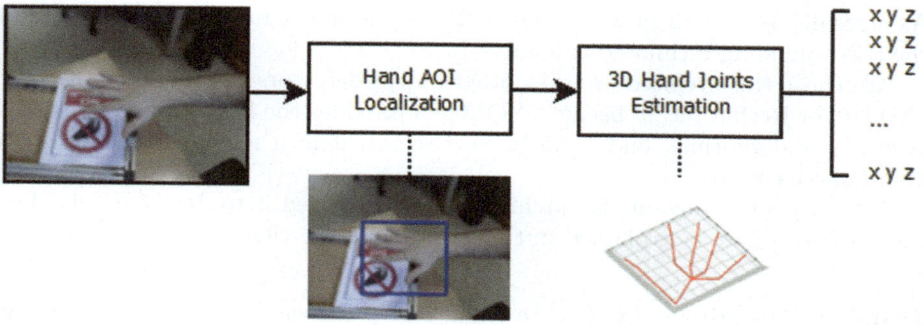

Fig. 1. Pipeline of the proposal.

in order to remove useless information. Namely, to estimate the position of the joints in a hand it is only needed the color information regarding the hand. Thus, we perform this step to remove the background. HAIL system is composed by a YOLO v3 [16] region convolutional neural network (R-CNN), that we custom trained to detect hands. Therefore, the HAIL system provides bounding boxes of the hands that are stated by its two corners in X and Y coordinates in the image plane.

Once the region of the image that depict the hand is extracted, the tridimensional HJE system uses it to estimate the tridimensional position of each joint in a hand. To do so, we used modified state-of-the-art convolutional neural networks. We put to test ResNet50, InceptionV3 and DenseNet. For all of them, we removed the last fully connected layer and appended a different one with 63 output neurons, matching the 63 values that define the tridimensional coordinates of the 21 joints in a hand.

Finally, the estimated positions of the 21 joints in the hand can be used to reconstruct an accurate and fine pose of the hand. A diagram of this pipeline is shown in Fig. 1. We used this prediction to teleoperate a low-cost robotic hand.

3.1 Hand Area of Interest Localization

The first piece of the pipeline is the Hand Area of Interest Localization system, which is in charge of detecting the hands in the images. Namely, to provide the area of interest (AOI) in which the hand is depicted within the image by stating its bounding box.

To carry out this task, we adopted the YOLO v3 R-CNN. This deep learning-based architecture is composed of a convolutional neural network backbone that is combined with custom detection layers. The custom detection layers operation is described next. The feature map as provided by the convolutional backbone is organized in a 2D grid of fixed size. Then, for each cell of the grid, a set of anchor boxes, centered in the grid, is used to extract different set of features for the different patches of the feature map. The anchor boxes are just a fixed set of different shapes. Each patch is then classified in either object or background.

As a result, YOLO v3 provides the AOI of each object in the input image and the corresponding estimated category.

We chose this architecture over other object detectors like Tiny-YOLO [16], SSD [10] or RetinaNet [9] because YOLO v3 provides the best trade-off between accuracy and runtime, and is fairly easy to run it in a real time fashion with affordable hardware.

We took advantage of this architecture and trained it to detect hands. The training procedure is explained in the following subsection.

Training Procedure. In order to train YOLO v3, we decided to use different datasets to give to the model generalization capabilities. The final dataset we have used to train YOLO is as follows.

First, 1000 random images were extracted of the Large-scale Multiview Hand Pose Dataset [5], which was also used used in the training procedure of the Tridimensional HJE system. More information about this dataset is given in Sect. 4. We manually annotated the hands in the mentioned images we extracted. Furthermore, another 1000 images were generated by applying random rotations. We also involved 2000 images of the Rendered Hand Pose dataset [25]. This dataset is formed by synthetic samples. The dataset provides the images, the depth map and the hand segmentation. To generate the bounding boxes we need, went through all the pixels of the segmentation mask, and took the largest and smallest X and Y. Then, 2000 images of the dataset FreiHand [1]. This dataset contains color images as well as its segmentation mask, so we did the same process that we have described above. Finally, additional 1960 images from the Dexter dataset [19] were also involved. In this dataset there is a sequence of images that changes the positions of the fingers, closing some, opening others, which we needed to improve the accuracy of the estimation. In addition, this dataset has different cameras for each gesture, so it has a great variability and can help us when changing the camera perspective. We manually annotated them, and also applied data augmentation just like in the first subset of data explained before.

We used this mix of data from different datasets to train the HAIL system. The system achieved great robustness as it works with different shades of light, both natural and artificial, and with different poses and perspectives of both the camera and the hand.

3.2 Tridimensional Hand Joints Estimation

The second part of the pipeline is the Tridimensional Hand Joints Estimation system. This piece takes the AOIs of the hands, as provided by the first step of the pipeline, and returns 63 values. Each triplet of values defines the X, Y and Z coordinates of a specific joint. In total, the hand is defined by 21 joints, which corresponds to 4 joints in each finger plus one more point to define the center of the palm.

To carry out this task, we considered different architectures that we put to test on the Sect. 5. The architectures of choice were ResNet50, InceptionV3 and DenseNet.

These deep learning-based models currently are the best performers in different challenges, such as ImageNet [3], so we adopted them for our Tridimensional HJE system. They all feature novelties that are behind the high accuracy and robustness they provide. For instance, ResNet introduced the residual connections within the architecture. Namely, the feature maps are forwarded to subsequent layers skipping certain blocks of the sequential architecture. This feature benefits the generalization capability and decrease the vanishing gradient problem. Then, InceptionV3 has different features, such as factorizing convolutions, that aim to reduce the number of parameters in the network without harming its efficiency. In addition, it also has an auxiliary classifier that enhances the accuracy on different scales. Finally, DenseNet follows the ResNet methodology, but instead of connecting a specific feature map to a following layer by skipping a block, DenseNet connects each feature map with all the following convolutional blocks.

These architectures are intended to perform classification, so we had to setup them to perform regression. To do this, we removed the last fully connected layer, and replaced it with a new fully connected layer with 63 output neurons to match the number of values that represent the 21 joints in a hand. We also used the *tanh* activation function to deal with negative values. We did this to every architecture mentioned before. We custom trained these architecture as explained in the following subsection.

Training Procedure. Before training we had to make a transformation of the points so that the reference system was relative to the hand, this process is explained in [4]. We do this to standardize the poses that are essentially the same, but the absolute position of the hand is not.

We train the Tridimensional HJE architectures by randomly grouping the images in batches in each epoch. We divided the dataset as follows: 60% training, 20% validation and 20% test. We do not decide on a particular number of epochs. Instead, we left it training in a loop and stopped it manually when we the learning stuck. Normally, the training needed about 100 epochs to converge, but in the experiments that involved data augmentation, it typically converged upon 50 epochs.

Validation loss values ranged from 3.55 to 2.42 and test loss values from 4.6 to 2.9, depending the architecture. The experiments that involved data augmentation the relative losses were higher. This is because the set is composed of a greater number of images. Anyway, these models worked better in terms of accuracy in spite of having a higher loss values.

3.3 Low-Cost Robotic Hand Actuator

The low-cost robotic hand of choice is part of inMoov. The inMoov project consist in the creation of an open source 3D printed life-size robot. This project is led by Gael Langevin, who is a French sculptor and designer. It begun in January 2012 as the first open source prosthetic hand but he eventually managed to create the first open source 3D printed life-size robot. The iMoov robot can be printed on any home 3D printer with a 12 × 12 × 12 cm area so it is conceived as a development platform for universities, laboratories and hobbyists.

We 3D printed the files for the right hand as provided by the author in its webpage[1]. The parts were printed by a Creality 3D Ender-3 Pro with PLA filament. It was assembled following carefully the guidelines provided in the inMoov webpage. The servos are MG946R brushless motor, 13 kg/cm torque, which are connected to an Arduino UNO. An external power supply was used to properly feed the 5 servos of the assembly, which is depicted in Fig. 2.

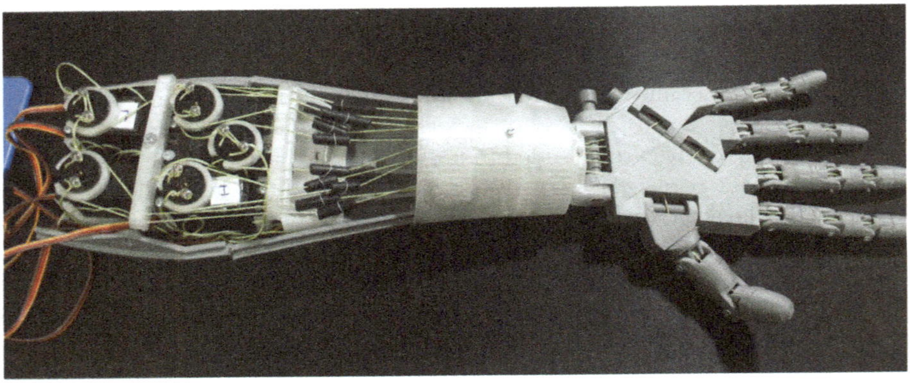

Fig. 2. Arm and hand assembly of the in Moov robot.

The robotic hand must replicate the hand poses depicted in the images by feeding rotation values to the servos. Nonetheless, the pipeline described in Sect. 3 provides the 3D position of the joints. Thus, a process to convert the 3D positions to rotation values for the servos must be carried out. To do so, first we empirically measure the working servo values limits from the finger completely relaxed to the finger completely flexed. Then, for estimated each hand pose, it is measured the distance from the center of the palm to the distal knuckle of each hand. Then, using this distance, the servo value is linearly interpolated using the limits we computed first. This approach whilst simple is also effective as the robotic hand of choice has a limited range of movements compared with a human hand.

4 Dataset Description and Data Balance Technique

The dataset of choice to train and test the systems is the Large-scale Multiview 3D Hand Pose Dataset [5]. It is composed of 21 sequences of different subjects

[1] http://inmoov.fr/hand-and-forarm/.

performing unconstrained hand poses. Each gesture is captured by 4 different points of view. It also contains the 21 point annotations (x,y,z) of each frame, as captured by a Leap Motion sensor.

One of the problems of the dataset is that the images are unbalanced, there are many more images of open or closed hands than with other gestures. Since we did not have the images grouped in classes, we decided to carry out a clustering process to group the images with similar poses in order to know the magnitude of the decomposition, and try and even the number of samples per pose.

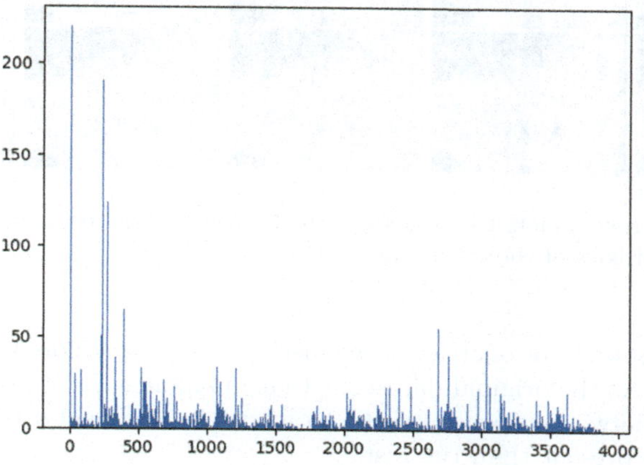

Fig. 3. Number of samples (Y axis) per cluster (X axis) distribution of the dataset we used to train and test the architectures. As it can be seen, there are a set of hand poses that are more present than others.

To do this clustering, we assumed two poses were the same if a similarity score between them was under a threshold. We iterate over all the poses of the dataset. We start with no clusters, so the first pose is set as a proxy of the first cluster. Then, we compute the similarity score between the current pose and the proxies of each cluster. If the similarity score is under a threshold, the current pose is appended to that cluster. If all proxies were tested and the similarity threshold had no match, a new cluster is created and the current pose is set as its proxy. All poses are iterated until they have been assigned to a cluster, either as a member of as the proxy. At the end, each cluster represents a set of similar hand poses. If the threshold is small, a larger number of clusters is obtained and more subtle is the difference between the poses of two clusters. We empirically set the similarity threshold to 5.

To calculate the similarity score between two poses, we used the Euclidean distance between the hands. We calculate the mean Euclidean distance between the pairs of 21 points corresponding joints of the hand. We use this similarity score to check if two hand poses are similar or not.

Once all the images of the dataset were clustered, we obtained the data distribution shown in Fig. 3. The X axis are the clusters and the Y axis show the number of samples in the particular cluster. As it can be seen, there are particular clusters, namely set of poses, that are more present that others, as shown in Fig. 4.

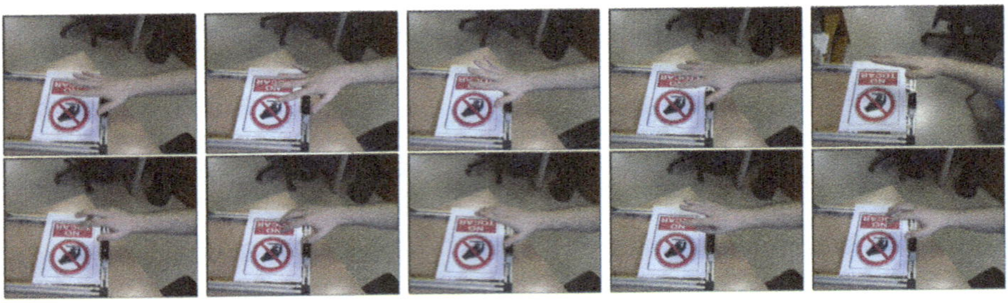

Fig. 4. The top row depicts the proxies of the 5 most present poses whilst the bottom row shows the proxies of the 5 least present poses.

Finally, we set a limit of 50 for each cluster, achieving a more balanced version of the dataset. As the original dataset is that unbalanced, this step is important to ensure the networks do not overfit to a certain, reduced set of poses which are much more present than the rest.

5 Experimentation

In this section we explain the experiments we carried out. First, we describe our first approach, which will serve as a baseline. Then, we show the results of performing the clustering described in Sect. 4 to reduce the unbalance factor of the dataset. Finally, the results we obtained by using different backbones for the Tridimensional Hand Joints Estimation system are also given. The best architecture is then chosen to be trained with an augmented version of the dataset.

5.1 Baseline Experiment

The first experiment we performed is training the Tridimensional HJE system, which architecture of choice is a ResNet50, with untreated images. Namely, without applying the HAIL system. Thus, the images are fed as they are in the dataset, including background and the rest of artifacts in them. No data augmentation or technique for balancing the data were applied.

The average test similarity error (ATSE) for this experiment was 3.599, which means that there is an average error of 3.599 mm for each predicted joint. Despite the fact that this error could appear small, it did not perform as expected when we tested the system with live data different from the samples present in the dataset.

The reason behind the noticeable difference of performance between the test and the live data relies in the background. The background is the same across all the samples in the training and test dataset, so the model learned it as a key feature of the poses.

To ameliorate this effect, we involved the HAIL system. We used it to extract the hands from the samples present in the dataset in order to isolate the hands from the background. Then, we retrained Tridimensional HJE system. This time, the average test similarity error for this experiment was of 4.607. This model performed better with live data, but it still failed often. The lack of generalization was likely introduced because the training dataset is highly unbalanced.

5.2 Experiments with a Balanced Version of the Dataset Using a Clustering Procedure

When we discovered that the dataset was highly unbalanced in terms of poses, we performed the clustering process explained earlier in Sect. 4. First, we limited the number of samples of all the clusters to 50 samples (200 images, since there are 4 cameras capturing the poses from 4 different points of view). The results obtained, using a ResNet50 as the backbone for the Tridimensional HJE system, threw an average test similarity error for this experiment of 3.641. Despite the low similarity threshold, this model did not show the same accuracy on live data.

5.3 Experiments with Different Backbones

In this experiment, the data setup was the same as explained in Sect. 5.2 but we tested different backbones for the Tridimensional HJE system. In this experiment, InceptionV3 and DenseNet were tested. The results obtained are shown in the Table 1.

Table 1. Average test similarity error (ATSE) achieved by different backbones for the tridimensional Hand Joint Estimation system

Architecture	Validation ATSE	Test ATSE
ResNet50	2.86	3.427
InceptionV3	**2.429**	**2.89**
DenseNet	3.41	2.994

In the light of this experiment, it can be seen that the InceptionV3 achieved the best results as for the testing split means. However, the lack of generalization to live data remained. It could be related to the fact that we removed a large quantity of samples as a result of the clustering procedure to balance the dataset.

5.4 Experiments Involving Data Augmentation

Since all the tests we carries out provided good results with the dataset itself, but did not show the same performance on live data, we decided to increase the number of samples of those clusters that had very few images. In this case, we took the InceptionV3 as the backbone for the for the Tridimensional HJE system as it was the best performer in the previous tests. The data augmentation procedure included rotations, fading, contrast changes and Gaussian noise. This data augmentation was done in the clusters that had less than 50 examples (200 images) and was done in an exponential way, data augmentation was done on the images that were in the cluster and if there were missing images it was done again on the new images.

In this case, the ATSE was of 4.649. This error seems higher than the one we achieved in the previous experiment, but it is only because the number of images noticeably increased as a result of the data augmentation process. As a result of the combined clustering process to balance the dataset and data augmentation to even the number of samples of each cluster, this model provided a similar performance in both the test set and live data.

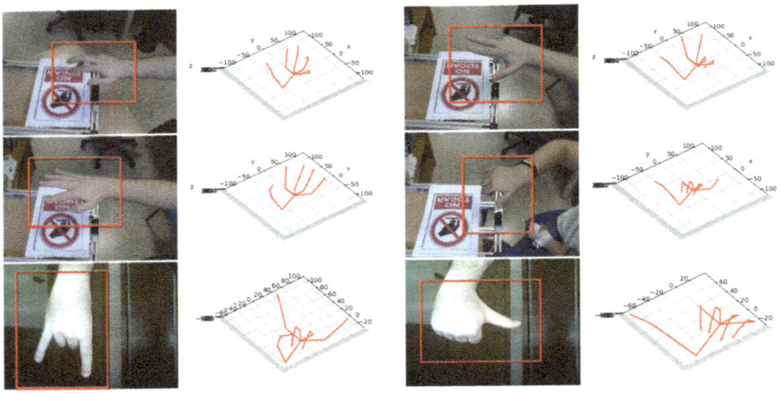

Fig. 5. The two first rows depict random images of hands from the dataset and predicted hand poses. The third row shows images from a live test alongside the corresponding predicted hand poses.

5.5 Experiments on the Full Pipeline

In this experiment, we used the hand pose estimation pipeline to control the robotic hand described in Sect. 3.3. It worked as expected, being robust and accurate. Some samples of this experiment are shown in Fig. 6.

A slight jittering can be noticed when the pipeline is run in a loop as a result of feeding it with live images from a video stream. This jittering is introduced by the hand pose estimation system as a result of the deviation discussed before.

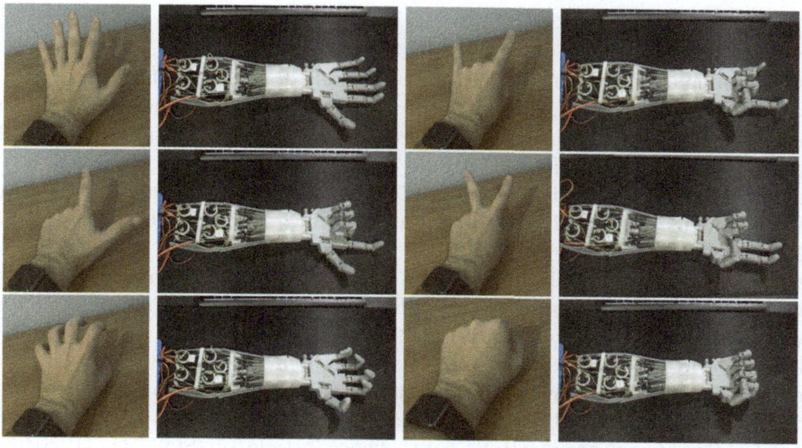

Fig. 6. Images of hands provided by a live video stream along the robotic hand being teleoperated by the estimated poses.

In addition, there are some poses that, despite being predicted correctly, the robotic hand cannot perform due its mechanical limitations. For instance, the "spock salutation" cannot be performed as then robotic hand cannot move the the fingers sideways. Likewise, the thumb cannot be moved next to the index finger. However, another robotic hand, like the Shadow Hand[2], could do so, as long as it has more degrees of freedom.

6 Conclusion and Future Work

In this work we studied a pipeline for hand pose estimation from color images of hands. We applied the prediction to teleoperate a low-cost robotic hand. The hand pose estimation pipeline is composed of two main pieces, the Hand Area of Interest Localization system and the Tridimensional Hand Joints Estimation system. The predicted hand poses yield high accuracy with an ATSE of less than 5 mm per joint. The robotic hand of choice performed as expected. It is worth noting that there are some poses that the robotic hand cannot perform due to mechanical limitations despite the estimated poses were accurate enough.

As a future work we plan to tackle the jittering we noticed when we fed the hand pose estimation pipeline with a live stream of images. In addition, we plan to further improve its accuracy by involving another datasets in the training stages and another backbones.

Acknowledgments. This work has been funded by the Spanish Government PID2019-104818RB-I00 grant, supported with Feder funds. It has also been supported by Spanish grants for PhD studies ACIF/2017/243 and FPU16/00887. Experiments were made possible by a generous hardware donation from NVIDIA.

[2] https://www.shadowrobot.com/products/dexterous-hand/.

References

1. Zimmermann, C., Ceylan, D., Yang, J., Russell, B., Argus, M.A., Brox, T.: Freihand: a dataset for markerless capture of hand pose and shape from single rgb images. In: IEEE International Conference on Computer Vision (ICCV) (2019). https://lmb.informatik.uni-freiburg.de/projects/freihand/
2. de La Gorce, M., Fleet, D.J., Paragios, N.: Model-based 3D hand pose estimation from monocular video. IEEE Trans. Pattern Anal. Mach. Intell. **33**(9), 1793–1805 (2011)
3. Deng, J., Dong, W., Socher, R., Li, L., Li, K., Fei-Fei, L.: Imagenet: a large-scale hierarchical image database. In: 2009 IEEE Conference on Computer Vision and Pattern Recognition, pp. 248–255 (2009)
4. Gomez-Donoso, F., Orts-Escolano, S., Cazorla, M.: Accurate and efficient 3D hand pose regression for robot hand teleoperation using a monocular rgb camera. Expert Syst. Appl. **136**, 327–337 (2019). https://doi.org/10.1016/j.eswa.2019.06.055. http://www.sciencedirect.com/science/article/pii/S0957417419304634
5. Gomez-Donoso, F., Orts-Escolano, S., Cazorla, M.: Large-scale multiview 3D hand pose dataset. Image and Vis. Comput. **81**, 25–33 (2019)
6. Grzejszczak, T., Kawulok, M., Galuszka, A.: Hand landmarks detection and localization in color images. Multimedia Tools Appl. **75**(23), 16363–16387 (2016). https://doi.org/10.1007/s11042-015-2934-5
7. Kumar, P.P., Vadakkepat, P., Loh, A.P.: Hand posture and face recognition using a fuzzy-rough approach. Int. J. Humanoid Rob. **7**(03), 331–356 (2010). https://doi.org/10.1142/S0219843610002180
8. Kuznetsova, A., Leal-Taixé, L., Rosenhahn, B.: Real-time sign language recognition using a consumer depth camera. In: 2013 IEEE International Conference on Computer Vision Workshops, pp. 83–90 (2013)
9. Lin, T., Goyal, P., Girshick, R.B., He, K., Dollár, P.: Focal loss for dense object detection. CoRR **abs/1708.02002** (2017). http://arxiv.org/abs/1708.02002
10. Liu, W., Anguelov, D., Erhan, D., Szegedy, C., Reed, S., Fu, C.Y., Berg, A.C.: SSD: single shot multibox detector. In: Leibe, B., Matas, J., Sebe, N., Welling, M. (eds.) Computer Vision - ECCV 2016, pp. 21–37. Springer, Cham (2016)
11. Marcel, S., Bernier, O.: Hand posture recognition in a body-face centered space. In: Braffort, A., Gherbi, R., Gibet, S., Teil, D., Richardson, J. (eds.) Gesture-Based Communication in Human-Computer Interaction, pp. 97–100. Springer, Berlin, Heidelberg (1999)
12. Marcel, S., Bernier, O., Viallet, J.E., Collobert, D.: Hand gesture recognition using input-output hidden markov models. In: Proceedings of the 4th IEEE International Conference on Automatic Face and Gesture Recognition 2000, p. 456, FG 2000, IEEE Computer Society, USA (2000)
13. Molina, J., Pajuelo, J.A., Escudero-Viñolo, M., Bescós, J., Sanchez, J.M.M.: A natural and synthetic corpus for benchmarking of hand gesture recognition systems. Mach. Vis. Appl. **25**, 943–954 (2013)
14. Oberweger, M., Wohlhart, P., Lepetit, V.: Training a feedback loop for hand pose estimation. In: 2015 IEEE International Conference on Computer Vision (ICCV), pp. 3316–3324 (2015)
15. Panteleris, P., Argyros, A.A.: Back to RGB: 3D tracking of hands and hand-object interactions based on short-baseline stereo. In: 2017 IEEE International Conference on Computer Vision Workshops (ICCVW), pp. 575–584 (2017)

16. Redmon, J., Divvala, S., Girshick, R., Farhadi, A.: You only look once: unified, real-time object detection. In: 2016 IEEE Conference on Computer Vision and Pattern Recognition (CVPR), pp. 779–788 (2016)
17. Simon, T., Joo, H., Matthews, I., Sheikh, Y.: Hand keypoint detection in single images using multiview bootstrapping. In: The IEEE Conference on Computer Vision and Pattern Recognition (CVPR), July 2017
18. Sinha, A., Choi, C., Ramani, K.: Deephand: robust hand pose estimation by completing a matrix imputed with deep features. In: The IEEE Conference on Computer Vision and Pattern Recognition (CVPR), June 2016
19. Sridhar, S., Oulasvirta, A., Theobalt, C.: Interactive markerless articulated hand motion tracking using RGB and depth data. In: Proceedings of the IEEE International Conference on Computer Vision (ICCV), December 2013. http://handtracker.mpi-inf.mpg.de/projects/handtracker_iccv2013/
20. Stenger, B., Thayananthan, A., Torr, P.H.S., Cipolla, R.: Hand pose estimation using hierarchical detection. In: Sebe, N., Lew, M., Huang, T.S. (eds.) Computer Vision in Human-Computer Interaction, pp. 105–116. Springer, Berlin, Heidelberg (2004)
21. Tang, D., Yu, T., Kim, T.: Real-time articulated hand pose estimation using semi-supervised transductive regression forests. In: 2013 IEEE International Conference on Computer Vision, pp. 3224–3231 (2013)
22. Tompson, J., Stein, M., Lecun, Y., Perlin, K.: Real-time continuous pose recovery of human hands using convolutional networks. ACM Trans. Graph. **33**(5), 1–10 (2014). https://doi.org/10.1145/2629500
23. Zhang, J., Jiao, J., Chen, M., Qu, L., Xu, X., Yang, Q.: 3D hand pose tracking and estimation using stereo matching. CoRR **abs/1610.07214** (2016). http://arxiv.org/abs/1610.07214
24. Zimmermann, C., Brox, T.: Learning to estimate 3D hand pose from single RGB images. In: The IEEE International Conference on Computer Vision (ICCV), October 2017
25. Zimmermann, C., Brox, T.: Learning to estimate 3D hand pose from single rgb images. Technical report, arXiv:1705.01389 (2017). https://lmb.informatik.uni-freiburg.de/projects/hand3d/, https://arxiv.org/abs/1705.01389

Author Index

A
Abellan, Daniel, 345
Aller, Aitor, 256
Almazán, Emilio J., 305
Alonso, Jose M., 30
Álvarez, Pedro, 30
Álvarez, Sergio, 256
Antunes, Miguel, 241
Araluce, Javier, 44, 241, 274
Arango, Felipe, 44, 274
Arroyo, Roberto, 256
Aumjaud, Pierre, 318

B
Bachiller, Pilar, 109, 167, 180
Baghel, Rishabh, 180
Bandera, Antonio, 94, 109
Bañuls, Alejandro, 345
Barea, Rafael, 44, 60, 241, 274
Barro, Senén, 79
Bergasa, Luis M., 44, 60, 241, 256, 274
Bernabé Murcia, José Manuel, 227
Bugarín, Alberto, 30
Bustos, Pablo, 109, 167

C
Calle, Alejandro de la, 305
Cañas, José-María, 291
Cardiff, Philip, 318
Casado, Fernando E., 79
Cazorla, Miguel, 345
Chapela-Campa, David, 30
Cilleruelo-Rodríguez, Carlos, 194
Cintas, Ramón, 109
Condés, Ignacio, 291

D
Del Egido, Javier, 44, 241
de-Marcos, Luis, 194
Díaz, Alejandro, 60
Duque Domingo, Jaime, 213

E
Escalona, Félix, 345
Evers, Christine, 151

F
Fabra, Javier, 30
Faria, Diego R., 167

G
García, Juan C., 109
García-Garrido, Miguel A., 30
García-Varea, Ismael, 18
Gil-Pita, Roberto, 332
Ginés Clavero, Jonatan, 3
Gomez-Donoso, Francisco, 345
Gómez-García-Bermejo, Jaime, 213
Gómez-Huélamo, Carlos, 44, 241, 274
Gutiérrez, Rodrigo, 60, 241
Guerrero Hernández, José Miguel, 3

H
Hernández, Noelia, 30

L. M. Bergasa et al. (Eds.): WAF 2020, AISC 1285, pp. 361–362, 2021.
https://doi.org/10.1007/978-3-030-62579-5

I
Iglesias, Roberto, 79
Inglés-Romero, Juan F., 94

J
Jorvekar, Ronit R., 167, 180
Junquera-Sánchez, Javier, 194

K
Kapoor, Aditya, 180

L
Lama, Manuel, 30
Lazkano, Elena, 137
Lema, Dylan, 79
Llamazares, Ángel, 30
López, Joaquín, 44
López-Guillén, Elena, 44, 60, 241, 274

M
Manso, Luis J., 167, 180
Marfil, Rebeca, 94
Martín Rico, Francisco, 3
Martínez-Barberá, Humberto, 227
Martínez-Colón, Antonio, 151
Martínez-Cruz, Jesús, 94
Martínez-Gómez, Jesus, 18
Martínez-Herráiz, José Javier, 194
Martínez-Otzeta, José María, 137
Martirena, Esteban, 109
Matellán Olivera, Vicente, 3
McAuliffe, David, 318
Mucientes, Manuel, 30

N
Naylor, Patrick A., 151
Núñez, Pedro, 109

O
Ocaña, Manuel, 30
Ortiz, Miguel E., 256

P
Perdices, Eduardo, 291
Pérez-Gil, Óscar, 60, 274
Perez-Lorenzo, Jose Manuel, 151

R
Regueiro, Carlos V., 79
Revenga, Pedro A., 30, 60
Ribas-Xirgo, Lluís, 123
Rodriguez, Igor, 137
Rodriguez-Criado, Daniel, 180
Rodríguez-Lera, Francisco Javier, 3, 318
Romero-Garcés, Adrián, 94
Romero-González, Cristina, 18
Rosa-Zurera, Manuel, 332

S
Salles De Freitas, Renan, 94
Sánchez-Hevia, Héctor A., 332

T
Tovar, Javier, 305

U
Utrilla-Manso, Manuel, 332

V
Vicente-Chicote, Cristina, 94
Viciana-Abad, Raquel, 151

Z
Zabala, Unai, 137
Zalama, Eduardo, 213

The manufacturer's authorised representative in the EU is Springer Nature Customer Service Centre GmbH, Europaplatz 3, 69115 Heidelberg, Germany. If you have any concerns regarding our products, please contact ProductSafety@springernature.com

Printed and bound by CPI Group (UK) Ltd, Croydon, CR0 4YY
30/09/2026
02225292-0005